Systems Engineering
for Microscale and
Nanoscale Technologies

Systems Engineering

for Microscale and Nanoscale Technologies

M. Ann Garrison Darrin and Janet L. Barth

CRC Press
Taylor & Francis Group
Boca Raton   London   New York

CRC Press is an imprint of the
Taylor & Francis Group, an **Informa** business

CRC Press
Taylor & Francis Group
6000 Broken Sound Parkway NW, Suite 300
Boca Raton, FL 33487-2742

© 2012 by Taylor & Francis Group, LLC
CRC Press is an imprint of Taylor & Francis Group, an Informa business

No claim to original U.S. Government works

Printed in the United States of America on acid-free paper
Version Date: 2011915

International Standard Book Number: 978-1-4398-3732-0 (Hardback)

Library of Congress Cataloging-in-Publication Data

Systems engineering for microscale and nanoscale technologies / editors, M. Ann
　　Garrison Darrin, Janet L. Barth.
　　　　p. cm.
　　Includes bibliographical references and index.
　　ISBN 978-1-4398-3732-0 (hardcover : alk. paper)
　　1. Nanostructured materials. 2. Systems engineering. I. Darrin, M. Ann Garrison. II.
Barth, Janet L. III. Title.

TA418.9.N35S983 2012
620'.5--dc23   2011034572

Visit the Taylor & Francis Web site at
http://www.taylorandfrancis.com

and the CRC Press Web site at
http://www.crcpress.com

Contents

Preface

Technology development on the microscale and nanoscale has transitioned from laboratory curiosity to the realization of products in the health, automotive, aerospace, communication fields, and numerous other arenas. As technology developers, the editors saw a need to bring together a multidisciplinary team to develop a handbook for product development managers, technology researchers, and systems engineers. This handbook is a first step in exploring the application of systems engineering to small-scale systems. The editors understand that *MNT* has multiple meanings. Here we use it to represent micro- and nanoscale technologies and not the often-used molecular nanotechnology. Also, it is intentional that we have not spent a great deal of time discussing microcircuits but rather have concentrated our efforts on MNTs as a less mature and advanced field.

Developing systems engineering methodologies that integrate stand-alone, small-scale technologies and interface them with macrotechnologies to build useful systems is critical to realizing the potential of micro- and nanoscale devices. A barrier to the infusion of the micro- and nanotechnologies into systems is a lack of insight into how to apply systems engineering principles and management processes to the integration of small-scale technologies. The result of this first-step book is the provision of practical guidance for systems engineers in the development of micro- and nanotechnologies. For nonconventional micro- and nanoscale systems, the systems engineer must also be knowledgeable about the roles of nonconventional disciplines, such as quantum mechanics, quantum chemistry, solid-state physics, materials science, and chemistry, in the development of small-scale systems. The results are also targeted toward small-scale technology developers who need to take into account systems engineering processes, such as requirements definition, product verification and validation, interface management, and risk management, in the concept phase of technology development to maximize the likelihood of successful, cost-effective micro- and nanotechnology to increase the capability of emerging deployed systems and long-term growth and profits. Contributors in this introductory first step include nanotechnologists, physicists, systems engineers, material scientists, chemists, electrical engineers, and futurists.

Contributors

Janet L. Barth
NASA Goddard Space Flight
    Center
Greenbelt, Maryland

Jason Benkoski
The Johns Hopkins
    University Applied Physics
    Laboratory
Laurel, Maryland

Jennifer Breidenich
The Johns Hopkins
    University Applied Physics
    Laboratory
Laurel, Maryland

M. Ann Garrison Darrin
The Johns Hopkins
    University Applied Physics
    Laboratory
Laurel, Maryland

Elinor Fong
The Johns Hopkins
    University Applied Physics
    Laboratory
Laurel, Maryland

Brian Jamieson
SB Microsystems
Columbia, Maryland

D.Y. Kusnierkiewicz
The Johns Hopkins
    University Applied Physics
    Laboratory
Laurel, Maryland

Bradley Layton
Department of Applied
    Computing and Electronics
University of Montana
    College of Technology
Missoula, Montana

Jeffrey P. Maranchi
The Johns Hopkins
    University Applied Physics
    Laboratory
Laurel, Maryland

Jennette Mateo
Virginia Commonwealth
    University
Richmond, Virginia

Bethany M. McGee
The Johns Hopkins
    University Applied Physics
    Laboratory
Laurel, Maryland

Timothy G. McGee
The Johns Hopkins
    University Applied Physics
    Laboratory
Laurel, Maryland

Robert Osiander
The Johns Hopkins
    University Applied Physics
    Laboratory
Laurel, Maryland

Stergios J. Papadakis
The Johns Hopkins
    University Applied Physics
    Laboratory
Laurel, Maryland

William Paulsen
The Johns Hopkins
    University Applied Physics
    Laboratory
Laurel, Maryland

Jennifer L. Sample
The Johns Hopkins
    University Applied Physics
    Laboratory
Laurel, Maryland

I.K. Ashok Sivakumar
The Johns Hopkins
    University Applied Physics
    Laboratory
Laurel, Maryland

John Thomas
The Johns Hopkins
    University Applied Physics
    Laboratory
Laurel, Maryland

Morgan Trexler
The Johns Hopkins
    University Applied Physics
    Laboratory
Laurel, Maryland

O. Manuel Uy
The Johns Hopkins
    University Applied Physics
    Laboratory
Laurel, Maryland

Acknowledgments

The editors have enjoyed this opportunity to provide an introductory exploration into concepts of systems engineering and the maturing world of micro- and nanoscale technologies. Both Ann Garrison Darrin and Janet Barth appreciate the continued opportunities and support from their respective organizations, The Johns Hopkins University Applied Physics Laboratory and the NASA Goddard Space Flight Center. In addition, Ann and Janet appreciate the opportunity to work together and to work with an outstanding team of contributors. As we are not micro- or nanoscientists or systems engineers, this has been a fascinating activity to pull together a wide range of technical fields. Special thanks to our husbands, Armond Darrin and Douglas Barth, and our other grown children along with little Ian. Finally, again it is a pleasure to work with Taylor & Francis and CRC Press.

PART 1

Systems Engineering Methodologies

A System Is ...
Simply stated, a system is an integrated composite of people, products, and processes that provide a capability to satisfy a stated need or objective.

Part 1 introduces the basic concepts of systems engineering in the context of enabling micro- and nanotechnology developments for systems. The systems engineering process is reviewed focusing on life-cycle stages and functions (Chart I.1) that are applicable to advanced technology developments. This part includes a comprehensive discussion on the special considerations that must be taken into account when incorporating new technologies (Chart I.2) and explores other system engineering methodologies that are applicable to developing and integrating micro- and nanotechnologies. Several systems engineering methodologies are reviewed, including traditional waterfall methodology and agile methods. Configuration management and risk management (Chart I.3) are discussed.

CHART I.1
Life-Cycle Functions

Life-cycle functions are the characteristic actions associated with the system life cycle. They are development, production and construction, deployment (fielding), operation, support, disposal, training, and verification. These activities cover the *cradle-to-grave* life-cycle process and are associated with major functional groups that provide essential support to the life-cycle process. These key life-cycle functions are commonly referred to as the eight primary functions of systems engineering. The customers of the systems engineer perform the life-cycle functions. The system user's needs are emphasized because their needs generate the requirement for the system, but it must be remembered that all of the life-cycle functional areas generate requirements for the systems engineering process once the user has established the basic need.

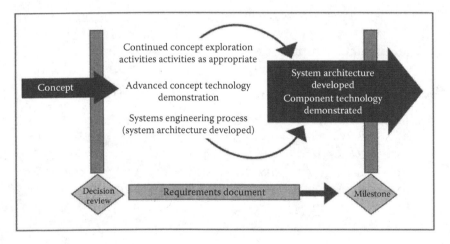

Chart I.2 Advanced Technology Development

CHART 1.3
System Level Risk Assessment

	Low Risk	Moderate Risk	High Risk
Consequences	Insignificant cost, schedule, or technical impact	Affects program objectives, cost, or schedule; however, cost, schedule, and performance are achievable	Significant impact, requiring reserve or alternate courses of action to recover
Probability of Occurrence	Little or no estimated likelihood	Probability sufficiently high to be of concern to management	High likelihood of occurrence
Extent of Demonstration	Full-scale, integrated technology has been demonstrated previously	Has been demonstrated, but design changes, tests in relevant environments required	Significant design changes required in order to achieve required/desired results
Existence of Capability	Capability exists in known products; requires integration into new system	Capability exists, but not at performance levels required for new system	Capability does not currently exist

1

Systems Engineering for Micro- and Nanoscale Technologies

Janet L. Barth
M. Ann Garrison Darrin

When mini meets macro.

Contents

Introduction

Applying systems engineering principles to the realm of micro-
and nanoscale technology (MNT) development has been recognized as key to solving the challenge of increasing the success
of the transition of MNT from the laboratory to operational
systems. In 2008 Yves LaCerte of Rockwell Collins, Cedar
Rapids, Iowa, addressed the International Council on Systems
Engineering (INCOSE) regarding systems engineering for
complex systems. In his presentation, he highlighted the key
role that systems engineering will play in developing micro-
and nanosystems by stating the following:

> Systems engineering will become a key enabler for the successful
> commercialization of multi-functional, micro and nano technolo-
> gies. Systems engineering delivers methodologies, processes, and
> tools to enable the efficient integration and exploitation of these
> disruptive technologies.

K. Eric Drexler responding to Richard E. Smalley in a now
famous debate in *Chemical and Engineering News*[1] wrote:

> To visualize how a nanofactory system works, it helps to consider
> a conventional factory system. The technical questions you raise
> reach beyond chemistry to *systems engineering*. Problems of con-
> trol, transport, error rates, and component failure have answers
> involving computers, conveyors, noise margins, and failure-tolerant
> redundancy. (p. 4)

In this *Systems Engineering for Microscale and Nanoscale Technologies* handbook, we provide guidance for applying system engineering methodologies to the development of micro- and nanotechnology-based devices and systems.* We meet this objective by providing a solid technical foundation for the systems engineer engaged in the development of MNTs and their integration into macrosystems. In their lead role of managing the technical specialists on their teams, this book provides systems engineers with an understanding of tools and methodologies used by disciplines involved in the development of micro- and nanotechnologies, including quantum mechanics, quantum chemistry, solid-state physics, materials science, and chemistry. This book is also a resource for micro- and nanoscale technology researchers and development teams who apply systems engineering processes, such as requirements development, key decision points, product verification and validation, interface management, and risk management. Finally, the book serves as a guide to technical and business program managers for developing and implementing robust micro- and nanoscale technology programs to increase the likelihood that new technologies will bridge the gap between the laboratory and applications.

What are Microtechnology and Nanotechnology?

Microtechnology is defined as systems with physical feature sizes near one micrometer (10^{-6} meter). In the late 1960s, researchers demonstrated that mechanical devices can be miniaturized and batch-fabricated, promising the same benefits to the mechanical domain as integrated circuit technology has given to the electronic world. The birth of microtechnologies, also known as Micro Electro Mechanical Systems (MEMS), began in 1969 with a resonant gate-field effect transistor

---

* In the context of this book, micro is used in reference to Micro Electro Mechanical Systems (MEMS) technologies. We have not attempted to address the vast field of microelectronics technologies.

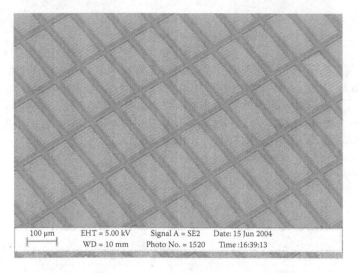

| 100 µm | EHT = 5.00 kV | Signal A = SE2 | Date: 15 Jun 2004 |
| | WD = 10 mm | Photo No. = 1520 | Time :16:39:13 |

Figure 1.1 Micro Electro Mechanical Systems (MEMS) microshutters on the National Aeronautics and Space Administration's (NASA) James Webb Space Telescope Near-Infrared Spectrograph (NIRSpec) instrument. (Courtesy of NASA.)

designed by Westinghouse. During the following decade, manufacturers began using bulk etched silicon wafer technology to produce pressure sensors, and technology breakthroughs continued into the early 1980s, creating surface-micromachined polysilicon actuators that were used in disc drive heads. By the late 1980s, the potential of MEMS devices was embraced, and widespread design and implementation emerged throughout the microelectronics and biomedical industries. In 25 years, MEMS had moved from the technical curiosity realm to the world of commercial potential.[2] Today, automobile air bags, ink-jet printers, blood pressure monitors, projection display systems, and space systems (see Figure 1.1) employ MEMS devices as key components, demonstrating their wide range of utility. It is conceivable that these devices will be as pervasive as microelectronics in the not too distant future.

Nanotechnology is the manipulation and control of matter at the scale of the nanometer, or one-billionth (10^{-9}) of a meter, roughly the diameter of a small molecule. Unlike its predecessor microtechnology that deals with the relatively gargantuan scale of amoebas, nanotechnology is engineering at the atomic and molecular levels. Nanotechnology demands more than just taking well-understood microtechnology engineering techniques down another step in size. It abruptly and vastly expands the limits of what is possible. Working with the basic material building blocks of nature, atoms and molecules, nanotechnology allows for an unprecedented level of engineering precision and control of matter. The nanometer scale (or nanoscale) is where the effects of "regular" Newtonian physics that governs everyday human experience and the "weird" quantum physics that governs the atomic and subatomic worlds begin to overlap. Working at the nanoscale permits engineers to take advantage of the benefits of both realms simultaneously.[3] The evolution of the critical dimension of technologies into the nanometer scale, together with the exploitation of completely new physical phenomena at the atomic and molecular levels, gives new momentum, creating opportunities for new solutions to current engineering problems in bioengineering, the environment, and human–machine interfaces.

Commercial nanotechnology is not as mature as microtechnology. Roco[4] describes four generations of nanotechnology products as demonstrated in Figure 1.2 and discussed in further detail in Chapters 7 and 15. From this list, one can see that the potential exists for the rapid emergence of nanotechnology products that will be enabling for intelligent integrated systems. Chad Mirkin, Director of the International Institute for Nanotechnology, observed that the field had transitioned in the last decade from "a lot of hype" to producing "real substantive science and engineering accomplishments."[5]

Characteristics of Systems Enabled by Microtechnology and Nanotechnology

Future product generations will be integrated systems enabled by micro- and nanoscale technologies. They will be systems of increasing complexity that use the convergence of a whole range of technologies for the improvement of the characteristics of the overall system. Features of future systems are

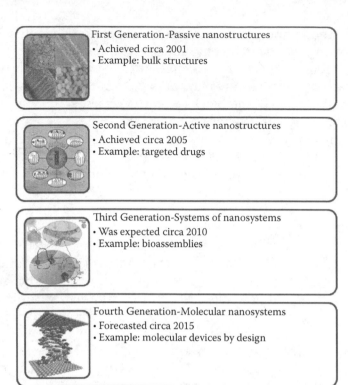

First Generation-Passive nanostructures
• Achieved circa 2001
• Example: bulk structures

Second Generation-Active nanostructures
• Achieved circa 2005
• Example: targeted drugs

Third Generation-Systems of nanosystems
• Was expected circa 2010
• Example: bioassemblies

Fourth Generation-Molecular nanosystems
• Forecasted circa 2015
• Example: molecular devices by design

Figure 1.2 The evolution of the field of nanosciences through four generations.

- Increasingly complex, involving quantum mechanics, quantum chemistry, solid-state physics, materials science, and chemistry principles, especially when considering micro- and nanoscaling
- Highly integrated systems of increasing complexity that use a range of technologies for the improvement of the overall system
- Networked, energy autonomous, miniaturized, and reliable for space, defense, medical, civil, and commercial applications
- Operating within larger systems in which they are embedded
- Interfacing with each other, with the larger system, the environment, and humans
- Easy to use
- Integrate mechanical, optical, and biological functions[6]

It is useful to consider Ottino's criteria[7] to identify complex systems to understand how to approach the application of systems engineering principles to developing micro- and nanoscale technologies and systems with embedded micro- and nanotechnology:

1. What they do—they display emergence
2. How they may or may not be analyzed—classical systems engineering approaches of decomposing/analyzing subparts do not necessarily yield clues of their behavior as a whole

Micro- and nanoscale technologies (MNTs) are complex systems. In addition, adapting these technologies to the human and environmental landscape requires that they be embedded within larger systems. This further increases system complexity, because the scale order between the macro and nano realm is very high (over 10^9). Integrating systems across macroscale to nanoscale regimes poses integration issues related to physical properties (e.g., physical, electronic, chemical) that do not scale as they would between differently sized macroscale objects. For example, van der Waals, surface tension, and frictional forces increase, and there are changes in fluid flow properties and melting point. These issues will be discussed throughout the book using examples, such as self-healing paint, nano-piezo-generators, polymer fullerene solar cells, MEMS accelerometer, and lab on a chip. The realization of systems based on micro- and nanoscale technologies is dependent on understanding their complexity, reproducibility, and interfacing with the systems within which they operate. These challenges were well summarized by Allhoff, Lin, and Moore: "It is sometimes very easy to get caught up in what is scientifically possible and ignore the engineering problems that come with it."[8]

Need for Systems Engineering Formalism

The traditional textbook approach for systems engineering is a top-down hierarchical approach of reductionism and discovery that is used to understand the system. In his appendix to the Rogers Commission Report on the Space Shuttle Challenger Accident, Richard Feynman pointed out that the National Aeronautics and Space Administration's (NASA) overreliance on the top-down systems engineering approach to design the Space Shuttle's main engine resulted in the inability of NASA to assess accurately the reliability of the engine. He also writes that another disadvantage of the top-down method is that if an understanding of a fault is obtained, a simple fix may be impossible to implement without a redesign of the entire system. Fortunately, systems engineering is agile enough to adapt to the bottoms-up or "coming into the middle" approach. Such is the case for space systems, where an instrument is designed first, and the spacecraft and mission are designed around it, or when a spacecraft is specified, and the other parts of the system have to modify their approach.[9]

In the case of nanotechnology, when development is often guided by bottom-up self-organization of molecules and super molecules, the less traditional bottom-up approach is more useful in synthesizing nanoscale technologies to determine technical feasibility and to drive out system enabling capability. In the realm of chemical engineering and nanotechnology, this bottom-up approach is often described as design synthesis. On the other hand, too much bottom-up engineering can lead to missed requirements and integration problems as noted by Graham Stoney.[10] Requirements flow down the system hierarchy, and the balancing force is feasibility that flows back up to ensure that higher-level design decisions do not result in downstream requirements that are excessively difficult or impossible to meet. He states that "good engineering has a balance between top-down and bottom-up design, but there should generally be a bias towards top-down because the ultimate goal is to meet the system requirements which flow in at the top level" (p. 1).

To be fully successful, development efforts for MNT-based systems must be system-centric, particularly in the concept development and feasibility phases where critical decisions must be made about interfaces in the multiscale system. In considering system integration on multiple scales, the systems engineer must address issues such as correlation between different scales, coupling between time and space dependencies, and identification of dominate mechanisms. This book explores the formalism of the "science of systems engineering" to gain insight on how to integrate discrete micro- and nanoscale components that exhibit scale-specific physical characteristics, the macrosystems with which they interface, and the overall system application or product.

Toward Common Terminologies and Taxonomy

Terms and definitions in the macro-micro-nano multiscale arena have been applied without strict definition or discipline. For example, the term *element* without any modifier, such as data, system, mission, machine, software, or support, has many meanings. Here are sample definitions of *element* in the context of systems engineering:[11]

1. A complete, integrated set of subsystems capable of accomplishing an operational role or function, such as navigation
2. A basic component of a system, typically controlled by a unique specification, such as a single line replaceable unit (LRU) or configuration item (CI)
3. A product, subsystem, assembly, component, subcomponent, or subassembly, or part of a physical or system architecture, specification tree, or system breakdown structure, including the system itself

Because so much of chemistry uses the term *element*, there can be confusion at the nanoscale. Element is commonly used in the microscale realm to indicate a structure such as a nand gate. At the micro- and nanoscale, a single entity may be referred to as an element, device, or structure.

Additional terms and definitions that are used throughout the book are described here:

Array: A large collection of nano- or microscale devices functioning together to achieve a common purpose [Note that *array* is also a term used for system of systems, but this use of the term is not typical in the MNT realm.]

Device or component: As in nanodevices or microdevices used to indicate several structures either integrated or interfaced

Structure: In nanostructures or microstructures used to indicate an entity

System: A complex collection of interactive units and sub-systems within a single product, jointly performing a wide range of independent functions to meet a specific operational mission or need. A system consists of many subsystems (and assemblies), each performing its own function and serving the system's major mission. Systems are capable of performing a complex mission, and their use involves considerable man–machine interaction.[12] In the MNT context, a system is defined as two or more components interacting together to achieve a common objective.

Subsystem: A building block of systems that may include device integrations/interfaces of nano/micro and macrodevices

Other frequently used terms related to systems hierarchy are[13]

Activity: A set of actions that consume time and resources and whose performance is necessary to achieve or contribute to the realization of one or more outcomes

Enabling system: A system that complements a system of interest during its life-cycle stages but does not necessarily contribute directly to its function during operation

Process: A set of interrelated or interacting activities that transform inputs into outputs

Project: An endeavor with start and finish dates undertaken to create a product or service in accordance with specified resources and requirements

Stage: A period within the life cycle of a system that relates to the state of the system description or the system itself

Overview and List of Contributors

The book is organized into four parts to guide the reader from an overview of standard systems engineering approaches to understanding the requirements for systems engineering methodologies that will need to be developed and applied for future generations of MNTs. The following provides a guide to the organization and introduces the authors who contributed chapters for the book.

- *Part 1: Systems Engineering Methodologies*: Concepts in the context of enabling micro- and nanoscale technology developments for systems.
- *Part 2: Technology Development Process:* Development on the micro- and nanoscale, including analyses that are important for the systems engineering process.
- *Part 3: Systems Engineering Process Elements*: Key technology development activities that support and run parallel to systems engineering verification and validation and risk management activities
- *Part 4: Systems Engineering Applications—Toward the Future:* Issues integral to the conduct of systems engineering efforts for future generations of micro- and nanoscale technologies

Part 1 introduces the reader to the realm of micro- and nanotechnology and will describe examples and applications from the space, military, and medical domains that will serve as a common thread throughout the book. The basic concepts of systems engineering will be reviewed in the context of enabling micro- and nanoscale technologies. This part contains three chapters in addition to this introductory chapter. Elinor Fong, Lead Systems Engineer at the Johns Hopkins Applied Physics Laboratory, introduces systems engineering principles used in the design and development of complex projects and systems. She discusses the dependence of the systems engineering approach to the system under development. David Y. Kusnierkiewicz, Chief Engineer of the Johns Hopkins University Applied Physics Laboratory Space Department, focuses on systems engineering processes as they relate to maturing new technologies. He contrasts technology development driven from the bottom-up versus top-down. Timothy G. McGee and Bethany M. McGee, both of the Johns Hopkins University Applied Physics Laboratory, discuss *agile* systems engineering concepts and their applicability to micro- and nanoscale systems.

Part 2 describes the process of technology development on the micro- and nanoscale and introduces concepts and analyses that are important to consider during the development of micro- and nanoscale systems. Stergios J. Papadakis, researcher at the Johns Hopkins Applied Physics Laboratory, gives an overview of the changes in fundamental operating characteristics which will occur as current generations of macroscale devices evolve to microscale and then nanoscale, or as entirely new device concepts are devised based on novel nanoscale materials. Technology development processes for MEMS and nanotechnologies are reviewed with the objective of providing the systems engineer with insight that can be used in the requirements analysis and concepts and system design phases of programs. Robert Osiander, researcher at the Johns Hopkins Applied Physics Laboratory, provides an overview of MEMS from design through applications. Examples and lessons learned are given to aid in the analysis of the design for critical interfaces between micro and macro functionality, define the requirements for design (macro and micro) and validation, and establish a management process. The objective is to assist the systems engineer with the technical and management process when designing an instrument using MEMS. Jennifer L. Sample, researcher at the Johns Hopkins Applied Physics Laboratory, discusses systems derived from nanotechnology in which diverse nanostructures work together to perform a specific function, as well as the application of systems engineering to ultimately integrate nanoscale devices with associated required architectures into systems and associated services. Jeffery P. Maranchi, researcher at the Johns Hopkins Applied Physics Laboratory, provides a detailed examination of emerging top-down assembly technologies that have significant promise to tackle the challenges for nanoscale electronic and opto-electronic systems. A second focus of this chapter is the top-down assembly of nanoscale composite materials that are not electronic or opto-electronic in nature, but rather fulfill other functionalities in larger engineered systems (e.g., structural, thermal, or optical functionalities). The chapter will conclude with a brief examination of the top-down assembly processes and potential scalability associated with Nano Electro Mechanical Systems (NEMS). Jason Benkoski, researcher at the Johns Hopkins Applied Physics Laboratory, reviews the principles that form the basis for the field of nanoscale self-assembly and the various types of forces involved and the conditions under which they are most important. He defines self-assembly versus directed assembly in practical terms from the perspective of carrying out bottom-up assembly, and in scientific terms to describe the underlying causes for the observed behavior. The chapter concludes with a description of various efforts in nanotechnology that rely on bottom-up assembly as a manufacturing technique.

Part 3 describes tools and techniques that are unique to complex technology development and run parallel to systems verification and validation and risk management. Their value in providing critical data for key decision points in micro- and nanoscale technology development programs is discussed. Morgan Trexler and John Thomas, researchers at the Johns Hopkins Applied Physics Laboratory, review modeling and simulation in the micro and nano world. They discuss the advances in computational theory for the nanoscale, as well as remaining challenges and prospects for the future. Jennifer Breidenich, formerly of the Johns Hopkins University Applied Physics Laboratory, reviews the interface control mechanisms that need to be integrated into MNT systems to allow the user to obtain feedback from the system. She describes the core function of interface controls within systems engineering, which are used to establish infrastructure that allows for monitoring and evaluation of the health of a given system, including establishing a work breakdown structure, determining the configuration of the system, auditing progress, employing trade studies, and measuring performance through metrics. O. Manual Uy, a senior researcher at the Johns Hopkins Applied Physics Laboratory, discusses system reliability including quantification of reliability for simple and complex systems, influences on system reliability, and how to analyze the influences. He also discusses statistical concepts and quantitative predictive analysis as they relate to complex systems. William Paulsen, a senior engineer at the Johns Hopkins University Applied Physics Laboratory, discusses test and evaluation principles for complex systems using experiences from the realm of very-large-scale integration (VLSI) devices to new developments in micro- and nanoscale technologies. Janet L. Barth, Chief of NASA Goddard Space Flight Center's Electrical Engineering Division, describes processes for managing technology development programs focusing on special considerations for micro- and nanoscale technologies. She discusses issues integral to the conduct of a systems engineering program from planning to consideration of broader management issues. Product improvement strategies, organizing and integrating system development, and management considerations are covered. Team culture, technology readiness assessment, staff development, inputs to systems engineering for program life cycle, and quality assurance and reliability are addressed from the micro- and nanoscale technology viewpoint.

The first three parts of this book reflect progress on developing the first two generations of nanotechnology and introduce the next generations. In Part 4, the exploration continues into third- and fourth-generation developments. M. Ann Garrison Darrin, Managing Executive of the Johns Hopkins Applied Physics Laboratory's Space Department and Janet L. Barth, Chief of NASA Goddard Space Flight Center's Electrical Engineering Division delves into the nature of multiscale systems and introduces emerging trends in system engineering that can be applied to the challenges of developing micro- and nanoscale technologies. Issues in scalability that inhibit the success of multiscale systems are reviewed. Brian Jamieson, President of Scientific and Biomedical Microsystems, and Jennette Mateo, Virginia Commonwealth University, discuss the current state of nano- and microscale systems in the biomechanical, biomedical realm. The appropriateness of this discussion is noted by the first demonstrations of third-generation nanosystems seen in bioassay and lab-on-a-chip technology. I.K. Ashok Sivakumar, a researcher at the Johns Hopkins University Applied Physics Laboratory, continues the discussion relating self-assembly to the biological evolutionary systems. Pushing into the uncharted fourth generation, Bradley Layton, the University of Montana College of Technology, Department of Applied Computing and Electronics, covers theory of the design and construction of complex nanosystems and introduces concepts in mechanoevolution.

Summary

This book is an introduction to a field that has yet to mature. Not being micro- or nanoscientists or systems engineers, the editors recognized the need for a handbook that brings together the thoughts and experiences of the multiple disciplines required for micro- and nanoscale technology developments. This book includes contributions from systems engineers, physicists, material scientists, electrical and mechanical engineers, and technologists. It is neither a "deep dive" into systems engineering nor an in-depth study of micro- and nanotechnologies, but is rather an introduction for an emerging field. The purpose is to introduce the nano- and microengineers and scientists to basic systems engineering precepts and systems engineers to the world of the small. And finally, this book serves as a guide to technical and business program managers for developing and implementing robust micro- and nanoscale technology programs.

References

1. Baum, R., Chemical and Engineering News, "Nanotechnology: Drexler and Smalley make the case against 'molecular assemblers,'" Volume 81, Number 48, CENEAR 81 48, pp. 37–42, December 1, 2003.
2. Osiander, R., Darrin, M.A.G., and Champion, J., *MEMS and Microstructures in Aerospace Applications*, CRC Press, Boca Raton, FL, 2006.
3. Vandermolen, T.D., "Molecular Nanotechnology and National Security," LCDR, USN, Source: *Air & Space Power Journal*, August 31, 2006.
4. Roco, M.C., "Nanoscale Science and Engineering: Unifying and Transforming Tools," National Science Foundation, Arlington, VA, and "Nanoscale Science, Engineering and Technology (NSET) Subcommittee," U.S. National Science and Technology Council DOI 10.1002/aic.10087.
5. *Chemistry World*, "US roadmap for nano development," www.rsc.org/chemistryworld/News/2010/October/06101001.asp, October 6, 2010.
6. "Towards a Vision of Innovative Smart Systems Integration," EPoSS (The European Technology Platform on Smart Systems Integration), Vision Paper, 2006, www.smart-systems-integration.org/public/documents/publications.
7. Ottino, J.M., "Engineering Complex Systems," *Nature* 427, 399 (29 January 2004), doi:10.1038/427399a.
8. Allhoff, F., Lin, P., and Moore, D., *What Is Nanotechnology and Why Does It Matter? From Science to Ethics*, John Wiley & Sons, New York, 2010.
9. Discussion with David Kusnierkiewicz, the Johns Hopkins University Applied Physics Laboratory, Laurel, MD.

10. Stoney, G., "Top-down vs Bottom-up Design," GreatEngineering.net, http://greatengineering.net/System-Engineering/Architectural-Design/Top-down-vs-Bottom-up-Design.html.
11. http://www.iso.org/iso/home/htm
12. Shenhar, A.J. and Bonen, Z., IEEE Transactions on Systems, Man, and Cybernetics—Part A: Systems and Humans, Volume 27, Number 2, The New Taxonomy of Systems: Toward an Adaptive Systems Engineering Framework, p. 141, March 1997.
13. Definitions were taken from ISO/IEC 15288.

2

Introduction to Systems Engineering

Elinor Fong

"What you mean the specs
are being changed again?!?"

Contents

Introduction

It is difficult to pinpoint when the discipline of systems engineering began. Different sources cite different dates and attribute the term to different persons or organizations. The discipline and the term began gaining broader acceptance sometime in the late 1940s or early 1950s. However, the principles of systems engineering were being practiced long before the discipline was formalized. In fact, it has been said that systems engineering is just doing what good engineers do. Of course, that does not help those who are trying to learn how to become good engineers. Before trying to discuss the principles and processes of systems engineering, a definition of both a *system* and *systems engineering* should be established. In our context, a *system* is defined as two or more components interacting together to achieve a common objective. Finding a concise definition for systems engineering that captures the breadth of systems engineering is not easy. Combining several accepted definitions, including those in Chapter 1, *systems engineering* is basically an interdisciplinary approach to realize a system, which meets the customer's needs and considers operations, performance, manufacturing, test, cost and schedule, support and maintenance, and disposal throughout design and development. Despite the many definitions of *systems engineering*, there are some fundamental properties that are common. These are

- Develop requirements based on customer or user *needs*. It may be the case that the customer or user may not fully understand what is needed or know what is achievable. Part of systems engineering is to work with the customer to gain this understanding and translate it into system requirements. It is important to focus on what is needed before determining how to realize it. Determine the *what* before the *how*.
- Consider the life-cycle of the system throughout the design and development. Considerations such as how will the system be tested, how will it be used, what will be its environment, how will it be maintained, how will improvements be made, and how will it be phased out and disposed, should be thought of throughout the systems development process.
- Establish a disciplined approach to system development. This disciplined approach should contain reviews that relate back to the requirements, configuration management, risk management, and a means to iterate between the phases or activities within the approach. It should also include planning for the entire system development from the start. Initially, the plans may be a bit vague for the latter stages of the development, but thought needs to be given to all the stages, and the plans will become more detailed and finalized as the development progresses.

Life-Cycle Models and System Engineering Methodologies

Consideration of the life-cycle is fundamental. For those who are new to systems engineering, likely the concept of life-cycle is also new. A life-cycle of a system can be defined as the stages of a system's existence from concept to disposal. The International Council of System Engineering (INCOSE)[1] states that the purpose of defining the system life-cycle is

> [T]o establish a framework for meeting the stakeholders' needs in an orderly and efficient manner. This is usually done by defining life cycle stages, and using decision gates to determine readiness to move from one stage to the next. Skipping phases and eliminating "time consuming decision gates can greatly increase the risks (cost and schedule), and may adversely affect the technical development.

INCOSE[1] succinctly describes the six life-cycle stages defined by the International Standard for Systems and Software Engineering—System Life Cycle Processes, ISO/IEC 15288. Table 2.1 from the INCOSE System Engineering Handbook[1] lists the six stages and their purpose along with the decision options.

TABLE 2.1
Description of Life-Cycle Stages

Life-Cycle Stages	Purpose	Decision Gates
Concept	Identify stakeholders' needs; explore concepts; propose viable solutions	*Decision Options* – *Execute next stage* – *Continue this stage*
Development	Refine system requirements; create solution description; build system; verify and validate system	– *Go to a preceding stage* – *Hold project activity* – *Terminate project*
Production	Produce systems; inspect and test (verify)	
Utilization	Operate system to satisfy users' needs	
Support	Provide sustained system capability	
Retirement	Store, archive, or dispose of the system	

Note: See the International Council on Systems Engineering (INCOSE), *Systems Engineering Handbook, A Guide for System Life Cycle Processes and Activities,* INCOSE-TP-2003-002-03.1, August 2007.

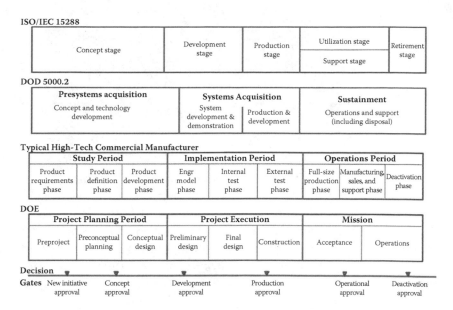

ISO/IEC 15288

Concept stage	Development stage	Production stage	Utilization stage	Retirement stage
			Support stage	

DOD 5000.2

Presystems acquisition		Systems Acquisition		Sustainment
Concept and technology development		System development & demonstration	Production & development	Operations and support (including disposal)

Typical High-Tech Commercial Manufacturer

Study Period			Implementation Period			Operations Period		
Product requirements phase	Product definition phase	Product development phase	Engr model phase	Internal test phase	External test phase	Full-size production phase	Manufacturing, sales, and support phase	Deactivation phase

DOE

Project Planning Period			Project Execution			Mission	
Preproject	Preconceptual planning	Conceptual design	Preliminary design	Final design	Construction	Acceptance	Operations

Decision						
Gates	New initiative approval	Concept approval	Development approval	Production approval	Operational approval	Deactivation approval

Figure 2.1 Life-cycle examples.

Some of these stages may overlap or be done in parallel. The system's exact life-cycle stages will vary based upon its application. For instance, if the technologies being employed in the system are well understood and have been used in other systems, the concept stage may be very short or nonexistent. However, if the technology is immature, such as a new nanotechnology, the concept stage may be lengthy, and the latter stages, such as support and retirement, may be ill-defined until much later in the process. In addition to identifying the life-cycle stages, decision gates need to be defined. At each decision gate, the decision options are the same, as listed in the Table 2.1. Figure 2.1 from the INCOSE *System Engineering Handbook*[1] compares the life-cycle stages of the ISO/IEC 15288 to other life-cycle formulations. There are similarities across the formulations, and the specific stages can be mapped to one another. Also included are typical decision gates, as identified by INCOSE.[1] The life-cycle most relevant to micro- and nanoscale technology (MNT) is the Typical High-Tech Commercial Manufacturer, but it still needs some modification. Figure 2.2 presents a suggested life-cycle and decision gates for a system based upon micro- and nanotechnologies.

Micro- and Nanotechnology System Life-Cycle

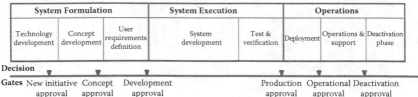

Figure 2.2 Micro- and nanotechnology life-cycle.

Note that stages within the system formulation may seem reversed. Ideally, user requirements definition (or establishment of the need) is first, followed by concept development, and then technology development. However, in the case of MNT, the technology is discovered and then concepts are developed to apply the technology to a customer or user need. This life-cycle formulation will serve as a guide to provide context for the systems engineering overview.

Based on the fundamental properties listed above, a system engineering process or methodology can be defined. There are several basic system engineering methodologies, such as the *waterfall*, the *vee*, the *spiral*, and the *defense acquisition* models. Each of these provides a general framework, but the methodology for any individual development needs to be tailored to the specific application. The waterfall methodology is well suited for introducing the principles of systems engineering. The other methodologies can be viewed as extensions or modifications to the waterfall methodology and will be introduced later in the chapter. Simplistically, the waterfall methodology can be broken into five activities:

- Requirements and conceptual design
- Design
- Implementation and integration
- Test and evaluation
- Deployment

Figure 2.3 shows how the systems engineering waterfall methodology aligns with the MNT system life-cycle. The traditional waterfall methodology does not have the iterations between phases or the feedback path. This could be viewed as

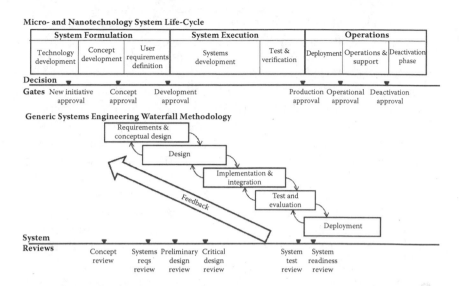

Figure 2.3 System life-cycle and systems engineering process relationship.

the "happy" path. However, over the years, the waterfall methodology has incorporated these to more appropriately reflect actual system development processes—the less "happy" but more realistic path. Also, note that the iterations are depicted to occur between adjacent activities. However, situations may occur that will necessitate going back more than one activity. For instance, an issue may arise in test and evaluation, which requires changes in the design or, perhaps, changes to the requirements. The further back one must go to resolve an issue, the larger the impact to cost and schedule. The same could basically be said for the iterations between adjacent activities. Generally, it is better to have the iterations occur earlier in the systems engineering process, for instance, iterating between the requirements and design rather than iterating later in the process, such as between implementation and test and evaluation. The system life-cycle and the systems engineering process have much in common and are even sometimes used interchangeably. However, the system engineering process and its activities support the different stages of the system life-cycle. As mentioned earlier, it is important to identify decision gates during the system life-cycle. The systems engineering methodology supports these decision gates through its system reviews. Figure 2.3 also shows the relationship between the decision gates and the system reviews. Typically, each of the system reviews has predefined entrance and exit criteria. These criteria should take into account the subsequent decision gates. The review should not be held prior to meeting all of the entrance criteria, and it should not pass until all of the exit criteria have been met. In some cases, where most criteria have been met but a few issues remain unresolved, a *delta* review may be held so that the entire review need not be repeated and closure can be achieved on the outstanding issues.

The technology and concept development stages of the system life-cycle are sometimes considered to be part of the requirements development of the system engineering process. However, in many cases, and likely for the applications of MNT, it occurs prior to a definition of sponsor or customer need or, perhaps, the conception of an actual system. Assuming that it does not hinder innovation, transitioning MNT or an emerging technology into an actual system is always easier if potential need is identified during the technology development. The technology development stage is exploratory and may consist of studies, experiments, and early prototyping or breadboarding. There may be competing or similar technologies that are attempting to quantify performance or capability of the technologies, and a "bake-off" or competition may decide which technology proceeds. As the technology matures and its performance and capabilities are better understood, the application of the technology can be explored, and transition to the concept development stage can begin. As noted earlier, although the diagram is drawn to show the life-cycle stages occurring serially, in reality, they overlap and some may be occurring in parallel with others. This is especially true for the technology development stage. The technology development and maturation may continue into the system development stage. Chapter 3 will discuss some commonly used standards for classifying the technology readiness or maturity.

The sponsor's or customer's need of the system must be defined prior to or during the concept development stage. The need could be a very broad, qualitative statement of the problem to be solved or could be very specific and quantitative. For example, the customer may ask for the best radar possible within certain size, weight, cost, and schedule constraints. Or the customer may specifically want radar that can detect and track particular threats by a given range and can support an existing weapon system with quantified requirements, in addition to having size, weight, cost, and schedule constraints. Reaching a clear understanding with the customer is extremely important. The customer need is what drives the subsequent system engineering process and decision gates; and if this is misunderstood, the consequences could be very costly and time consuming. It is also the job of the system engineer to ensure that the customer understands the implications of any constraints or potential conflicting requirements. For example, if the customer has strict size and weight constraints, the system engineer needs to make clear the impact of these constraints to the performance of the system. As depicted in Figure 2.3, the traditional systems engineering waterfall methodology typically begins during the concept development stage of the system life-cycle. Early in the process, a system concept review is held to determine the feasibility of the system and the risks and issues of meeting the customer's objective. This review will be part of the concept approval decision gate. The decision to proceed essentially starts the systems engineering process for the development of the system beginning with the requirements and conceptual design.

Description of Waterfall Methodology

Requirements and Conceptual Design

The system requirements are developed based upon the definition of the customer need, along with any standards that must be followed or project constraints. Initially, the system requirements may be broad and qualitative. Depending upon the application, a concept of operations may need to be developed in order to better define the system requirements.

Trade studies will likely be needed to better quantify the system requirements or to determine the feasibility or performance of technologies. For requirements where trade studies are needed, it is good practice to specify as much as possible and use "to be determined" (TBD) as an indication that quantitative results are needed for the requirement to be fully specified. As mentioned earlier, a principle of systems engineering is to consider the system life-cycle throughout development. Requirements development is no exception. Certainly, technical performance requirements must be developed, but any operational requirements along with maintenance and support requirements must also be considered. An assessment of technical, schedule, and cost risks should also be performed. Risk management will be discussed in more detail later in the chapter. It is important to begin the risk management (identification and mitigation of risks) early in the systems engineering process. During this phase, conceptual designs are developed. At a minimum, the system is decomposed into its subsystems or components; their interactions, performance, and functional allocations must be defined. Part of the requirements development is also to determine how the requirements will be verified—whether the requirement will be verified by inspection, analysis (includes simulation), demonstration, or test. The Department of Defense (DoD)[2] defines these verification methods as

Inspection: The visual examination of the system, component, or subsystem. It is generally used to verify physical design features or specific manufacturer identification.

Analysis: The use of mathematical modeling and analytical techniques to predict the compliance of a design to its requirements based on calculated data or data derived from lower-level component or subsystem testing. It is generally used when a physical prototype or product is not available or not cost effective. Analysis includes the use of both modeling and simulation.

Demonstration: The use of system, subsystem, or component operation to show that a requirement can be achieved by the system. It is generally used for a basic confirmation of performance capability and is differentiated from testing by the lack of detailed data gathering.

Test: The use of system, subsystem, or component operation to obtain detailed data to verify performance or
to provide sufficient information to verify performance
through further analysis. Testing is the detailed quantifying method of verification, and it is ultimately
required in order to verify the system design.

It is important to properly identify the verification method.
As stressed in the DoD *System Engineering Fundamentals
Guide*,[2] choice of verification methods must be considered an
area of potential risk. "Use of inappropriate methods can lead
to inaccurate verification. Required defining characteristics,
such as key performance parameters (KPPs) are verified by
demonstration and test. Where total verification by test is
not feasible, testing is used to verify key characteristics and
assumptions used in design analysis or simulation. Validated
models and simulation tools are included as analytical verification methods that complement other methods."[2]

It is worth a digression to discuss the difference between
verification and validation. The terms are used interchangeably by many, but in systems engineering there is a difference.
Verification is the process of showing that the system meets its
requirements. Verification typically occurs after components,
subsystems, and the system have been built. It answers the
question, "Was the system built right?" Validation is the process of showing that the system to be built will meet the customer's need and that the built system meets the customer's
need. Part of validation is the review and approval of the system requirements. Validation also occurs once the system is
built and installed in its operating environment. Validation
answers the question, "Was the right system built?"

Once the requirements are established, a systems requirements review (SRR) is held. The requirements review should
evaluate the system requirements and ensure that they are
complete, unambiguous, not in conflict with one another, feasible, verifiable, and flow from the customer's need. That being
said, many system requirements fail to meet one or more of
these. For instance, it is not uncommon to see a requirement
that states something shall "optimize" or "minimize." Although
such terms may be used in the customer's need or objective,

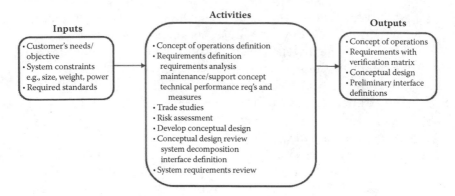

Figure 2.4 Requirements and conceptual design phase.

this is not a verifiable requirement, and terms such as these should be avoided when writing requirements. To do so, studies or experiments may be necessary to quantify the requirement. Ideally, at the end of the requirements development process, the systems requirements and the verification matrix will be complete. Also ideally, a conceptual design with functional, performance, physical decompositions, and interface definitions will be completed. For most applications, these will not be in their final version but will be well enough defined to proceed to the design phase. Iterations or modifications to the requirements, decompositions, and interface definitions will likely occur, but it should be recognized that the impact and risk increase if these get modified later in the design phase. Figure 2.4 summarizes the activities typically performed during the requirements and conceptual design phase along with the inputs and outputs.

Design

The next phase of the system engineering process is the design phase. The system requirements are flowed down to the subsystems and components. The conceptual design is further defined and the subsystems further broken down into smaller components. During this phase, trade studies that are assessing feasibility or evaluating technologies are complete, and their results are used to influence the design and update the requirements. Depending upon the maturity of the

technologies, prototyping of some of the subsystems or components may be done to reduce risk or as a proof of concept. Risk
assessment and mitigation will be discussed in greater detail
later in this chapter. Evaluation of commercial off-the-shelf
(COTS) items should also be done. When evaluating COTS
items, not only should the performance and technical specifications be considered, but also availability, replacement cycle,
and impact of technology refresh (improvement) need to be
considered. As mentioned earlier, life-cycle considerations are
made throughout the process. The design must not only meet
its performance requirements, but also meet any maintainability, reliability, and supportability requirements. COTS
items are no exception and often warrant additional scrutiny.
If the system under development must have an extended life,
any COTS items must support the system's life-cycle. How
long the COTS item will be available, how often it will need
to be upgraded, and its expected backward compatibility (the
ability of the future item to work with the older versions) are
all issues that need to be assessed. Hardware and software
requirements are defined in this phase. Depending upon the
complexity of the system, it may be necessary to have a preliminary design review (PDR) prior to completing the detailed
hardware and software specifications. The PDR examines the
design approach for the system and subsystems. The focus is
on the requirements compliance and ensuring the appropriate
flow-down of the requirements to the subsystems and hardware and software. Preliminary performance assessments
are evaluated through analysis or simulation to ensure that
the design can support the system performance requirements.
Once the preliminary design has been reviewed, the detailed
drawings for the hardware, the code specifications for the software, and the interface specifications can be completed. A critical design review (CDR) is held to determine that the system
can proceed to fabrication. It is very important to ensure that
all of the lower-level requirements, such as the hardware, software, and interface requirements all have traceability to the
system requirements and that all system requirements flow
down to the lower-level requirements. This helps to ensure
that the design of the system is appropriate and not over- or
underdesigned. The drawbacks for underdesigning a system

Figure 2.5 Design phase.

are obvious, but overdesigning a system also can cause issues. It can lead to additional costs in many areas throughout the system life-cycle, including fabrication, testing, maintenance, and support, and it can add additional and unnecessary risk. Certainly, growth and margin are appropriate design considerations, but it is good systems engineering practice to have them explicitly accounted for in the requirements. The results of SRR and, depending upon the complexity and risks of the system, the results of the PDR and CDR are used in the development approval decision gate. An important activity that should be done in parallel with the design development is planning for the implementation and integration of the system. The planning must ensure that the implementation and integration is properly sequenced and sufficient time is allotted for the procurement, fabrication, and component testing. Figure 2.5 summarizes the activities for the design phase.

Implementation and Integration

Once the design is complete and reviewed and the planning for implementation and integration is complete, then implementation can begin. Implementation includes fabricating the hardware, purchasing the COTS items, and coding the software. Testing of individual components is a key activity in system development and must be completed before integration of the system. Prior to formal component testing, it is usually beneficial to do some informal integration testing with the components or subsystems. Tests to ensure the compatibility of the hardware and software and initial interface testing, which verify the message exchange and timing, are examples of tests that can uncover potentially costly issues early in the program development. As components or subsystems are completed and tested, they are delivered along with the component test results for system integration. It is good practice to have at least one other person not involved in the design and implementation of the component participate in the component testing and the review of the test results. The test results should show that the component meets all of its requirements, and if there were any problems or issues, that they have been resolved and subsequently tested. As a result of any problem resolution, the design may need to be modified or, less likely, the requirements may need to be modified. Configuration management of the design and requirements, which will be discussed in more detail later, must be followed regardless of how small or minor the change may be. As components are delivered, preliminary integration of these components allows another level of testing of interfaces to verify the correct flow and use of the data. For complex systems, there will inevitably be some misinterpretation of the interface document that seemed unambiguous to developers on both sides of the interface, yet causes problems once integrated. Usually, the confusion is not in the interpretation of the data format, but rather in the behavior or functionality between the two interfaces. For instance, a common omission in the interface document is a description of the behavior when the component or subsystem receives unexpected data.

Planning for the system test and evaluation needs to be finalized during this phase. The method for verifying each of the system requirements should have been identified prior to the review and approval of the requirements. Each of the requirements and its verification method is examined, and a plan for verification of each of the requirements is determined. It is likely that many, if not most, of the requirements will be verified by test. Initial planning for the testing may begin as the verification methods are being developed, as early as during the requirements and concept design phase. Addressing any long-lead items, such as field test site arrangements or special test equipment procurement, also begins in this phase.

Figure 2.6 Implementation and integration phase.

The detailed planning of the system testing occurs during the implementation phase. Test plans, if necessary, are developed. The test plans should include the objective, state which requirements will be tested, and describe each test. Wherever possible, multiple requirements should be evaluated with one test, otherwise the time and cost for the testing could become excessive. Many times the schedule makes the assumption that the testing and the results of the testing will be as expected, because it is so difficult to plan for the unknown. However, for the schedule to be realistic and achievable, it needs to assume that all will not go as planned and to allow time to address any shortfalls discovered during the testing. The activities for the implementation and integration phase are summarized in Figure 2.6.

Test and Evaluation

During the test and evaluation of the system, two goals should be accomplished. One goal is the testing of all the requirements and verifying that they are met. The other goal is the evaluation of the end-to-end performance of the system. The tests and the collection of the data and metrics need to support both of these goals. Detailed test procedures must be defined prior to conducting the test and evaluation. They should include any system setup, environment requirements, data collection and storage requirements, and details of the test implementation. It is valuable to have both system designers and system testers involved in defining the test procedures. The system designers

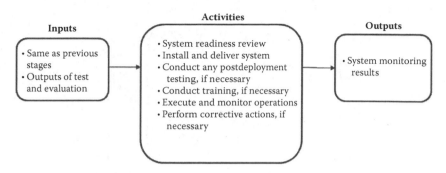

Figure 2.7 Deployment phase.

can ensure that the proper system setup is identified and that the appropriate functionality and components are stimulated. The system testers can ensure that requirement is fully evaluated and the appropriate metrics and data are collected. Once defined, the test procedures must be approved and put under configuration management. During the execution of the system test, problems will be identified and corrective action will need to be taken. If any problem discovered prevented the verification of a requirement, the test that uncovered the problem will need to be repeated after the corrective action is performed. Previous tests, which may include component testing, will need to be evaluated to determine if their results may have been impacted by the change. Some systems develop a set of regression tests that must always be performed after any changes have been done to ensure the change did not cause another problem. Any corrective action needs to be reflected in the design. Ideally, the corrective action should not impact the requirements, but it can happen, and if it does, the requirements must also be updated. One of the final reviews for a system is the system test review (STR). During this review, compliance with the test procedures, completeness of the test, the test results, and corrective actions are reviewed. A final review of the requirements verification matrix is also performed. For some systems, a few of the requirements may not be able to be verified until the system is deployed; for instance, some specific environmental conditions may not be able to be created until the system is installed and deployed. A system readiness review is also held prior to deployment (Figure 2.7).

Figure 2.8 Test and evaluation phase.

The purpose of this review is to ensure that all open issues from previous review have been resolved, and the system risks have been mitigated and closed. In addition, review of any operations procedures, user manuals, training materials, and shipping plans may be included. The results of the STR and the system readiness review results are the basis of the operational approval decision gate. Figure 2.8 contains the test and evaluation diagram.

Deployment

Although this is probably the most exciting and rewarding phase of the system development, there is not much that can be described in a general overview. The activities that occur in this phase are very system specific. In general, the system is installed, and postdeployment verification is conducted. For instance, if a requirement states that the system, installed on a ship, must have a certain level performance in bad weather with high sea-states, verification of this requirement (by queasy testers) may not be possible until after deployment. Required training of persons operating the system also occurs. Once the system is deployed, its operation is monitored, and its performance is assessed. Hopefully, by following the principles of system engineering, which include consideration of the system life-cycle, minimal or no corrective actions will be necessary during this phase. Figure 2.7 summarizes the activities of the deployment phase.

System Engineering Support Processes

Configuration management and risk management are two critical processes that support systems engineering that should be considered. They are part of the rigor underlying systems engineering and are essential in the development of any large, complex system.

Configuration Management

Typically there are several teams when developing a large system: hardware design teams for the various subsystems or components, software development teams, and test and evaluation teams. Each of these teams works in parallel, making decisions and contributing to the evolution design. Trying to discern what changes were made, when they were made, and their impact rapidly becomes quite untenable. There are two components to configuration management. One component is the creation and maintenance of a record of changes throughout the entire system life-cycle. The other component is establishment of a process to control the changes, typically referred to as *change control*. Change control, which generally includes an approval process, ensures that impact of the change is fully evaluated before the change is made. Part of the approval process is the demonstration that the technical, schedule, and cost impacts of the change are understood. At the beginning of a system development, a configuration management plan should be established. The configuration management plan should identify the items, such as requirements, design, interfaces, control documents, and products, which will be placed under configuration management. It also establishes the change process by addressing the following questions:

1. When will the items be controlled? At what point will the item be ready to be placed under configuration management? It does not make sense to place the item under configuration management until it is mature

enough to establish a baseline that has been reviewed and approved.

2. How will changes be initiated? Is a trouble report written? Is a meeting or review convened?

3. Who will approve changes? This will likely depend upon the item being changed and the extent of the change. If the change is confined to one component, perhaps only the lead engineer of that component and the lead system engineer need to approve the change. For some systems, an Interface Control Working Group is formed to evaluate any changes that interface between subsystems. Usually, the lead system engineer (and program manager, if cost or schedule is impacted) must approve any change.

4. How will these changes be documented and tracked? It is important that the changes be tracked so that the entire team is aware of them. A process for disseminating the change information and releasing an updated revision is established. It is also good practice to track the changes in a database containing the item being changed, a description of the change and its impacts, the approval date, and notation of who approved the change.

5. Is there a means or a mechanism to revert back to a previous configuration? This can be particularly useful when trying to isolate a problem or evaluating a solution.

Obviously, there is a balance between having the flexibility to allow for the evolution of the system and maintaining the rigors of configuration management. As the system design matures, the impact of a change tends to increase the technical risk and cost and pushes out the schedule. These must be weighed carefully. The configuration management process should enable well-informed decisions to be made and help to mitigate the impacts.

Risk Management

In developing any new system, there are uncertainties that pose risk to performance, cost, or schedule. Throughout the life-cycle of the system, risk management consisting of

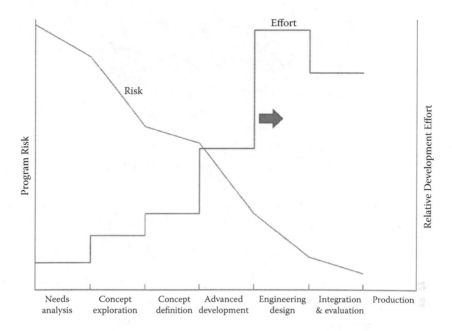

Figure 2.9 Variation of program risk and effort throughout system development. (Kossiakoff and Sweet, *Systems Engineering, Priciples and Practice,* John Wiley & Sons, New York, 2003. With permission.)

identification, tracking, and mitigation of the risks is performed. In general, there are more risks at the beginning of system development, but their resolutions have less impact on schedule and cost. Further into the system life-cycle, it is expected that there are less risks, but their potential impacts are greater and more costly to address. Figure 2.9 from *Systems Engineering, Principles and Practices* by Kossiakoff and Sweet,[3] illustrates this.

The two main components of risk management are risk assessment and risk mitigation:

Risk assessment: Periodically, assessments of the risks to the system are performed. At the start of a system development, there are many uncertainties regarding the technologies and capabilities, so there will likely be several technical risks. As the system development progresses and modifications or changes are proposed, risks to cost and schedule may increase. For each risk identified, the

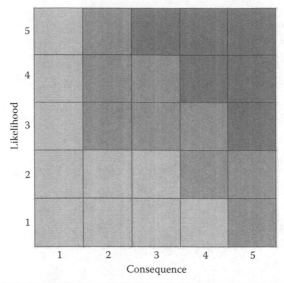

Risk Index	Risk decision criteria
High Risk	Unacceptable, implement new process or change baseline
Medium Risk	Aggressively manage; consider alternate process
Low Risk	Monitor

Figure 2.10 Risk Assessment Matrix example.

likelihood of occurrence and the impact (technical, schedule, and cost) of the risk should be assessed and then categorized by severity. Figure 2.10 is a common risk assessment chart used by many DoD programs.

The chart helps to assess the overall severity of the risk. The exact classification of each box will be program specific, but the idea is well illustrated by the figure. An item would be a very high risk if the likelihood was probable and the impact or consequence was significant or catastrophic. An item would be low risk if the likelihood was negligible and the impact insignificant or minor. As the system development progresses, it is expected that the risks should move to the left and downward. Some programs use arrows to indicate the trend of the risk. Items that are in the two darker gray regions or have an upward trend should have a risk mitigation plan associated with them.

Risk mitigation: For each risk identified, determination of how the risk will be mitigated should occur. For low-risk items, it may be as simple as program management oversight. For high- or very-high-risk items, a risk mitigation plan is warranted. The plan outlines actions to be taken to reduce the likelihood of occurrence and also actions that lessen the impact of the risk if it occurs. Note that the mitigation need not necessarily address both the likelihood and impact, but the expected outcome should be to reduce the overall risk.

Periodic reviews to track the current risks and identify any new risks are performed. Evaluation of the risk mitigation for each item should also be reviewed. Obviously, if the risk mitigation is effective, the risk should be decreasing in likelihood and consequence. A successful system development will have closed most of its risks and have only a few low risks remaining toward the end of its development.

Other Systems Engineering Methodologies

As mentioned earlier, there are other basic system engineering methodologies, as illustrated in Figures 2.11 to 2.13.

These methodologies contain many of the same steps or phases as the waterfall methodology, and much of the difference lies within how much iteration within or between steps is expected. Among the methodologies the phases are relatively similar, but the feedback mechanisms and how the process may be iterated are different. More detailed discussions on the Vee and Spiral methods[4] and on the defense acquisition systems engineering methodology[2] can be found in the literature. The methodology that best applies to a given application will depend upon many factors, but the Spiral and defense acquisition methodologies tend to be better suited for situations where the system objectives are less well defined, such as for proof-of-concept or prototype developments. Although all of these models accommodate iterations between activities, they are basically top-down approaches, meaning that

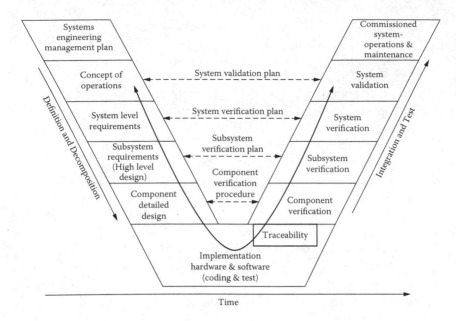

Figure 2.11 The Vee systems engineering methodology.

the system objective or need is the driver or initial input to the process. There are cases in which the top-down approach is not best suited for the application. However, it is good practice to start with a top-down perspective and then modify or constrain the process based on the application. For instance, due to cost or time constraints, a system may need to use existing components and technologies. This may lend itself better to a bottoms-up approach, because the system capabilities and performance will be driven by these existing components and technologies. But usually, it is still possible to define the system objective and requirements while constraining it to the existing components and technologies. This maintains the rigor of systems engineering by ensuring that those portions of the system that can be are driven by the top-down approach while accommodating those that are driven by a bottoms-up approach. This also encourages the developers to evaluate the system as a whole, rather than as individual components, from the start. It is important to note that all of the systems engineering methodologies have

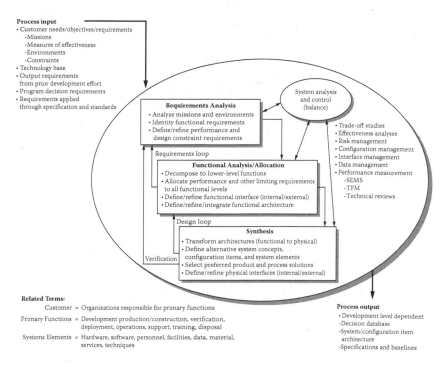

Process input
- Customer needs/objectives/requirements
  - Missions
  - Measures of effectiveness
  - Environments
  - Constraints
- Technology base
- Output requirements
  from prior development effort
- Program decision requirements
- Requirements applied
  through specification and standards

System analysis and control (balance)
- Trade-off studies
- Effectiveness analyses
- Risk management
- Configuration management
- Interface management
- Data management
- Performance measurement
  - SEMS
  - TFM
  - Technical reviews

Requirements Analysis
- Analyze missions and environments
- Identify functional requirements
- Define/refine performance and
  design constraint requirements

Requirements loop

Functional Analysis/Allocation
- Decompose to lower-level functions
- Allocate performance and other limiting requirements
  to all functional levels
- Define/refine functional interface (internal/external)
- Define/refine/integrate functional architecture

Design loop

Synthesis
- Transform architectures (functional to physical)
- Define alternative system concepts,
  configuration items, and system elements
- Select preferred product and process solutions
- Define/refine physical interfaces (internal/external)

Verification

Related Terms:
Customer = Organizations responsible for primary functions
Primary Functions = Development production/construction, verification,
  deployment, operations, support, training, disposal
Systems Elements = Hardware, software, personnel, facilities, data, material,
  services, techniques

Process output
- Development level dependent
  - Decision database
  - System/configuration item
    architecture
  - Specifications and baselines

Figure 2.12 Defense acquisition systems engineering methodology.

paths back to previous activities. These iterative paths are there to accommodate changes and corrective actions and to provide a mechanism to incorporate lessons learned. The traditional waterfall process did not have these feedback mechanisms, but updated versions have incorporated them. The feedback or iterations gives the system engineering process the flexibility to evolve and to address issues that arise. For some systems, particularly those that are explorative in nature, the iterations are expected and, therefore, planned for in terms of cost and schedule. For other systems, the iterations may be a result of an unanticipated design change or issue and may impact cost and schedule. Managing the feedback and iterations using the other elements of systems engineering principle, such as reviews, configuration management, and risk management, will help maintain the proper balance.

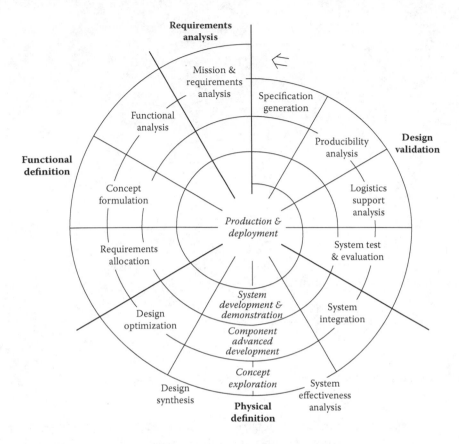

Figure 2.13 Spiral systems engineering methodology.

Development of Systems Engineers

One final note is an observation about how systems engineers are developed. Prior to having a formal discipline called *systems engineering*, system engineers studied and were trained to be another type of engineer, such as an electrical or mechanical engineer. Typically, they grew their expertise in a somewhat focused technical area and over time became exposed to broader areas. For instance, a radar systems engineer may have started by being a hardware design engineer and was involved with the testing and integration of the hardware component of the radar. On a later project, the same engineer may have become the lead hardware engineer and would have been involved in the flow down of the system requirements to the hardware design giving the engineer the experience with requirements development. The lead hardware engineer would also need to interact with the lead software engineer, gaining an understanding of the issues and concerns involved with software development. In addition, in this role, the lead hardware engineer would have been involved in the overall system integration and testing. If field testing occurred, the lead hardware engineer would have also gained experience in how the system would be deployed and would have interacted with the users. These experiences were instrumental in developing the hardware design engineer into a radar systems engineer. Some challenges in this approach are ensuring the engineer takes the broader perspective and also applies the systems engineering rigor. Systems engineering is now a formal discipline, largely offered as a master's degree but also offered as an undergraduate degree at some universities. An early-career systems engineer will undoubtedly be familiar with the systems engineering methodologies and rigor, but may not be likely to have the in-depth technical knowledge about the specific system or component. Their technical knowledge may be developed through the requirements analysis or system verification process. The challenge in this approach is gaining the technical knowledge and judgment so that risks can be appropriately identified, trade-offs can be made, and the performance can be assessed. The first approach can be considered as a bottoms-up approach and the second as a top-down approach. In this case, the terms are used for the development of a systems engineer rather than the system engineering process. This will not be the last time the reader will see these well-worn, yet appropriate, terms applied. As with the other applications, different strengths and weaknesses may result depending upon whether the bottoms-up approach or the top-down approach is chosen. For example, the systems engineer developed following the bottoms-up approach will have the technical knowledge to judge risks and system performance but may not develop the system as efficiently or well managed, as possible. On the other hand, the system engineer developed by the top-down approach may be too focused on the process and may not appropriately balance risk and performance. Regardless of the approach taken, these weaknesses can be mitigated by gaining experience and adhering to the system engineering principles.

References

1. International Council on Systems Engineering (INCOSE), *Systems Engineering Handbook, A Guide for System Life Cycle Processes and Activities*, INCOSE-TP-2003-002-03.1, August 2007.
2. Department of Defense Systems Management College, *Systems Engineering Fundamentals*, Defense Acquisition University Press, January 2001.
3. Kossiakoff, A. and Sweet, W., *Systems Engineering, Principles and Practice*, John Wiley & Sons, New York, 2003, Chapter 4.
4. Blanchard, B. and Fabrycky, W., *Systems Engineering and Analysis*, 4th ed, Pearson Prentice Hall, Upper Saddle River, NJ, 2006.

3

Systems Engineering in Technology Development Phases

D.Y. Kusnierkiewicz

When Capabilities don't meet Requirements.

Contents

Introduction

Systems engineering has evolved into a formal discipline with the flexibility to adapt to various modes of system development. But some specific core functions are always present: define, control, validate, and verify requirements; determine system architecture; define and control interfaces; and so forth. When introducing something truly new into a development effort, such as micro- or nanoscale technologies (MNT), the natural tendency is to manage the technology and the overall system developments in the usual, established way. This may result in a "forced fit" at first. As experience accrues, it may become evident that a different approach is needed. This chapter will look at the potential impacts of using early-generation MNT in a system and the application of traditional technology development strategies.

Incorporating new technology into a system development can be a double-edged sword. New technology can be enabling, allowing realization of the system, and often making more efficient use of technical resources (e.g., mass, power) while achieving higher performance than is possible using an existing, mature technology. However, depending on its maturity, incorporating new technology may present considerable programmatic risk. Can the technology be brought to the required level of maturity in time for integration into the system or system component? Will it perform to expectations? Delays in technology development can be extremely costly. At a minimum, the overall system development schedule is adversely affected—sustaining support of the "marching army" to complete and deliver the system increases cost. At the worst extreme, the development fails to achieve the desired end result, requiring re-design of the system or a part of the system, consuming additional technical and programmatic resources, and perhaps compromises in system performance as well.

Technology incorporation can come from the bottom-up (technology push) or the top-down (technology pull). Whatever their origins, technologies that are developed with a targeted system application have the advantage of established requirements, or at least goals, for the new components to guide the technology development. Not only can performance specifications be provided, but also targets for technical resource allocations and operating environment parameters can all be given consideration from the start. For example, President Ronald Reagan's Strategic Defense Initiative for missile defense required the development or significant advancement in multiple new technologies: target identification and tracking in multiple phases of intercontinental ballistic missile (ICBM) flight, discrimination of target versus decoy, directed energy weapons and kinetic kill vehicles, and so forth. Strawman system architectures were developed that provided context for the development of sensors to perform the target discrimination and tracking functions.

On the other hand, invention may yield technology that presents opportunities for applications not foreseen during development. But if a specific end use is not considered during the development phase, additional work may be required. For instance, the performance of the electronics used in spacecraft does not match state-of-the-art commercial devices. Commercially developed devices are not aimed at application in the space environment, which can be extremely harsh in many respects, such as the radiation or launch environment. Also, many commercial devices are meant for application in mass-produced units that are replaced in the field upon failure and do not have the required reliability for a long duration space mission where repair or replacement (and failure) is not an option.

Assessment of Technology Maturity

The organization that first employs a new technology for a given application provides new options for subsequent projects. But the first project pays a price for the benefits they and subsequently others realize. They assume the burden of qualifying the technology and must assume and mitigate the risk that the technology may fail to achieve its promise. The National Aeronautics and Space Administration (NASA) and the Department of Defense (DoD), to try and address the problem of incorporating new technology into systems, have established a common taxonomy for classifying technology maturity, or *readiness*. The Technology Readiness Levels (TRLs) are defined in Table 3.1 from least to most mature.

In order to minimize the risk of a project becoming derailed by immature technology, new technology is typically expected to achieve TRL 6 by the time of the project preliminary design review (PDR). Technology development plans are established at the outset, and progress against them is assessed throughout the preliminary phases. Alternate methods may be identified to mitigate any risks associated with new technology. Additional resources required to pursue alternatives are usually identified, and "trigger dates" are established for decision points. Preliminary design review is chosen as the gate, because the monetary investment in a project is (relatively) minimal until then. After PDR, the spending profile increases dramatically. If a technology has not matured to the appropriate level by then, the project may be terminated, or alternative implementations that can be implemented with existing, proven technology may be pursued instead.

The synthesis of MNTs and macrosystems presents new challenges to this technology maturation model. Among them are the "critical scaling" issues referenced in the "Hardware Description" entry of TRL 6 of Table 3.1, and potential difficulties in conducting "tests in relevant environments" to validate analytical models and simulation results. (Scaling issues are discussed in Chapter 5; modeling and simulation are discussed in Chapter 10.)

TABLE 3.1
Technology Readiness Levels (TRLs)

TRL	Definition	Hardware Description	Software Description	Exit Criteria (to Next TRL Level)
1	Basic principles observed and reported	Scientific knowledge generated underpinning hardware technology concepts/ applications	Scientific knowledge generated underpinning basic properties of software architecture and mathematical formulation	Peer reviewed publication of research underlying the proposed concept/ application
2	Technology concept or application formulated	Invention begins, practical application is identified but is speculative, no experimental proof or detailed analysis is available to support the conjecture	Practical application is identified but is speculative, no experimental proof or detailed analysis is available to support the conjecture; basic properties of algorithms, representations, and concepts defined; basic principles coded; experiments performed with synthetic data	Documented description of the application/ concept that addresses feasibility and benefit
3	Analytical and experimental critical function or characteristic proof of concept	Analytical studies place the technology in an appropriate context and laboratory demonstrations, modeling, and simulation validate analytical prediction	Development of limited functionality to validate critical properties and predictions using nonintegrated software components	Documented analytical/ experimental results validating predictions of key parameters

TABLE 3.1 (*Continued*)
Technology Readiness Levels (TRLs)

TRL	Definition	Hardware Description	Software Description	Exit Criteria (to Next TRL Level)
4	Component or breadboard validation in laboratory environment	A low fidelity system/component breadboard is built and operated to demonstrate basic functionality and critical test environments, and associated performance predictions are defined relative to the final operating environment	Key, functionally critical, software components are integrated, and functionally validated, to establish interoperability and begin architecture development; relevant environments defined and performance in this environment predicted	Documented test performance demonstrating agreement with analytical predictions; documented definition of relevant environment
5	Component or breadboard validation in relevant environment	A medium fidelity system/component brassboard is built and operated to demonstrate overall performance in a simulated operational environment with realistic support elements that demonstrate overall performance in critical areas; performance predictions are made for subsequent development phases	End-to-end software elements implemented and interfaced with existing systems/simulations conforming to target environment; end-to-end software system, tested in relevant environment, meeting predicted performance; operational environment performance predicted; prototype implementations developed	Documented test performance demonstrating agreement with analytical predictions; documented definition of scaling requirements

6	System/subsystem model or prototype demonstration in an operational environment	A high fidelity system/component prototype that adequately addresses all critical scaling issues is built and operated in a relevant environment to demonstrate operations under critical environmental conditions	Prototype implementations of the software demonstrated on full-scale realistic problems; partially integrated with existing hardware/software systems; limited documentation available; engineering feasibility fully demonstrated	Documented test performance demonstrating agreement with analytical predictions
7	System prototype demonstration in an operational environment	A high fidelity engineering unit that adequately addresses all critical scaling issues is built and operated in a relevant environment to demonstrate performance in the actual operational environment and platform (ground, airborne, or space)	Prototype software exists having all key functionality available for demonstration and test; well integrated with operational hardware/software systems demonstrating operational feasibility; most software bugs removed; limited documentation available	Documented test performance demonstrating agreement with analytical predictions

TABLE 3.1 (Continued)
Technology Readiness Levels (TRLs)

TRL	Definition	Hardware Description	Software Description	Exit Criteria (to Next TRL Level)
8	Actual system completed and "flight qualified" through test and demonstration	The final product in its final configuration is successfully demonstrated through test and analysis for its intended operational environment and platform (ground, airborne, or space)	All software has been thoroughly debugged and fully integrated with all operational hardware and software systems; all user documentation, training documentation, and maintenance documentation completed; all functionality successfully demonstrated in simulated operational scenarios; verification and validation (V&V) completed	Documented test performance verifying analytical predictions
9	Actual system flight proven through successful mission operations	The final product is successfully operated in an actual mission	All software has been thoroughly debugged and fully integrated with all operational hardware/ software systems; all documentation has been completed; sustaining software engineering support is in place; system has been successfully operated in the operational environment	Documented mission operational results

Source: From NASA Procedural Requirements 7120.8, NASA Research and Technology Program and Project Management Requirements, Effective Date: February 05, 2008, Appendix J.

New Technology Risk Mitigation and System Development Methodologies

"Successful" system development, viewed from a technical perspective, would be concerned with how the completed system performs to specification, with the ease expected by its users, within the resources allocated. However, the sponsor who funds the system development effort will also consider "success" to entail on-time delivery, for the agreed-upon cost (or less). Note that the "sponsor" may be the system developer's own organization, in the case of an internal development to bring a new system to market. Successful program management then must manage programmatic risks, as well as risks to technical performance.

The largest risk associated with incorporating new technology into a system may be more programmatic than technical. The new technology may enable higher system performance. In the best case, this higher performance would also consume fewer system resources and take less time and money to produce. But, the path to such success may be filled with time- and money-consuming false starts and dead ends. How does one *manage* such risks (as opposed to merely *monitoring*)? Closing the technology readiness gap between low TRLs and TRL 6 remains a murky challenge to the underfunded technologist, who may not understand all the requirements of a potential system application. Understanding these requirements is key to ensuring the invention meets the needs of the system. The system engineer and project manager, anxious to realize the potential system benefits of a new technology, also need to manage the development to ensure successful incorporation into the system, in a timely manner.

One of the basic tenets of system engineering focuses on *interfaces* between system components (see Chapter 11). This is where the system engineer "lives." Interfaces between components are established, and then *controlled*. If each system component is regarded as a "black box" with defined performance requirements and interfaces, then the contents of the black box may change, as long as the performance and interface requirements are met.

Classical system engineering is largely a requirements-driven activity (see Chapter 2), although this process may also work in reverse. This reverse approach, sometimes referred to as "capabilities-driven" development, may start with an existing component of known *performance and interface requirements*; the system can then be designed around this component.

A similar approach may be taken when having to develop a system component based on a new technology. *If* the performance and the interfaces to the new technology component can be defined, and "frozen" at the appropriate time in the development life cycle, the classical waterfall system development approach may work with an acceptable level of programmatic and technical risk. Alternate designs that do not rely on the new technology may also need to be identified, depending on the level of confidence that the new technology will work as advertised. The system engineer and project manager also need to determine what events may trigger the abandonment of the technology development in favor of implementing an alternate approach to avoid protracted schedule delays and cost overruns. Another approach, if confidence in the technology development is low is as a technology development (or risk reduction) phase that *precedes* the rest of the system development. The development and evaluation of prototypes, models, and simulations are typical risk reduction activities (see Chapter 10), whether a project uses a waterfall development, or an iterative "agile" approach.

Waterfall System Development

Real-world system development rarely follows faithfully any one idealized, textbook methodology. The linear "waterfall" system development methodology discussed in Chapter 2 is common, although it rarely proceeds in such an orderly, monotonic fashion. However, it suffices for large systems with a significant hardware component employing mature technologies and with well-defined requirements at the outset. It may even be employed when the requirements are initially not well defined. In such cases, successive iterations through the process result in requirements refinement. But this methodology

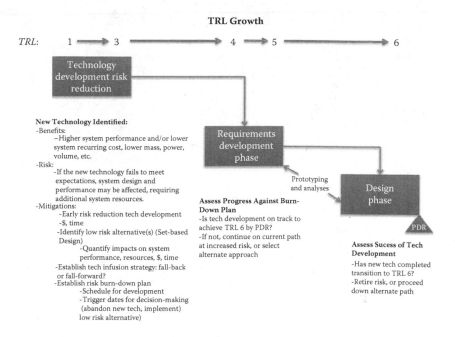

Figure 3.1 Traditional technology growth in waterfall life cycle.

may not be well suited to system development that plans to incorporate immature technology, although adjustments to the process may be made to address such risks. Figure 3.1 depicts the desired TRL growth through the early phases of a project.

For instance, consider the example of a system for a space mission. A common system trade involves mass and power. The power available comes from solar arrays. Increased power demand by the spacecraft may result in growth to the solar arrays that then require more mass. If sufficient mass is not available to support solar array growth, the system engineer will need to constrain growth in the system power demand. One power-saving option the engineer may consider is the use of flash memory technology for on-board data storage. The advantage of such memory is that it is nonvolatile; the memory contents are not lost if power is interrupted or removed. However, the flash memory is more susceptible to the space radiation environment. Therefore, if the project wishes to take advantage of newer technology, higher-density devices that have never flown before in space, the technology will need to be qualified.

The project then must qualify the devices, demonstrating TRL 6 by the time of the project PDR. From Table 3.1, TRL 6 requires a system/subsystem model or prototype demonstration in a relevant environment. This would typically require that a prototype of the memory circuit board intended for flight be built and "demonstrated in a relevant environment." The principal relevant environment is the space radiation environment, characterized by a variety of different types of radiation: the total dose of ionizing radiation (usually measured in kilorads, Silicon), neutrons, protons, and other heavy ions. The operating prototype would be subjected to a variety of radiation sources on the ground to demonstrate an acceptable incidence of induced bit errors and the lack of any destructive latch-up. Because the relevant environment also includes the mechanical stresses induced by launch and the thermal environment, such qualification may also involve subjecting the prototype to "shake-and-bake"—tests on a vibration table, followed by thermal cycling in a thermal-vacuum chamber that simulates the thermal and/or radiation environment of space where heat is transferred by conduction as opposed to convection.

At the successful conclusion of such tests, the technology has achieved TRL 6. Note that deficiencies may be exposed in such testing, which may require mitigation. Such mitigations may involve adding additional shielding mass around the devices to limit the total ionizing radiation dose during the mission, or latch-up protection circuitry to limit the input current (preventing device destruction) and to cycle power to the device if it does latch up when hit with a heavy particle.

Now, the system engineer will need to plan for the contingency that his first-choice option, the flash memory mass storage, may not succeed. He or she may consider alternative designs and their impact on spacecraft resources and performance. Prudence may dictate that several parallel paths be pursued until a decision can (or *must*) be made.

Testing MNT or macrosystems employing MNT may not be as straightforward as with traditional systems. Producing and testing a high-fidelity prototype that successfully addresses all "critical scaling" issues (Reference Table 3.1, TRL 6) may be difficult or very costly to realize. Modeling and simulation may prove a more cost-effective approach. Even though this may not completely satisfy the traditional criteria for TRL 6, it may provide sufficient confidence to proceed with system development at an acceptable level of risk (see Chapter 10).

Agile Development

The discipline of software engineering has given rise to several *agile* development models, such as the Spiral (or Boehm) and Scrum models.[2] These models address difficulties encountered using the waterfall approach when undertaking a major software development. Even though development using the waterfall method is regarded as being more easily managed, it is also regarded as being less efficient. Each phase must be considered complete (i.e., as bug-free as possible) before moving to the next phase, so that flaws do not propagate forward.

Agile models progress through multiple iterations of requirements development and analysis, design, prototyping, test, and evaluation (Figure 3.2). (The team structure has been described as cross-functional, closely knit, and *self-organizing*.[2]) Where agile development models are applied to a "system" development (hardware and software, or hardware only), they tend to be small systems, developed by small, close-knit teams. In these cases, requirements may not be well defined at the outset, so prototypes might be constructed and evaluated, leading to a refinement of requirements, back to a new prototype for evaluation, and so forth. The agile development model may be appropriate when incorporating immature technology in general and MNT in particular into a system, but the actual execution may need tailoring. For instance, can the MNT be iterated as quickly as the other system elements? If so, then the process may execute with little to no modification.

However, the development of new technology is often not "rapid." It is very likely that the system component using the new technology will not be available for evaluation until the later passes through the spiral. If it is not known until late in the development cycle that the new technology does not meet all the expectations and requirements, changing to an alternate implementation may be very costly. If the MNT cannot be

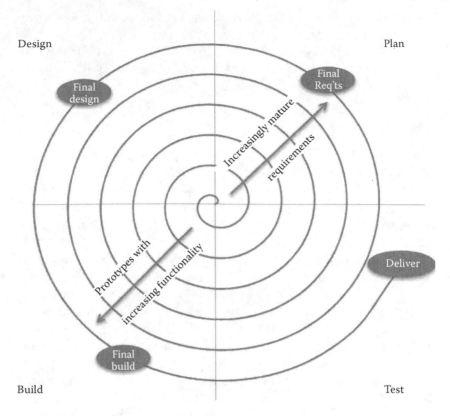

Figure 3.2 Agile (Spiral) development model.

iterated as quickly as the other elements, then the risk with this approach increases. If the element with the MNT cannot be modeled as a black box, with *bounded performance characteristics* and *well-defined interfaces that can be frozen*, then proceeding with the rest of the system development may result in wasted effort and cost. In this case, a hybrid development model may be necessary where the main system development using the agile model should be preceded by a technology development risk reduction phase using the waterfall model, as previously discussed. Regardless, a point (or interval) in the development effort where the new technology must be sufficiently mature needs to be identified. Technology maturation should be monitored so that alternatives can be exercised in a timely manner, if necessary.

Evaluating the Risk of Immature Technology

Even though the risks and benefits of reliance on an immature MNT technology must be evaluated case by case, some general guidelines may be beneficial to both the MNT technologist and the system engineer for deciding when to apply the various risk mitigation techniques. These guidelines are provided in Table 3.2.

TABLE 3.2
Guidelines for Evaluating Alternate Approaches for Incorporating an Immature Micro- and Nanoscale Technology (MNT) System Component

Consideration	Guideline
1. What is the preferred development methodology? (Waterfall versus Agile)	If agile is preferred, make sure the technology development iterations are consistent with the system development iterations; otherwise, consider a waterfall or hybrid model with a dedicated technology development phase that precedes the system development.
2. What is the Technology Readiness Level (TRL) of the MNT?	For TRL < 6, develop Technology Development Plan. TRL 1 to 3 may necessitate a technology development phase that precedes the main development effort. In some cases, additional mitigations may be needed (e.g., identify an alternate design to replace the immature MNT, and identify impacts on technical and programmatic resources).
3. Can the interfaces between the MNT and the system components be defined and "frozen"?	If they can, and if the risk is low of timely MNT maturation, then it may be low risk to proceed concurrently with system development. If not, a precursor risk-reduction technology development phase may be warranted.
4. Can the performance of the MNT component be bounded?	If not, a precursor technology development phase may be warranted.

Case Study: Self-Healing Paint

Consider the application of an existing first-generation nanotechnology self-healing paint. The paint contains, in effect, nanoscale sealed "cans" of wet paint within the dry paint coat applied to a surface. When the surface of the paint is scratched, these "cans" of paint rupture, sealing over the scratch, and restoring the integrity of the painted surface. The paint was invented with the notional application to naval ships. The self-healing nature of the paint has the potential to reduce the frequency of new paint applications, thus lowering maintenance costs and downtime. A set of firm requirements was not generated for the nanotechnologist inventor, but the invention was based on existing paint for such an application. Therefore, the expectation is that the paint is compatible with existing techniques and materials such as primers.

The new properties of this paint may be of interest to other applications that were not considered from the outset. For instance, might this type of paint also be useful in a space application for a thermal-control surface, such as a radiator or reflector? Such a coating may be black (radiator) or white (reflector). The self-healing nature of the paint holds promise that the thermal properties of the treated surface do not degrade over time as much as existing paints. What properties are important in such an environment? The following illustrates the properties that would require evaluation before incorporating the paint into the baseline system design.

Requirements on Self-Healing Paint for Space Application

Two potential applications are as follows:

1. White paint for parabolic antenna
2. Black paint for radiator/external electronic box

(Another potential application might be for optics baffles. For an optical baffle, many of the same properties below must be met. The resulting painted surface is to be "flat" without specular reflections. Self-healing properties would be very useful in that handling damage would be self-repairing.)

Assume an eclipsing low earth orbit application, 100-minute period, 5,256 orbits/year.

Paint must be compatible with surface treatment for aluminum (alodine, iridite) or composite (e.g., graphite epoxy) structure.

Unless stated otherwise, properties are to be met after exposure to the mission environment. Therefore, in order to qualify the use of the paint, a test program would be conducted to demonstrate compliance with requirements. Properties such as emissivity and absorptivity, surface resistivity, adhesion, and compatibility with surface cleaning agents would be measured prior to environmental exposure, and remeasured afterward. Environmental exposure would consist of thermal cycle tests over the operational temperature range and number of cycles, with margin added. Typical requirements for this application are shown in Table 3.3. The risk reduction needed for this application amounts to a *qualification* of the paint. The evaluation program might look like this:

1. Prepare N samples of alodined aluminum and carbon composite.
2. Apply black and white paint to samples of each material.
3. Test adhesion (e.g., via tape lifts).
4. Perform vacuum bakeout at 100°C; measure total mass loss and collected volatiles.
5. Measure optical properties (emissivity/absorptivity).
6. Measure surface resistivity.
7. Subject samples to ~10,000 temperature cycles (demonstrating margin over operational requirement) from −100 to 100°C; illuminate samples with ultraviolet light during thermal cycling.
8. Perform ground-based test for atomic oxygen reactivity.
9. Repeat adhesion tests and optical property and surface resistivity measurements.

TABLE 3.3
Requirements for Self-Healing Paint for Space Application

Parameters/Properties	White Paint	Black Paint
Temperature range	−100 to 100°C	−40 to 60°C (for radiator application; optical baffle would typically be colder)
Number of cycles	5,256/year of mission life	5,256/year of mission life
Thermal emissivity[a]	0.8 to 0.9	>0.9
Thermal absorptivity[b]	Beginning of life: 0.2 or less End of life: 0.4 or less	>0.9 >0.9
Surface resistivity (to minimize spacecraft charging)	10^5 to 10^7 ohms/square (the lower the better)	10^5 to 10^7 ohms/square (the lower the better)
Darkening due to exposure to ultraviolet light	Minimal; thermal properties must stay within specifications	N/A
Reactivity with atomic oxygen	Minimal; thermal properties must stay within specifications	Minimal; thermal properties must stay within specifications
Outgassing (after vacuum bake-out, if required to meet these specifications)	<1% total mass loss (TML), <0.1% collected volatile condensed material (CVCM)	<1% TML, <0.1% CVCM
Adhesion	No degradation over mission life	No degradation over mission life
Painted surface cleaning	Compatible with isopropyl alcohol (IPA) and deionized water	Compatible with IPA and deionized water

[a] Thermal emissivity: ratio of energy radiated by a surface to energy radiated by a blackbody at the same temperature.
[b] Thermal absorptivity: fraction of incident radiation absorbed by a material.

In this particular example, risk to the system development ment is low if the new paint fails to meet requirements. Qualified paints for these applications already exist that the system engineer can fall back on and employ. The self-healing paint represents a potential performance improvement but is not necessarily system *enabling*. Also, this example lends itself well to the traditional technology maturation process; later-generation MNT may not be as straightforward (consider the MNT that exhibits emergent behavior). There are, however, numerous examples of programs that have suffered significant developmental delays and cost overruns because of unexpected difficulties in qualifying an immature enabling technology (e.g., James Webb Space Telescope[3]).

Summary

The introduction of new technology into a system development presents system engineering challenges and several risks. There is the technical risk—will the new technology live up to expected performance? Then there is programmatic risk—will the new technology meet performance expectations but take longer to develop than planned? If so, delivery schedules and system cost may be adversely affected. A judgment must be made based on the maturity of the new technology and the effort required to develop the technology to the appropriate level for inclusion in the system.

Technologies at the earliest TRL levels (1 to 3) may warrant a focused effort that precedes the main system development. More mature technologies (4 to 5) may be developed within the main system development effort with acceptable risk, with the goal of demonstrating TRL 6 by PDR. However, these judgments must be made on a case-by-case basis with the approach tailored to the specific system development methodology. Other considerations include the likelihood of successful technology development and incorporation into the system and the consequence of failure. Alternative paths for risk mitigation should be considered as well as the associated impacts on system performance and resources.

Finally, MNT challenges traditional processes associated with technology maturation. A heavier reliance on modeling and simulation may be more cost effective at the risk of deferring definitive performance verification to the full-up system, once realized. Such an approach requires acceptance of this risk and places a heavier burden on the early validation of models and simulations that replace tests.

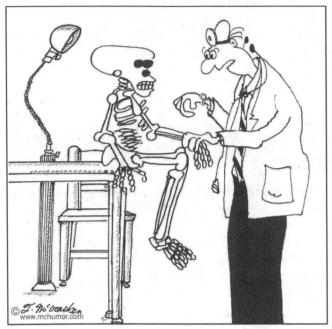

Locking down requirements
later in the life cycle

References

1. From NASA Procedural Requirements 7120.8, NASA Research and Technology Program and Project Management Requirements, Effective Date: February 5, 2008, Appendix J.
2. Buzzle.com, Intellegent Life on the WEB, "Waterfall vs. Agile," www.buzzle.com/articles/waterfall-model-vs-agile.html, created on October 24, 2010, last updated on October 24, 2010.
3. Space Science Telescope Science Institute, "James Webb Space Telescope," www.stsci.edu/jwst/overview/history/hist05.html.

4

Agile Systems Engineering

Timothy G. McGee
Bethany M. McGee

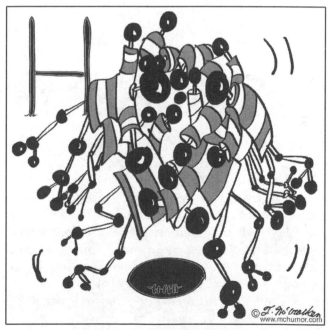

Getting ready for the SCRUM.

Contents

Introduction

At an abstract level, all product development requires at least two steps. The first step is the analysis to define what work needs to be done, and the second step is the actual execution of the work defined by the analysis step.[1] In practice, this process becomes much more complicated, especially in common situations when the work requirements are not clearly understood and well defined before the execution begins or when the individuals performing the work are neither funding, nor acting as the end users, of the product. The interplay between the work definition and work execution stages and how the various parties involved in these stages interact can have major impact on project cost, with the cost of changing the work definition historically increasing as the execution stage progresses. A variety of methodologies exist that formalize the processes for defining what work is desired by the customer and how this work is to be carried out and verified. The previous chapters discussed classical methods in which the customer requirements are clearly laid out at the beginning of the effort. These methods address the cost of change by making a large initial investment in requirements analysis so that the desired work is well enough understood so that it should not change. This chapter will briefly review these classical methods and then discuss in detail more recent agile methods that manage requirements that will change and continuously evolve. These agile methods attempt to directly reduce the cost of change by implementing more testing and continuous feedback. This chapter will also discuss how these agile methodologies are applicable to the development of systems based on micro- and nanoscale technologies (MNTs).

Review of Classical Techniques and Waterfall

The rise in use of classical project management techniques can arguably be traced back to the initial study of "scientific management" by Frederick Taylor at the turn of the nineteenth century.[2-4] Taylor turned his attention to studying management techniques in order to reduce production inefficiencies, specifically in the steel industry in which he worked. In his 1911 work, *The Principles of Scientific Management*, he outlined what he believed were the three main causes of these inefficiencies. First, there existed a misbelief among the workers that an increase in productivity would result in a reduced need for workers and a rise in unemployment. Second, Taylor observed poor management and work planning systems under which it was in the best interests of the workers to work at a slow, inefficient pace. He also believed that people have a natural tendency to work at a minimum effort. Third, most manufacturing knowledge at all levels of detail was nominally passed along using inefficient "rule-of-thumb" techniques. In order to address these problems, Taylor performed

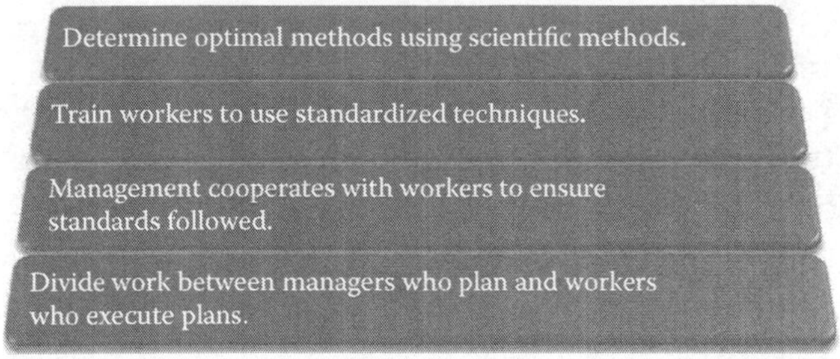

Figure 4.1 Taylor's four principles of scientific management.

experiments to measure and compare the productivity of various approaches of performing each step of the production process. As a result of his work, Taylor converged on four duties of management that can be viewed as the core principles of his scientific management system. First, a scientific study should be performed to determine the optimal method to perform each work task. This optimal approach should become the uniform standard that replaces all "rule-of-thumb" approaches. Second, management should select and train workers in the standardized approaches, replacing the previous system of the workers growing and passing knowledge down among themselves. Third, management should supervise the workers to ensure that the developed scientific methods are being followed. Fourth, there should be an equal division of work between the management and the workers under which the management frees the workers from the burden of planning. Taylor's principles of scientific management are illustrated in Figure 4.1. The end result of Taylor's principles was to create a system in which all planning and work definition was done up front by management with little or no feedback from the workforce. Under this system, "the establishment of many rules, laws, and formula replace[d] the judgment of the individual workman."[2]

Overall, Taylor's scientific management encouraged a hierarchical top-down management approach in order to eliminate more inefficient bottom-up approaches that he observed. Even though many of the manufacturing processes and worker skill sets have changed during the century since Taylor outlined his ideas, the core principles of the hierarchical

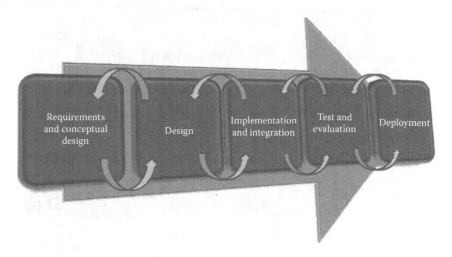

Figure 4.2 Classical waterfall.

top-down management style and rigorous processes of his scientific management are still very apparent in many of the project management frameworks of today, including the waterfall method.[5] Under the waterfall framework, the execution of each project follows a very linear path consisting of several distinct phases that are performed serially. As discussed in Chapter 2 and illustrated in Figure 4.2, in its most basic form, these phases include requirement and conceptual design, design, implementation and integration, test and evaluation, and deployment. Incidentally, these phases also correspond to the elements of most current instantiations of the systems engineering process. Even though these phases can potentially overlap, feedback in the waterfall process is usually confined between sequential steps.

The waterfall approach, along with many traditional management techniques, has many underlying assumptions that seem to coincide with the mechanized, assembly line view of Taylor's day.[6] The clearly defined initial, one-time phase of requirements and planning assumes that projects are predictable and well enough understood at the start to break down into clearly defined tasks that can be distributed among the workers. While many organizations that utilize waterfall style frameworks will involve the more experienced and knowledgeable workers in the requirements definition phase, most will maintain clearly defined hierarchies. During planning and resource allocation within these hierarchies, employees are often assumed to be interchangeable with little consideration for specific skill sets or past team dynamics. Finally, because introducing change to a project once the waterfall has gone into motion typically requires revisiting the initial requirements phase and disrupting the linear flow of the process, change can be difficult and inefficient to implement.

Agile Methods

Frederick Taylor performed his production experiments in order to address inefficiencies of the existing management techniques as applied to manufacturing during the late 1800s and early 1900s. Similar to Taylor's recognition of flaws in the predominant management techniques of his day, there has been a growing movement over the last few decades to address the inefficiencies of rigid classical management applied to various industries in the modern era. This movement has coincided with the growth of the knowledge-based sector of the economy relative to the manufacturing sector and the large increase in the number of highly educated workers as opposed to the larger number of manual laborers in the past. The most widely visible and organized effort to explore alternatives to the waterfall method began in the software development industry. With the rise of lower-cost computing platforms, many software developers found that the top-down, linear development strategies that originated in manufacturing and construction were not a good fit for software development. In a paper that is often cited as the first formal description of the waterfall method,[1,5] Winston Royce described the waterfall method as a poor fit for software development. His major objection focused on the location of testing and evaluation at the end of the development chain. If problems or flaws are found during the testing phase, the development chain may have to revisit the design or even the requirement phases. For a linear development process, this can have major cost and schedule consequences. Two other potential objections to top-down development processes in the software industry are their poor ability to respond to change and their inability to allow ideas to flow from the bottom up. In more traditional industries such as construction or large manufacturing, the end products can often be well understood and the technologies usually change slowly, allowing top-down methods to produce desirable results. In the software industry, however, the rapid changes in computer and network technologies over the last few decades have increased the rate at which old technologies become obsolete and new possibilities grow. Thus, the ability to continuously respond to change and encourage bottom-up creativity is an advantage in the software industry.

In response to the constantly changing environment of software development, various grassroots efforts to explore better suited incremental development frameworks have emerged. Although references to the use of incremental software development can be found as early as 1957,[7] the formalization and more widespread implementation of the most prevalent incremental development frameworks of today occurred during the 1990s. In February of 2001, a group of 17 software developers met at a ski resort in Utah to discuss their common goals. This group included "representatives from Extreme Programming, Scrum, Dynamic Systems Development Method (DSDM), Adaptive Software Development, Crystal, Feature-Driven Development, Pragmatic Programming, and others sympathetic to the need for an alternative to documentation driven, heavyweight software development processes."[8] Even though these developers differed in the specifics of their respective methodologies, the meeting resulted in the drafting of the Manifesto for Agile Software Development, a brief statement (coining the use of the word *agile* for iterative development methods) that outlined the core values shared by the various new approaches. Figure 4.3 is an overview of the Agile Manifesto.

In addition to the manifesto, the group expressed a set of principles that form the foundation of the manifesto. In order to maximize customer satisfaction, technical teams should be in continuous, even daily, contact with the business teams and make frequent delivery of working software, which is the primary measure of progress. Working on short time frames

"We are uncovering better ways of developing software by doing it and helping others do it. Through this work we have come to value:

Individuals and interactions over processes and tools
Working software over comprehensive documentation
Customer collaboration over contract negotiation
Responding to change over following a plan

That is, while there is value in the items on the right, we value the items on the left more."

Figure 4.3 The Agile Manifesto.

of a few weeks or months allows changing requirements throughout the project timeline to be harnessed for competitive advantage. Projects should be built around talented and motivated individuals who will do the best work when trusted and allowed to form self-organizing teams. These teams work best when they have face-to-face communication and regularly reflect on how to improve productivity. The technical teams and sponsors should work at sustainable paces and avoid "fire drills" whenever possible.

At first glance, the four values listed in the Agile Manifesto may appear to advocate overthrowing the rigor and proven methodologies that began with Taylor's four principles for scientific management in favor of a return to ad hoc unstructured projects and potentially chaos. Although the Agile Manifesto and various agile development frameworks advocate more trust and responsibility in individual workers, they do not advocate the lack of rigor and methodology. Jim Highsmith, one of the original signers of the manifesto noted, "We want to restore a balance. We embrace modeling, but not in order to file some diagram in a dusty corporate repository. We embrace documentation, but not hundreds of pages of never-maintained and rarely-used tomes. We plan, but recognize the limits of planning in a turbulent environment."[8] The various formalized agile frameworks have their own sets of rules and methods that are based upon empirical evidence of what worked for the teams that implemented them rather than idealized methods imposed by a hierarchical organization. The implementation of agile techniques requires both consensus among the team members to implement the agreed upon agile methods and a high level of discipline to follow them.

Examples of Agile Techniques

There are currently a number of well-established and popular agile methods including Scrum, Crystal, Extreme Programming, Adaptive Software Development, Feature Driven Development, Dynamic Systems Development Method (DSDM), and Pragmatic Programming.[9] Many of these methods began in the field of software development, but in recent years, a variety of industries including pharmaceuticals, software, automobiles, and integrated circuits have begun to explore agile product development.[10] A comprehensive survey of the many agile methodologies is beyond the scope of this chapter, although summaries of three popular methods: Scrum,[11-15] Extreme Programming,[16-18] and Kanban,[19-23] will provide a good picture of how the principles of agile product development can be implemented in practice.

Scrum

The term *scrum* was initially applied to product development in an article by Hirotaka Takeuchi and Ikujiro Nonaka in the *Harvard Business Review* in 1986.[12] The concept of Scrum evolved from the original idea of relating product development to the game of rugby where a self-organizing team moves collectively down the field toward a goal. These initial ideas were further developed and formalized in the mid-1990s by Ken Schwaber and Jeff Sutherland.[13] The resulting project management approach known as Scrum is a framework of methods used to obtain incremental functional products from small self-organizing teams on a cyclical, iterative basis. Figure 4.4 illustrates the Scrum development methodology.

A Scrum team consists of a product owner, a Scrum master, and the team members. The product owner is responsible for determining the most desirable product features while minimizing the cost-to-benefit ratio, which can be literally applied

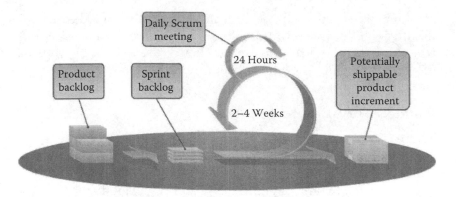

Figure 4.4 The Scrum development methodology.[15] (Copyright ©2005, Mountain Goat Software).

to commercial products or more abstractly applied to balancing different demands within research environments. The product owner will frequently interact with the Scrum master and the Scrum team to help prioritize features as scope or demands change. The Scrum master is the "coach" of both the Scrum team and the product owner, in that he or she is responsible for educating the product team (owner and members) about Scrum, keeping the entire team aligned and on task, and ensuring on-time deliveries of the iterative products. The team typically consists of five to nine[24] dedicated, cross-functional, self-organizing members who work well together and are responsible for developing and delivering the product specified by the product owner.

The Scrum process is built around iterative product development cycles called *sprints*. A sprint is a focused period of development, usually ranging from 1 to 4 weeks, during which a set number of tasks is to be completed in order to deliver a potentially usable product at the end of the sprint. At the beginning of each sprint, the product owner brings a prioritized list of desired product features to the Scrum master and the team during the sprint planning meeting. The team, facilitated by the Scrum master, tells the product owner how many of the higher-priority features they are confident they can complete by the end of the sprint. The team and Scrum master, with input from the product owner, then convert the list of product features into specific achievable tasks. Each day during the sprint, the team and Scrum master meet for brief updates on progress that cover three questions: (1) What was accomplished yesterday? (2) What is the goal for today? and (3) Are there any impediments to progress? The Scrum master will maintain a burn-down chart tracking the team's progress toward completing all of their tasks within the sprint. If there are impediments (e.g., illness, expired software licenses, unforeseen development hurdles), the Scrum master facilitates removal of the impediments, and if the impediments have a large impact or emergency tasking comes during a sprint, the Scrum master works with the product owner to reprioritize and, if necessary, remove some of the initial features. At the end of the sprint period, the team demonstrates the incremental product for the product owner. A retrospective is also held among the team as a feedback method to determine what worked well and what needs to change in the next sprint. The process then repeats.

Traditional product development methods are framed by the requirements placed on the final product, but Scrum is structured using time-boxing allowing the requirements to potentially vary. The planning meeting, the task breakdown meeting, the daily meetings, the product review/demo, and the retrospective are deliberately limited in time. The length of the sprints should also be rigidly maintained. For example, if 2 weeks is chosen as the duration, it should not vary between 8 workdays and 12 workdays instead of 10. The time limits force the team to focus on what is most important; otherwise, even just the breakdown of the features into specific achievable tasks could drag on for several days. If the task breakdown meeting is limited to only a few hours, the team members have to make the highest-impact breakdowns and decisions first, leaving specific details to work out when they are encountered and needed. This also raises the importance of communication and coordination among team members as well as communication between the product owner and the team.

Extreme Programming (EP)

Extreme Programming was created by Kent Beck based on the project management techniques he developed as a software lead working on the Chrysler Comprehensive Compensation

System between 1996 and 2000. During this project that explored the use of object-oriented programming to improve payroll systems, Beck asked his team to "to crank up all the knobs to 10 on the things [he] thought were essential and leave out everything else."[16] At a high level, Extreme Programming is similar to Scrum in many ways, including organizing the customers, managers, and developers as partners in the development process; encouraging open work spaces to foster communication; and holding short daily meetings during which each team member states what was accomplished the previous day, what is planned for the current day, and what obstacles are in their way. Developers are encouraged to become involved in as many aspects of the project as possible to become knowledgeable in multiple areas to avoid knowledge loss, allow balancing of workloads, and foster creativity. Although the formalized Extreme Programming method has a large number of rules, it also encourages modifying the rules as necessary to meet the needs of the team. Figure 4.5 shows the flow of the Extreme Programming process.

The complete Extreme Programming system specifies more specific technical software engineering practices than some agile methodologies such as Scrum, although its five core values of communication, simplicity, feedback, respect, and courage apply to many disciplines. Face-to-face communication is encouraged using daily meetings, open work

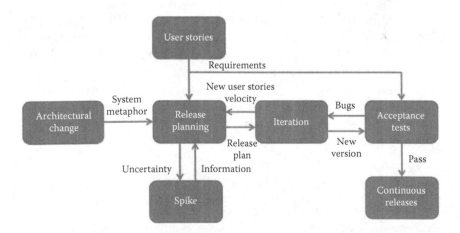

Figure 4.5 Extreme Programming.

areas, and frequent meetings between developers and customers. Simplicity seeks to maximize the return on investment by only doing what is needed and no more. Creating simple designs that are easy to understand and maintain requires in-depth knowledge; therefore, Extreme Programming stresses refactoring to clean up existing code. Feedback is encouraged at all levels of the project from release meetings that occur on the order of several months to iteration meetings that occur every 1 to 3 weeks to daily meetings and constant feedback between pairs of programmers working together to create code. Extreme Programming holds that processes should be adapted based on feedback to meet the product needs as opposed to the other way around. Respect is reflected by the shared ownership of the project by all team members: trust of the managers and customers in the developers to share in responsibility, and openness of the developers to new ideas from customers. Courage is required to maintain truthfulness in all progress reports and estimates so that problems can be identified and resolved early without fear of blame.

The planning process in Extreme Programming begins with user stories provided by the customer. Each story consists of a few jargon-free sentences and explains a feature that the customer would like in the product. The work required to implement a story should be achievable in 1 to 3 weeks. At the beginning of the project, a release meeting is held in which customers work with the team to prioritize stories and decide what features should be incorporated in the next release. At the beginning of the project or after any major architectural change, the team also formulates a system metaphor that is a simple narrative explaining how the system works. This includes agreed-upon naming methods to make functionality of software components apparent from names. The actual work is performed during iterations. The length of iterations can be between 1 and 3 weeks, although the length should be fixed over the project to encourage rhythm and allow improved planning and progress measurement. Detailed planning is done only one iteration at a time at the beginning of each iteration. The highest-priority user stories and uncompleted or unsuccessful stories from the last iteration must be chosen first for the next iteration. During planning, developers decompose user stories that are in the customer's language into more detailed tasks of 1 to 3 days that are in the developers' language. The amount of work allowed during each iteration is determined by the velocity of the team, which is the empirically determined amount of work completed in past iterations. Modifying the team's work estimates to add more work into iterations is not allowed because it violates honesty and ignores the "cold reality of consistent estimates."[16] The user stories may need to be reformulated every three to five iterations as understanding of the system evolves. The highest-priority stories are always implemented first; therefore, the process should work to maximize value to the customer.

Other notable properties of Extreme Programming are its use of spikes, its emphasis on testing, and its use of pair programming. A spike is a short program or study that is used to explore new ideas or potential solutions. One is started when the project encounters a risk such as a large uncertainty in required work during release planning. Any code developed during spikes is meant to be informal and disposable. Extreme Programming also has rules to develop tests before the work is done. The tests can then act as a definition of the required functionality of the product. There are two primary levels of tests: acceptance tests and unit tests. Acceptance tests are black box tests defined with customer feedback to determine if the product satisfies the user stories. Products that pass acceptance tests are included in the next small release. Unit tests are smaller tests to evaluate the correct functionality of individual components. For software programs, all tests are incorporated into an automated testing framework that allows all tests to be regularly run on the system in order to ensure that new features do not negatively impact any functionality of the existing system. The third unique aspect of Extreme Programming is its use of pair programming. During pair programming, coding is done by two developers who sit at one computer and take turns writing code. Although it is counterintuitive to have two individuals working on the same code, advocates of Extreme Programming have shown that it can often increase productivity overall by reducing errors in the code. By rotating pairs, team members can also share knowledge across the team.

Kanban

Kanban originated in manufacturing with Toyota in the 1950s and is a main component of just-in-time and lean manufacturing processes. The basic tenet is that as parts are used on the manufacturing floor, a *Kanban* card is sent to the supply location to replenish the parts. The supply location then replenishes their supply from the parts manufacturer. This system allows the rate of finished products supplied to meet customer demand to echo all the way back to the parts manufacturer.

Kanban can be applied to practices outside of manufacturing and has become popular in agile development platforms, similar to Scrum. Kanban processes and rules are fewer than Scrum with only three rules needed: visualize the workflow, limit work in progress, and measure the lead time.

A big part of Kanban is the highly visual Kanban board. Each task is described on a card and posted on the board. As work progresses, the cards are moved across the board into subsequent sections until the task is completed. However, each intermediate step is limited by the number of tasks allowed to be placed in that category. This prevents bottlenecks caused by incomplete tasks or processes piling up in any given step. An example of a Kanban board is shown in Figure 4.6.

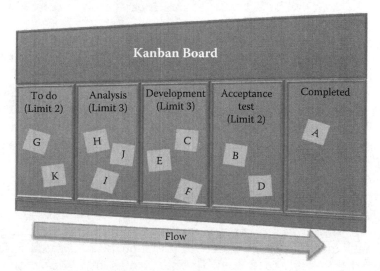

Figure 4.6 Example of a Kanban board.

The team leader or product owner prioritizes the needed tasks and indicates the highest-priority items for work. The team works on the items moving the cards across the board as progress is made. If an impediment arises, stalling a task at a given point, the team must work to clear the roadblock before progress can be made on the next task. Kanban is not time-limited like Scrum but limits the work-in-progress tasks instead. If a high-priority task comes up, it is moved to the top of the waiting task area to be the next item tackled by the team once enough room is available in the work-in-progress area. In order to provide estimates of how much time future tasks will require to move through the process, each task should have the amount of time it takes to complete recorded.

The Kanban board can be used to simultaneously track multiple projects or products as well. Each product can be tracked using different colored cards or separate designated paths along the board. The backlog can be mixed among the products, or the team can work on the projects or products in order.

Semiagile Methods

The classic linear waterfall and agile methodologies should be viewed as opposite ends of a continuum of potential project management methods as opposed to a dichotomy. Between these extremes, there are a variety of management techniques that preserve much of the preplanned aspects of the waterfall while introducing the iterations and progressive elaboration of agile techniques. Such methods can be viewed as semiagile when compared to full agile methods such as Scrum or Extreme Programming. Semiagile methods include iteration and adaptability to change and uncertainty but do not necessarily include all of the principles laid out in the Agile Manifesto. For example, the length of iterations in semiagile methods is typically on the order of 6 months to 2 years rather than the 2-week to 2-month time frames of agile methods. Semiagile methods also may not necessarily focus on placing as much responsibility in individual workers and small self-organizing teams. Two examples of semiagile methods are spiral and rolling wave development.

Spiral Development

In the spiral development process,[25] development work is done in multiple iterations working toward the final deliverable product. Each cycle typically consists of four main steps: determining objectives, identifying and resolving risks, developing and testing the current iteration of the product, and planning for the next iteration. During each iteration, a prototype system with increasing functionality is produced, and the requirements for the system become more detailed with each successive prototype. The delivery of functioning prototypes and inclusion of testing and revisiting requirements with each cycle is similar to many of the agile methods, and it could be argued that many of the specific agile methodologies can be viewed as spiral methods taken to the extreme. The spiral development process is illustrated in Figure 4.7.

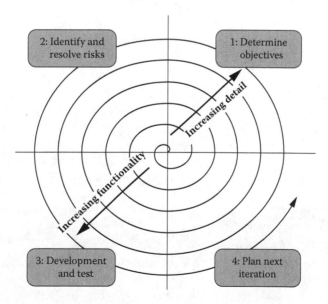

Figure 4.7 Spiral development.

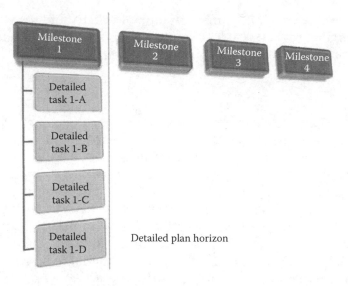

Figure 4.8 Rolling wave development.

Rolling Wave Development

In rolling wave development,[26,27] the high-level requirements are assumed to be known at the start of the project, but it is acknowledged that detailed planning for long time frames into the future is difficult to do accurately. Thus, the project deliverables and milestones are initially only defined at a high level. Detailed planning is only done for a shorter fixed time into the future, such as 1 year, over which there is reasonable confidence in the ability to accurately predict events. The definition of milestones over a long timescale, detailed planning beyond a few weeks or months, and the potential to perform planning in a top-down approach are the main differences between rolling wave techniques and full agile methods. The rolling wave development process is shown in Figure 4.8.

Applicability of Waterfall versus Agile Methodologies

No project management or development methodology is a panacea. Various methodologies have their own strengths and weaknesses that must be understood in order to match the right techniques to the right projects and development environments.

The Iron Triangle

The iron triangle of project management consists of time, cost, and scope. One clear way to illustrate the fundamental differences between classical waterfall and agile methods is to consider their respective approaches to the iron triangle and how they prioritize and balance risk between time, cost, and scope.[28,29] The scope can include both the quantity and quality of the deliverables. The time or schedule is when they will be delivered, and cost or budget refers to the resources required to produce them. There is an old adage in project management that you can only pick two of the three, and the third depends on the other two. In classical methods such as the waterfall, the scope is typically the dominant of the three and is rigorously set by the requirements of the project. The cost and schedule required to achieve the desired scope are then estimated. In most agile methods, this triangle is inverted. The schedule and cost are usually the rigorously defined factors as defined by the set time frame of the iteration cycles. With fixed schedule and cost, the deliverables or scope must be regularly reprioritized and reestimated in order to define reasonable objectives for each cycle. When evaluating the applicability of agile methods to a specific project, one important consideration is how much the scope of that project can vary. Many products will have a minimum level of scope that must be achieved to function. For example, an aircraft or computer system that is missing a major component such as an engine or processor is likely of minimal value to the end user. Figure 4.9 compares the iron triangle for classical and agile processes.

Factors Not Explicitly Addressed by Agile Methods

The agile methods outlined above deal primarily with how to organize teams of technical individuals and sponsors on

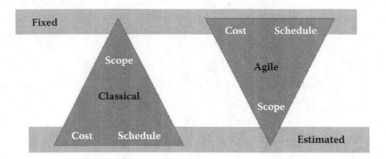

Figure 4.9 The Iron Triangle.

existing projects in order to produce products. When viewing product development in a larger context, however, there are factors that are not explicitly addressed by agile methodologies.[30] These factors include project initiation, infrastructure development and training, and other organizational factors such as human resources. The defined agile methodologies typically have the implicit assumption that there is an existing relationship between a sponsor and the technical team. The values outlined in the Agile Manifesto can also require that the sponsor has a high level of trust that the technical team will help define and develop the best products for the defined time and cost without clearly predefined contractual agreements on the final product. In practice, developing this type of relationship will take time and can be complicated by factors such as the desire of the development team to maximize profit. Without this type of relationship, methods such as waterfall that predefine requirements can provide the sponsor and customer confidence because they have a rigorously defined document stating what they should receive. Agile methods can also impact the types of contractual agreements that are most appropriate between those who fund and execute the work. For example, if a project prioritizes time and schedule over product scope, a contract that defines the price per unit of time worked by the team could be more appropriate than a contract that clearly defines a predetermined price for a clearly defined final product.

Agile methods also do not necessarily address infrastructure development. Although they may provide the best way for an existing team with the right tools to efficiently produce innovative products, they may not provide the best way to develop the infrastructure needed to build and maintain this team. This infrastructure can include physical buildings and work areas, technical tools, and development of personnel. Constructing and configuring the physical workspace for new teams will likely take longer than the several week cycle times used by agile methods and, therefore, require more classical planning. The required time and investment to incorporate new technical tools and to maintain existing tools is also not explicitly addressed by most agile methods. These tools could include new software development environments and architectures, laboratory equipment, or manufacturing capabilities to construct prototypes of physical systems. Another important investment is the training of staff. Agile methods encourage small teams where individuals can become involved in multiple aspects of the product. This will work best with a team of experienced members with a wide variety of existing skills. Developing and maintaining these skills in a field with rapid technology advancement will require some level of training that can take away from development time. With their emphasis on rapid and continuous delivery of viable products at each development iteration, agile methods do not explicitly address how to handle these long-term investments.

Running a technical organization will usually require more than just the small team of technical contributors. Agile methods can provide a way to organize small technical teams within a larger organization, but they do not address the details of organizing the larger organization. This can include legal team members or human resource staff who handle the logistics, such as payroll and hiring of new staff.

Important Questions for Comparing Traditional and Agile Methodologies

When comparing the pros and cons of waterfall, agile, and semiagile project management techniques in order to determine what is the best fit for a given project, there are several important questions that should be asked:[31]

How well understood and stable are the requirements?

All projects will have some risk that the requirements will change during the development effort. If the requirements are believed to be well enough understood that the risk and potential cost from them changing is below an acceptable level, a more traditional waterfall-style method could be a reasonable approach. Agile methods, in contrast, harness uncertainty and allow continuous updating of requirements at each iteration cycle.

Who are the end users of the product?

If there is a small group of known end users who know exactly how they will use the product, the uncertainty in the requirements is more likely to be low enough to make waterfall methods applicable. If there is a larger, more distributed population of end users that may not be well characterized early in the development cycle, agile methods can be more appropriate.

What is the timeline of the product?

Agile methods emerged to develop time critical products. When delivering a product with less functionality in a short time frame (possibly in multiple stages) is more valuable to the customer than delivering a final fully functioning product at a later date, agile can be advantageous over linear development systems. Agile methods could also be used on projects with a longer development timeline.

What is the size of the project?

Most agile methods advocate developing new products using small teams on the order of 10 members. One approach to apply agile methods to larger projects is to form teams of teams, such as Scrums of Scrums, where the teams interact with each other on a regular basis. Some organizations have reported successful use of agile methods on projects with over 500 members, but the majority of projects that have reported successfully applying agile methods have much smaller team sizes.[24] One benefit of the clearly defined requirements of linear projects is that the requirements can act as interface definition

between groups. As long as the requirement specifications are met, the products from many different teams can be merged together into larger, more
complex systems.

What is the physical location of the team members?

In addition to advocating for small teams, agile methods stress continuous communication. This is much
easier to achieve when team members work together
in the same physical location. Coordinating teams
in different geographic regions of the country or
world and across time zones can make it difficult to
maintain the constant communications to achieve
all of the benefits of agile methods. As communications technologies such as videoconferencing
improve, close coordination of noncolocated teams
may become more feasible.

What resources are required to implement the project?

As previously discussed, agile methods do not necessarily address how to deal with the development of
infrastructure such as laboratory space and equipment, software infrastructure, or personnel training. The need to procure materials with long lead
times that span multiple iterations of agile methods
can also create difficulties.

Agile for Micro- and Nanoscale Technologies

In its most general terms, nanotechnology refers to technologies that manipulate molecules and systems on the order of
1 to 100 nanometer scale in order to exploit characteristics
not achievable on the macroscopic scale. Thus, nanotechnology could cover a wide variety of fields and product types
from purely passive coatings to futuristic nanoscale robots for
medical applications. For applications such as passive coatings
of nanoscale particles, where the properties are well understood and the particles can be generated in large quantities,
waterfall-type development may be a good fit for assembly-
line-style development. As nanotechnology advances into the

development of custom molecules and actively controlled nano-systems, agile methodologies may provide many advantages. As Ralph Merkel noted, nanotechnology can someday allow us to "snap together the fundamental building blocks of nature easily, inexpensively and in most of the ways permitted by the laws of physics."[32] Early generations of equipment to manufacture custom molecules of the future will likely be large and costly, similar to early room-sized computers made of vacuum transistors. However, as technology advances, smaller, cheaper systems may allow more small teams to "program" new molecules and nanosystems similar to the programming of software today. Although technology that allows nanosystems to be developed as rapidly as software is today may be years into the future, management of research and development of nanosystems today could take advantage of many of the properties of agile methodologies.

Bottoms-Up Approach and Short Time Frames

When designing systems at the nanoscale level, engineers and scientists push closer and closer to the fundamental laws of physics. The material properties and phenomena at the nanoscale can be much more difficult to control and design than those on the macroscopic level. With fundamental physical limits constraining the development process, developers will often be forced to optimize solutions within these limits rather than design to specifications that may not be physically realizable. As nanotechnology is incorporated into more and more viable products with "new science," there will likely be more and more product development teams with scientists who are experts in the technology and its limitations playing lead roles. In order to maintain a competitive edge, the time frames from concept inception to product release will also likely be short, possibly less than 12 months.[32] This type of product development environment lends itself to evolving requirements and organizing projects around fixed time constraints and development iteration cycles rather than fixed product scope.

Evolutionary Development and Testing

Agile development processes can be viewed as a form of evolutionary development. Each development cycle allows the creation of a new, more developed system. The testing of the product at each iteration allows the new designs to be evaluated, and replanning meetings on a regular basis allow new ideas to be incorporated. Agile methods such as Scrum and Kanban use prioritized lists of new features and properties. These lists and the explicit prioritization steps allow new ideas to be tested and feasible concepts advanced while infeasible designs are identified early and abandoned. As Jim Highsmith, one of the original signers of the Agile Manifesto notes, "New technologies such as combinatorial chemistry and sophisticated simulation are fundamentally altering the innovation process itself. When these technologies are applied to the innovation process, the cost of iteration can be driven down dramatically, enabling exploratory and experimental process to be both more effective and less costly than serial, specification-based process"[10] (p. xix). This reduction in the cost of iterations directly reduces the cost of change in a project.

Communication and Small Teams

Product development involves teams of people working together to solve a problem; therefore, successful projects require strong communications between team members. One of the strengths of agile methods is their focus on small teams and continuous communication. Similar to Micro Electro Mechanical Systems (MEMS), many technical aspects such as thermal, structures, and electrical that can be separated into clearly defined subsystems at the macroscopic level are much more tightly coupled at the small scales of nanosystems. The use of small colocated teams with short daily meetings where individuals discuss what work they have recently done, what obstacles they encountered, and what work they are working on next can help teams developing nanosystems to optimize communications between subject area experts. The use of team idea boards used by many agile methods to organize new ideas can also foster strong team communications. Organizations with multiple independent teams working on nanosystems could also foster idea sharing by implementing "Scrum of Scrum" type meetings on a regular basis to encourage cross-pollination of ideas between the groups.

Interaction with Customer

As Chapter 9 will discuss in more detail, the majority of nanomaterials that have currently been turned into viable products consist of spherical or cylindrical molecules. Other more complex nanoscale molecules have been developed in research settings but have not yet been incorporated into viable products because the research was not need driven. Agile methodologies that encourage continuous communications between the technology development teams and the customer can serve as a bridge between the research environment and the market. This bridge will connect the individuals with technological expertise and knowledge of nanomaterial phenomena with the business teams that have expertise in the marketplace and in how to turn the new ideas into products.

Conclusions

Both classical waterfall methods and more recent agile methods share the same goals of optimally organizing people and resources to produce working measurable products. When Frederick Taylor first formalized his principles of scientific management, he was seeking to improve upon the existing methods to better match the manufacturing environment of his day. In the same way, the early advocates of agile methods were seeking to find improved management approaches to better fit the highly technological and knowledge-based software development environment of today. In both cases, the respective methods survived because they worked. Given the similarities between software and nanosystems of high technology, highly educated technical employees, and rapid product development, many of the agile methods that have proven effective for software development offer strong promise for developing MNT-based systems. Ultimately, the best approaches for managing MNT system development will be determined by the teams doing the development as they experiment with different approaches and let the empirical results speak for themselves.

References

1. Royce, W., "Managing the Development of Large Software Systems," *Proceedings of IEEE WESCON*, August 1970.
2. Taylor, F. W., *The Principles of Scientific Management*, New York, and London, UK: Harper & Brothers, 1911.
3. "Frederick Taylor & Scientific Management." *NetMBA Business Knowledge Center*. November 20, 2010. www.netmba.com/mgmt/scientific/.
4. "Scientific Management." *Wikipedia, The Free Encyclopedia*. November 20, 2010. http://en.wikipedia.org/wiki/Scientific_management.
5. "Waterfall Model." *Wikipedia, The Free Encyclopedia*. November 20, 2010. http://en.wikipedia.org/wiki/Waterfall_model.
6. "Agile Project Management." 2003–2008. CC Pace Systems, Inc. November 20, 2010. www.ccpace.com/Resources/documents/AgileProjectManagement.pdf.
7. "Agile Software Development." *Wikipedia, The Free Encyclopedia*. November 20, 2010. http://en.wikipedia.org/wiki/Agile_software_development.
8. *Manifesto for Agile Software Development*. November 20, 2010. http://agilemanifesto.org/.
9. Boehm, B. and Turner, R. *Balancing Agility and Discipline: A Guide for the Perplexed*. Boston, MA: Addison-Wesley, 2004. Appendix A, 165–194.
10. Highsmith, J. *Agile Project Management: Creating Innovative Projects*. Upper Saddle River, NJ: Addison-Wesley, 2010.
11. "Scrum (development)." *Wikipedia, The Free Encyclopedia*. 20 Nov. 2010 http://en.wikipedia.org/wiki/Scrum_(development).

12. Takeuchi, N. and Nonaka, I., "The New Development Process," *Harvard Business Review*. January-February 1996. pp. 137-146.
13. Schwaber, K. "Scrum Development Process," *OOPSLA Business Object Design and Implementation Workshop*. J. Sutherland, D. Patel, and J. Miller, eds. Austin, TX 1995.
14. DeGrace, P. and Stahl, L. *Wicked problems, righteous solutions*. Prentice Hall, New York, 1990.
15. "Learning Scrum—Free to Use Figures and Wallpapers about Scrum." *Scrum Training Agile Training from ScrumMaster Mike Cohn*. 20 Nov. 2010 http://www.mountaingoatsoftware.com/pages/18-learning-scrum--free-to-use-figures-and-wallpapers-about-scrum.
16. Wells, D. *Extreme Programming: A Gentle Introduction*. 20 Nov. 2010 http://www.extremeprogramming.org.
17. *Development Resource*. 20 Nov. 2010 http://xprogramming.com.
18. "Extreme Programming." *Wikipedia, The Free Encyclopedia*. 20 Nov. 2010 http://en.wikipedia.org/wiki/Extreme_Programming.
19. Making the Most of Both." 21 Dec. 2009. 20 Nov. 2010 http://www.infoq.com/minibooks/kanban-scrum-minibook.
20. Hiranabe, K. "Kanban Applied to Software Development: From Agile to Lean." 14 Jan. 2008. 20 Nov. 2010 http://www.infoq.com/articles/hiranabe-lean-agile-kanban.
21. Hudgik, S. "What is Kanban?" *Graphic Products—Introduction to Kanban*. 20 Nov. 2010 http://www.graphicproducts.com/tutorials/kanban/index.php.
22. "Kanban." *Wikipedia, The Free Encyclopedia*. 20 Nov. 2010 http://en.wikipedia.org/wiki/Kanban.
23. Salloway, A. "An Overview of Lean-Agile Method." 5 Sept. 2010. 20 Nov 2010. http://www.agilejournal.com/articles/columns/column-articles/3222-an-overview-of-lean-agile-methods.
24. Amber, S, "Agile Practices Survey Results: July 2009", 09 Jan. 2011.http://www.ambysoft.com/surveys/practices2009.html.
25. "Spiral Model." *Wikipedia, The Free Encyclopedia*. 20 Nov. 2010 http://en.wikipedia.org/wiki/Spiral_model.
26. Reiling, J. "Rolling Wave Planning and Progressive Elaboration" 11 Feb. 2008. 20 Nov. 2010. http://pmcrunch.com/project_management_process/rolling-wave-planning-and-progressive-elaboration/.
27. Githens, G.D. "Rolling Wave Project Management." *Proceedings of the 29th Annual Project Management Institute*. Long Beach, California, October 1998.
28. Seaver, D. "Respect the Iron Triangle." 20 Nov. 2010. http://www.ittoday.info/Articles/IronTriangle.htm.

29. Cottmeyer, M. "Inverting the Iron Triangle." 9 Apr. 2008. 20 Nov. 2010 http://www.leadingagile.com/2008/04/inverting-iron-triangle.

30. Walters, K. "Agile Management is Not Enough." 23 Mar. 2009, 20 Nov. 2010. http://www.pmhut.com/agile-project-management-is-not-enough.

31. "Which Life Cycle is Best for Your Project?" 22 Oct. 2008. 20 Nov. 2010. http://www.pmhut.com/which-life-cycle-is-best-for-your-project.

32. Cleland, D. and Bidanda, B., eds. *Project Management Circa 2025,* Project Management Institute, Newton Square, PA. 2009.

PART 2

Technology Development Process

Design Synthesis ...
Synthesis begins with the output of functional analysis and allocation (the functional architecture).

The functional architecture is transformed into a physical architecture by defining physical components needed to perform the functions identified in functional analysis and allocation. The objective is to identify the functional, performance, and interface design requirements; it is not to design a solution...yet!

Part 2 describes issues that are unique to the process of design synthesis on the micro- and nanoscale. The purpose is to show how the design process (Charts II.1, II.2) affects systems engineering functions during system formulation and development. To that end, an overview of the importance of scaling in microworld is given. Then technology development processes for Micro Electro Mechanical Systems (MEMS) and nanotechnologies are reviewed with the objective of providing the systems engineer and technology developer with insight that is critical to assess impacts on overall system development in the requirements analysis, concepts, and system design phases.

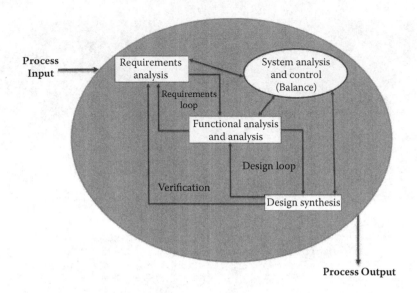

Chart II.1 The Systems Engineering Process

CHART II.2
The Phases in the Systems Engineering Process

Requirements Analysis

The first step of the systems engineering process is to analyze the process inputs. Requirements analysis is used to develop functional and performance requirements; that is, customer requirements are translated into a set of requirements that define what the system must do and how well it must perform. The systems engineer must ensure that the requirements are understandable, unambiguous, comprehensive, complete, and concise. Requirements analysis must clarify and define functional requirements and design constraints. Functional requirements define quantity (how many), quality (how good), coverage (how far), time lines (when and how long), and availability (how often). Design constraints define those factors that limit design flexibility, such as environmental conditions or limits; defense against internal or external threats; and contract, customer, or regulatory standards.

CHART II.2 *(Continued)*
The Phases in the Systems Engineering Process

Functional Analysis/Allocation

Functions are analyzed by decomposing higher-level functions identified through requirements analysis into lower-level functions. The performance requirements associated with the higher level are allocated to lower functions. The result is a description of the product or item in terms of what it does logically and in terms of the performance required. This description is often called the functional architecture of the product or item. Functional analysis and allocation allow for a better understanding of what the system has to do, in what ways it can do it, and to some extent, the priorities and conflicts associated with lower-level functions. It provides information essential to optimizing physical solutions. Key tools in functional analysis and allocation are functional flow block diagrams, time line analysis, and the requirements allocation sheet.

Requirements Loop

Performance of the functional analysis and allocation results in a better understanding of the requirements and should prompt reconsideration of the requirements analysis. Each function identified should be traceable back to a requirement. This iterative process of revisiting requirements analysis as a result of functional analysis and allocation is referred to as the requirements loop.

Design Synthesis

Design synthesis is the process of defining the product or item in terms of the physical and software elements that together make up and define the item. The result is often referred to as the physical architecture. Each part must meet at least one functional requirement, and any part may support many functions. The physical architecture is the basic structure for generating the specifications and baselines.

Design Loop

Similar to the requirements loop described above, the design loop is the process of revisiting the functional architecture to verify that the physical design synthesized can perform the required functions at required levels of performance. The design loop permits reconsideration of how the system will perform its mission, and this helps optimize the synthesized design.

Verification

For each application of the system engineering process, the solution will be compared to the requirements. This part of the process is called the verification loop, or more commonly, *verification*. Each requirement at each level of development must be verifiable. Baseline documentation developed during the systems engineering process must establish the method of verification for each requirement.

5
Scaling

Stergios J. Papadakis

Gravity can get you when you
move to the macro scale.

Contents

Introduction

For the design and engineering of large systems, testing scale models is often an integral part of the process. For many systems, this is a relatively straightforward process. For example, early in the design process of new ships, aircraft, spacecraft, and vehicles, progressively larger models from about 1/20th to 1/4th are often employed. Even for such a narrow range of scaling ratios, some physical parameters change significantly, which is why there is often a progression of scale model testing. For example, a physical structure that is 1/20th scale in linear dimension is only 1/8000 or 1.25×10^{-4} times the mass of the full-sized structure. Such scale models are nevertheless extremely useful in engineering endeavors where building multiple full-scale prototypes is overly expensive and impractical, as there is a well-developed methodology for the testing of scale models that yields valuable information used in design modifications before the next-larger model is built.

In designing systems incorporating micro- and nanoscale technologies (MNTs), some of these procedures and much intuition about scaling needs to be reconsidered in light of the many orders of magnitude in length scale that are spanned by the components of the system. The example above demonstrates that even 1/20th scale ratios result in a factor of 1.25×10^{-4} change in mass. The differences in behavior between macroscopic (meter-scale) and micron-scale devices are even more dramatic. For a ratio of 10^6, as is seen when scaling from 1 m to 1 µm, mass changes by a factor of 10^{18}. When nanoscale structures are considered, a length ratio of 10^9 yields a mass ratio of 10^{27}. To put these ratios in context, note that the mass of a proton is roughly 10^{-27} kg.

Such large changes in the mass/length ratio put micro- and nano-scale mechanical devices into very different operating regimes from their macroscopic counterparts. For example, the effective spring constant k of a beam bending under a load is proportional to I/l^3, where l is the length and I is the second moment of the area. I scales as area squared, or l^4. Therefore, stiffness scales as l, mass m scales as l^3, so resonance frequencies scale as $\sqrt{k/m} = 1/l$. Thus, microscale mechanical devices can easily have mechanical resonance frequencies much higher than is typical for macroscopic systems. This phenomenon is taken advantage of in radio-frequency Micro Electro Mechanical Systems (RF-MEMS), now a well-developed field of applications development enabling on-chip frequency standards that can replace conventional stand-alone quartz-crystal oscillators in the GHz regime [2,3]. Conversely, if relatively low resonance frequencies are desired, sophisticated engineering and processing techniques are often required. The lowest frequencies obtained with Micro Electro Mechanical Systems (MEMS)-scale devices have been in the kHz regime, while macroscopic systems can easily be made which resonate at sub-Hz frequencies [4].

The preceding example of changes in mechanical stiffness and response frequency as a function of size requires a shift in intuition when considering the responses of the structures to loads. Micro- and nanoscale structures appear to be remarkably spindly compared to macroscale counterparts of similar function, yet they are very stiff in response to applied vibration and shock loads, all because of their reduced mass. Much larger shifts in intuition are often required when considering other responses of micro- and nanoscale systems. For mechanical, fluidic, electronic, optical, and molecular systems, as the length scale of the system is reduced, different regimes of fundamental physical behavior can be crossed requiring wholesale modifications of analysis techniques and tools. This chapter gives examples of some of these regimes and of the types of phenomena that can be harnessed or must be compensated for when designing systems incorporating such micro- or nanoscale components.

Mechanical Effects in the Micro and Nano Regime

In the introduction to this chapter, the example of RF-MEMS is given as one where scaling devices to the micron-scale allows for devices with mechanical resonance frequencies in the RF regime. This increase in mechanical resonance frequency is a continuous evolution. As devices are reduced in size, their resonance frequency increases continuously until it reaches the RF regime. There are other phenomena that cannot be so simply understood as an obvious consequence of scaling down.

Surface Stresses

Following the discussion above, it is also self-evident that the surface area to volume ratio increases as $1/l$. As the characteristic length scale of devices approaches the micron scale, changes at the surface of devices become more important. Of course, surface features and surface conditions are very important in macroscopic structures for issues such as crack initiation and propagation, corrosion and corrosion resistance, and friction and wear properties. These phenomena are distinctly micro- and nanoscale phenomena that have important effects on the performance of macroscale objects. These and other surface phenomena have more dramatic effects on devices that are micro- or nanoscale, and can be harnessed to create new microscale devices with functionality not possible at the macroscale. A good illustration of this is MEMS sensors that rely on monolayer or submonolayer adsorption of molecules on surfaces to induce surface stresses that, due to the small size of the devices, cause significant deformations or resonance

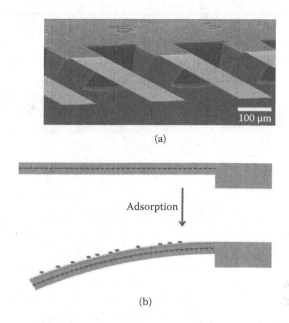

Figure 5.1 (a) Scanning electron microscope (SEM) image of microcantilever array. (See Fritz, J., et al., *Science,* 288(5464), 316–318, 2000.) (b) Microcantilever beam bending due to analyte adsorption on one surface. In this example, the adsorption causes compressive stress and downward deflection, but either direction is possible depending on the adsorbate and surface characteristics.

frequency shifts [1, 5–11]. Such devices rely on changes in surface stress to induce bending in a device component, typically a cantilevered beam (Figure 5.1). Thicknesses of the cantilevers range from 600 nanometers (nm) to 1 micrometer (µm), which demonstrates that the effects of submonolayers of adsorbates are significant even for structures three to four orders of magnitude thicker than the adsorbate layer.

These devices have been tested in a wide variety of configurations and sensing modes. They can be operated in air (or other gases), vacuum, or liquid. The analyte interaction with the surface can range from physisorption where the interaction is primarily van der Waals [9], to chemisorption in which the interaction is a stronger chemical bond, as in the case of thiols binding to gold surfaces [8]. More sophisticated chemistries at the surface can be used to make the cantilevers selective for particular analytes. In these cases, one surface of the cantilever is typically precoated with "probe" molecules that selectively bind with analytes of interest [1,10,12–14]. Typical sensitivities for these sensors are in the parts per billion range.

Casimir Forces

Micro- and nanoscale electromechanical systems often have nanoscale gaps between components. Casimir forces arise between any two surfaces that are separated by a very small distance. They are a prime example of a phenomenon that exists only at the nanoscale. They arise from a fundamental quantum mechanical property of space and have dramatic effects. Casimir forces are part of the same phenomenon as van der Waals forces [15–17]; they arise from random fluctuations of the vacuum. In Casimir's initial prediction of the effect, he considered perfectly conducting plates separated by a small gap. It is a quantum mechanical principle that there are quantum fluctuations in the electromagnetic field of a vacuum. Photon/antiphoton pairs spring into existence randomly and then annihilate following the Heisenberg uncertainty principle of $\Delta E \Delta t > h$, where E is energy, t is time, and h is Planck's constant. In this context, the uncertainty principle states that in a vacuum, in the absence of any sources, photon/antiphoton pairs of energy ΔE may spring into existence for periods of time $\Delta t \leq h/\Delta E$.*

If there are two uncharged parallel infinite perfectly conducting plates in the vacuum, only quantum fluctuations (photons) that meet the boundary conditions imposed by the plates can exist between them (Figure 5.2).[†] Outside the plates, the full spectrum of quantum fluctuations can exist. Thus, there is a radiation pressure that acts to push the plates together. The pressure scales as $1/d^4$, where d is the distance between the plates. This is a small pressure for large separations, but at a gap of 10 nm, it is 10^5 N/m² (~ 1 atmosphere), and at 1 μm,

---

* In principle, any particle/antiparticle pair can be created and annihilated, but photons dominate this process.
† For perfect conductors, these boundary conditions lead to the familiar allowed standing waves with an integer number of wavelengths between the plates.

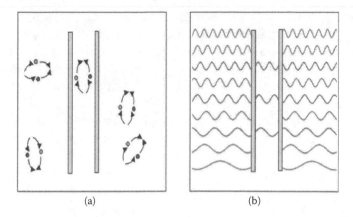

(a)   (b)

Figure 5.2 Quantum fluctuations leading to the Casimir force. (a) Photon–antiphoton pairs springing into existence and annihilating. (b) Photon wavelengths allowed between the plates are dark gray. Wavelengths not allowed between the plates are light gray. The larger number of allowed wavelengths outside the plates leads to radiation pressure pushing the plates together.

it is 10^{-3} N/m². The more familiar van der Waals forces are also due to such quantum fluctuations, but the description is arrived at by considering the interactions between atoms separated by a small distance rather than infinite plates [16–18].

While the above discussion assumes perfectly conducting plates for simplicity, only small corrections are required for real metals, and even dielectrics exhibit a Casimir force. Furthermore, a true vacuum is not required. Many measurements have been done at moderate vacuum [19–22], and experiments in liquids with appropriately selected polarizabilities have demonstrated both positive [23] and negative (repulsive) [24,25] Casimir and van der Waals forces. Finally, the Casimir force also exists for nonflat surfaces. In all of the above-referenced works, the force was measured between a sphere and a flat surface. This eliminates the experimental challenge of ensuring that two flat plates are parallel to nanometer precision.

The first good quantitative measurement of the Casimir force was done in 1996 with a torsion balance of the type used for gravitational experiments [26], and gaps between 10 nm and 1 μm are common in MEMS devices. A range of excellent experimental measurements of the Casimir force have been made using MEMS devices [21,22] and atomic force micro-scopes, which use MEMS cantilevers as force sensors [19,20]. Conversely, the Casimir force is also a challenge for MEMS device design as it is a major contributing factor to stiction, which can prevent MEMS components from moving or stick them in place at the extremes of their travel range [27,28].

Casimir/van der Waals forces have also been harnessed by evolution. There is strong evidence that the ability of a gecko to climb vertical surfaces is due to remarkably small hair-like structures on their feet. These structures, called setae, split at their ends into many nanoscale structures called spatulae that stick to surfaces via van der Waals interactions (Figure 5.3) [29–31]. There are about 5000 setae/mm², and as Figure 5.3 shows, each seta is split into many spatulae [32]. At the nanoscale in complex outdoor environments like those in which geckos live, there are other effects that could cause adhesion, the most common of which is the thin layer of water molecules on the surfaces, which leads to thin-film capillary adhesion. This capillary adhesion is operative in a number of animals that have glands on their feet to secrete a fluid, but geckos do not have such glands. Furthermore, experiments showing that the adhesion does not depend on the degree of hydrophobicity (or hydrophilicity) of the surface provide robust evidence that the adhesion is dominated by van der Waals

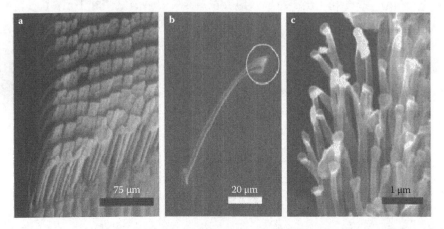

Figure 5.3 Gecko setae and spatulae. (a) Rows of setae on a gecko foot. (b) A single seta. The circled region is the end of the seta where it breaks into spatulae. (c) Spatulae.

forces [29]. In essence, the gecko relies on a vast number of microscopic setae to ensure a huge area of contact on both smooth and rough surfaces creating enough van der Waals force to support the animal's entire weight. In order to unstick its feet, the gecko simply peels its feet up from the heel, thereby progressively loading individual setae at the edge of the contact patch beyond their release load until the entire foot is free. Discovery of this phenomenon has spurred a variety of biomimetic adhesive efforts aimed at duplicating this mode of adhesion and locomotion [31,33–36].

Nanoscale Effects on Material Properties

Much of this chapter is focused on changes that occur in scaling from macroscale to nanoscale objects, but where material properties are concerned, the range of relevant length scales is much smaller. For example, typical structural metals and ceramics have grain sizes in the few microns to tens of microns range. Composite materials such as fiber-reinforced plastics typically have fiber diameters of similar length scale. However, if the individual grains in a metal or ceramic or the fiber diameters in a composite are in the range of a few nanometers rather than a few microns, the mechanical properties of the material can be very different. Much of the interest in this field was generated by an early observation of relatively ductile nanocrystalline ceramics CaF_2 and TiO_2 [37] and by an excellent review of the nanocrystalline materials in general [38].

For monolithic nanocrystalline materials (i.e., not composites), quite a few properties can be different from their bulk counterparts [38–40], including mechanical properties such as ultimate strength, hardness, ductility, and toughness. Other properties such as magnetism, diffusivity, and thermal expansion coefficient have also been observed to be different. A material is typically considered nanocrystalline if about half or more of the atoms in the material are in a defect in the crystalline lattice [38]. This typically results from grain sizes less than 10 nm. Note that even the main criterion for classifying nanocrystalline materials essentially makes use of a scaling argument. The surface-to-volume ratio of the individual crystalline grains dominates the behavior of the material.

It is important to distinguish nanocrystalline materials from glassy materials. Glasses are formed when a melted material is rapidly quenched, so the thermally induced disorder of the melt is frozen into the solid. Thus, in a glass, there are no recognizable crystalline regions. In contrast, nanocrystalline materials have such recognizable regions. In a glass, the density everywhere is a few percent smaller than the density of a perfect crystalline material, whereas in nanocrystalline materials, the density in the defect regions is as much as 15% to 30% smaller than in the crystalline regions [38] (Figure 5.4).

The most dramatic differences in the response during mechanical deformation of a nanograined material, compared to a microcrystalline material, occur in regimes where the response is dominated by grain boundary effects. Yield stress, plastic deformation, creep, failure stress, and derivative qualities such as hardness, toughness, and work hardening are examples of such properties. In many different metals, nanocrystalline grains result in a significant improvement of such qualities for most applications. Unfortunately, for many real-world applications where bulk structural properties are

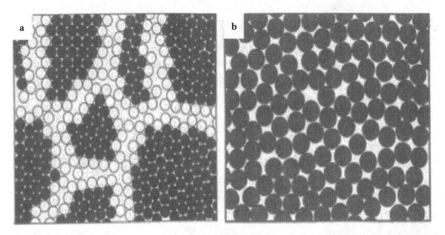

Figure 5.4 (a) Nanocrystalline material. The white circles represent atoms that are associated with the defects; the black circles represent atoms associated with crystalline regions. (b) Glassy material. (Reprinted with permission from Gleiter, H., *Progress in Materials Science,* 33(4), 223–315, 1989.)

concerned, such measurements are complicated by the fact that some of the processing techniques required for creating the nanocrystalline materials also result in the creation of voids and pores, which significantly negatively impact the mechanical properties [40–42]. Many measurements of mechanical properties are done with microscale test samples or by nanoindentation in order to minimize the impact of voids and pores on the fundamental materials properties being measured or because the materials are fabricated in such small quantities in the laboratory [42–44]. Thus, when considering the use of nanocrystalline materials, one must take into account that the properties of a macroscale structure may differ somewhat from those that would be projected from the microscale measurements if the measurements do not include the effects of voids and pores. For example, it is more straightforward to apply such experimental results to wear-resistant coatings than to structural members.

An example of material behavior that is different when it is composed of nanoscale grains is the well-known Hall-Petch relationship, which describes the yield stress of a material across a few orders of magnitude of grain size. The Hall-Petch relationship begins to break down as the grain size reaches a regime where the generation and mobility of dislocations become limited by the density of grain boundaries [45]. In some experiments, the slope of the yield strength as a function of grain size was seen to become negative, and some sophisticated atomistic simulations suggest that such behavior results when the deformation mechanism changes from dislocation motion to grain-boundary sliding [46–48]. For the purposes of this chapter, it is sufficient to note that the Hall-Petch relationship cannot be simply assumed once grain sizes become smaller than a few tens of nanometers. There is a large body of experimental data demonstrating deviation from the conventional Hall-Petch predictions and considerable debate about whether or not there is a maximum in the yield stress for some particular materials as a function of grain size. The main point is that the yield stress of a nanocrystalline material is typically not as high as the conventional Hall-Petch relationship would predict, and that the behavior of a particular nanocrystalline material is not predictable from the behavior of its microcrystalline counterpart. Similar deviations from expected behavior occur with other mechanical properties, such as ductility. In general, one must be aware that the properties of the bulk material can be unpredictable using conventional materials science methods if the grain sizes are in the nanometer regime making empirical measurements essential.

Fluidic Scaling

Fluidics is a field where scaling creates dramatic differences in behaviors. The different fluid-flow phenomena observable in microscale systems have enabled entirely new classes of fluidic devices that are not possible with larger-scale systems. It is in fluidics, also, where the different behaviors of macro- and microscale systems are most obviously driven by scaling. One of the first things considered in the engineering of any system where fluid flow is involved is the Reynold's number of the flow,

$$Re = \frac{\rho V L_0}{\eta}$$

where ρ is the fluid density, V is the velocity, L_0 is a characteristic length, and η is the fluid viscocity. Re is the dimensionless parameter that determines whether a flow is laminar or turbulent or, equivalently, whether or not the inertia of the fluid plays any appreciable role in the flow behavior. In external flows (for example, around an aircraft or ship), the turbulence occurs for Re greater than approximately 50,000, and for pipe flows, the transition to turbulence occurs for Re of a few thousand. In the vast majority of macroscopic systems, turbulent flow is the norm. For example, the transition to turbulence is expected to occur within centimeters of the bow of the ship under normal cruising speeds. This transition is delayed by the presence of a beneficial pressure gradient, but this discussion is beyond the scope of this chapter.

In microscale systems, turbulence is impossible in most cases. This difference leads to truly dramatic differences in behavior. These strictly laminar flows can be harnessed to create devices that are impossible on a larger scale. Due to fabrication technology restraints, microfluidic channels typically have square or rectangular cross sections. The smaller of the two dimensions is what governs the Re, and it is at most a few hundred microns. Thus, unlike most of the micro- and nanoscale systems discussed in this chapter, microfluidic devices are relatively large. It is worth remembering, however, that the length/volume scaling relationships result in the total volume of these devices where channel lengths can easily be many centimeters to a few meters and volume in the nanoliter (nL) range, such that microfluidic devices are incapable of processing more than very small volumes in a reasonable period of time.

For those more comfortable with macroscopic fluidic systems, the lack of turbulence at any point in the system leads to some unintuitive results. A complete absence of inertial effects and, therefore, the dominance of viscous effects, means that mixing of fluids is difficult. Microfluidic flows are often reversible, meaning that if the flow through a structure is run in one direction for a period of time and then reversed to run in the other direction for the same period of time, the system returns to its original state. Another manifestation of this phenomenon is the parallel flow of two liquids in the same channel with minimal mixing (Figure 5.5a), which is one of the earliest and most straightforward demonstrations of the unintuitive nature of microfluidics.

The inertial effects that would make this impossible in a macroscopic system are simply absent for microfluidics, but the effects of diffusion often are not. Diffusion allows for microfluidic devices that have no macroscopic analogue. A seminal example of an application of this phenomenon is the H-filter [49,50]. This device allows separation of fast-diffusing from slow-diffusing molecules in a solution (Figure 5.5b). In this schematic, the gray molecules diffuse through the liquid significantly faster than the black molecules. The connecting channel length and flow rates are chosen such that the gray molecules diffuse out to fill the full width of the channel, while the black molecules diffuse very little during their residence time in the connecting channel. Thus, the lower left exit channel in the schematic contains only the fast-diffusing species.

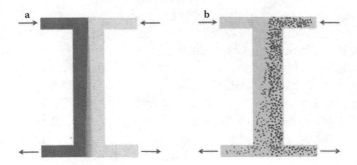

Figure 5.5 Microfluidic systems examples. Flow rates into the devices at the top and out of the devices at the bottom are equal. (a) Two liquids in the same channel flow side-by-side with very little mixing. (b) H-filter schematic. The gray dots represent fast-diffusing species; the black dots represent slow-diffusing species. The lower left exit channel contains only the fast-diffusing species.

For equal flow rates, it is at half the concentration of the input, which is excellent performance for such a separation.

In many fluidic systems, there is a need for mixing, which we have just shown to be absent from most microfluidic systems. Some innovation was required to develop microfluidic structures where mixing could occur in a reasonable period of time, or stated more appropriately, with a reasonable channel length for a given flow rate. The solution to this problem is to force the two streams that must be mixed to have a very large interfacial area for their volume, which allows diffusion (rather than turbulence or shear as in large systems) to mix the two products. This is impractical to do simply by making a channel cross section of very high aspect ratio, but this can be done with a variety of grooves or protuberances in the channel into which the two streams have been flowed (Figure 5.6) [51–53]. The flow pattern generated by the fluid flowing over the grooves without turbulence generates thin layers of each of the two fluids, which can then become homogenous by diffusion.

Other phenomena that become very important in the fluidic behavior in systems with few- to hundred-micron length scales are those controlled by surface tension and hydrophobicity. In this case, there is not a fundamental change in behavior as in a transition from turbulent to laminar flow, but the importance of both surface tension and hydrophobicity (or

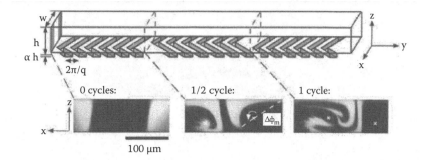

Figure 5.6 Herringbone mixer and images of flow streams being mixed. (See Stroock, A.D., et al., *Science,* 295(5555), 647–651, 2002.) Flow is in the *y*-direction. After a small number of cycles of the herringbone structure, the two flows have been forced into enough thin layers with large interfacial area that diffusion rapidly homogenizes the liquid.

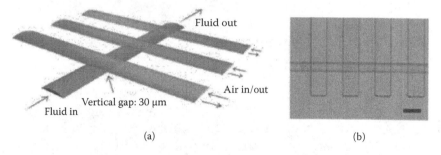

Figure 5.7 (a) Polydimethyl siloxane (PDMS) microfluidic valve design. By operating the valves sequentially, these can also pump fluid. (b) Optical microscope image of the valves. The scale bar corresponds to 200 um. (Reprinted with permission from Unger, M.A., et al., *Science,* 288(5463), 113–116, 2000.)

hydrophilicity) becomes larger as fluidic systems are scaled down until they both can be harnessed to create fluidic driving forces in ways that are not possible in macroscale systems. Because the surface-to-volume ratio of microfluidic channels is much larger than that of their macroscale counterparts, surface effects including the surface energies and liquid surface tension become important factors both in driving fluid flow and in creating novel valving approaches (Figure 5.7).

Many of the earliest microfluidic devices were based on polydimethyl siloxane (PDMS), a soft polymer that can easily be cast into structures of the appropriate dimensions [54,55].

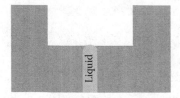

Figure 5.8 Capillary burst valve. Surface tension holds the liquid at the sharp expansion of channel width. Above a threshold pressure, the liquid starts flowing.

This material allows for microfluidic valves whose operating principle is similar to conventional valves in that one microfluidic channel is pinched closed by pressurizing and expanding a crossing control channel [56]. Many other valve designs are also based on conventional macroscale valves [57]. There are also valve designs that can only operate in a microfluidic system as they rely on surface tension effects that are far too small for macroscopic systems (Figure 5.8) [58–60].

As is evident from the above discussion, scaling fluidic systems allows for devices that harness phenomena that do not occur in larger channels. There is another scaling advantage for microfluidic systems that is much more prosaic. Very small quantities of fluid are required because of the very small volumes of the channels. A cube 100 μm on a side has a volume of 1 nL. Reasonable flow rates through typical microfluidic channels are 1 to 10 nL/s. Stated more intuitively, that flow rate range would require 28 to 280 days to consume 1 mL. The characteristics described in this section have resulted in a huge number of potential and realized applications for microfluidic systems, including DNA sequencing and separation [61–65], protein crystallography [66–68], many forms of chromatography [69–72], and various biological and medical detection devices [49,55,73,74]. The phrases *lab-on-a-chip* and *micro-total-analytical-system* (μTAS) refer to such systems [69].

The microfluidic devices we have been discussing up to this point all have a minimum dimension of at least a few microns. As channel diameters approach the nanometer regime, some phenomena become much more dramatic, and some new molecular phenomena emerge. For many fluidic sensor applications, the liquid of interest is an aqueous solution consisting of water, various salts, and often proteins or polymers. One critical parameter in nanoscale fluidics is the Debye length in a liquid. The Debye length, formally, is the distance from a charged surface at which the electric field created by the surface drops to $1/e$ of its value at the surface. More intuitively, it can be thought of as the distance beyond which the free charges in the liquid screen out other charges. When a dimension of a channel becomes similar to or smaller than a Debye length (typically 1 to 100 nm for salt concentrations of interest in biological or chemical systems), the properties of flow through the channel change [75]. Most importantly, significant net flow can be driven with electric fields through electroosmosis [76–78]. The basis of this phenomenon is as follows. Any surface in a liquid will have a slight charge due to chemical effects. Free ions in the liquid will screen that charge forming an electrical double layer. If an electric field is applied parallel to the surface, flow will be driven within that double layer. In principle, in a channel of any diameter, an electric field parallel to the walls will drive a net flow, because the flow within the charged double layer pulls the neutral bulk with it due to viscosity. In the case of a nanofluidic channel where the diameter is less than the Debye length, the electric field is acting on the entire volume of the fluid. Such flow has a relatively flat velocity profile with most of the slip occurring very near the surface, and the velocity is a function only of electric field, not of radius. This is in contrast to pressure-driven flows where a smaller radius results in a dramatic reduction in flow velocity (as the square of the radius) due to viscous drag.

Throughout the many orders of magnitude of length scale described above, liquid is reasonably approximated as continuous such that a continuum fluidics model applies. Scaling down yet further, eventually pore diameters can approach the dimensions of interatomic spacing in water, at which point treating the water (or gas) as a continuous fluid is inappropriate. In this regime, behavior depends very strongly on the details of the materials involved. For example, when using single-walled carbon nanotubes with diameters in the 2-nm range as channels for water, pressure-driven flow can travel at a rate a few orders of magnitude larger than would be expected from viscous drag predictions in the continuum fluid approximation [79,80]. This occurs because of the atomic smoothness of the walls and because the water molecules are forced to assume a "water-wire" configuration that has less drag. In contrast, polymer membranes with similar pore diameters do not show this enhanced water flow rate because of surface effects. While the carbon nanotube membranes require considerable effort to fabricate and may serve only as testbeds for fundamental physics investigations, flow through nanoscale channels is ubiquitous in living cells. For example, aquaporins are channels that allow water to flow in and out of cells [81], and bacteriorhodopsin controls proton flow [82]. In both cases, the formation of water molecule chains in the nanopore plays a significant role. Thus, fluidics in this extreme nanoscale regime turns out to be very important to life and could be very powerful in future devices that harness these phenomena.

Quantum Effects

The quintessential example of a new phenomenon that arises as the length scale of systems of interest is reduced is the emergence of quantized behavior. The differences between Newtonian (and classical relativistic) mechanics and quantum mechanics are so large they require entirely different languages, both in mathematics and in written description. A description of the mathematical differences is beyond the scope of this chapter, so this section will simply describe some of the qualitative differences that arise as the dimensions of various objects are reduced to a regime where quantum mechanical effects cannot be ignored.

It is, of course, impossible to explain many of the properties of solids without quantum mechanics. For example, to explain why some materials are insulators, some are semiconductors, and some are metals requires a quantum mechanical description of the electrons in the lattice. That quantum mechanical description leads to electronic band structure. Band structure is a strictly quantum mechanical construct, as is the fact that electrons carrying current can scatter from impurities or dislocations but cannot scatter from lattice atoms. Understanding the effective mass of electrons and holes in semiconductors also requires quantum mechanics. Nevertheless, many problems in electron behavior in solids can be qualitatively understood without quantum mechanics if the band structure and effective mass of the carriers are simply accepted. Also, many systems involving precise sensors have to deal with some aspects of quantum mechanics. For example, shot noise in photodetectors is a manifestation of the fact that the quantum of light is a photon. Similarly, shot noise in a current sensor is evidence that the electron is the quantum of charge. Generally, such manifestations of quantum mechanics do not require any deeper understanding of the theory than the knowledge that photons and electrons are indivisible. In this section, some areas where quantum effects play a fundamental role in the macroscopic properties of a system or in its interaction with the environment are outlined.

The development of quantum mechanics as a theory (mostly in the early 1900s) was driven by inconsistencies in the classical theories of electromagnetism and blackbody radiation, by the development of atomic theory, and, experimentally, by the invention of the cathode ray tube, which allowed for experiments with beams of electrons. Its development remained confined mostly to fundamental scientific questions until the development of modern silicon-based electronics. Once the silicon metal-oxide semiconductor field-effect transistor (MOSFET) was developed in the 1960s, the quantitative study of the behavior of electrons and holes in quantum wells became a very technologically relevant field leading to confirmations of many of the predictions of the theory of quantum mechanics to remarkable accuracy. These devices offered something very rare in experimental physics—model systems whose governing equations required very few or no approximations to completely solve mathematically. The electrons or holes in the channel of a MOSFET are trapped in a triangular quantum well. Their energy levels and phenomena related to them (such as temperature dependence and optical response) are easily calculated using very simple quantum mechanics (Figure 5.9) [83].

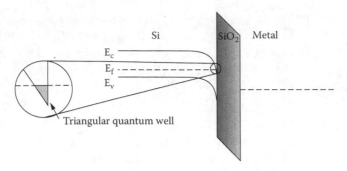

Figure 5.9 Band structure of a metal-oxide semiconductor field-effect transistor (MOSFET). The inversion layer consists of electrons trapped in a triangular quantum well.

Similarly, the electrons or holes in semiconductor hetero-structures such as $Al_xGa_{1-x}As/GaAs/Al_xGa_{1-x}As$ or Si/Ge/Si are an equally simple quantum mechanical model system with electrons trapped in a square quantum well. The analysis of these systems extends directly to nanoparticles; therefore, they are reviewed in more detail here. Such quantum wells are created when a very thin semiconductor layer of one type is sandwiched between appropriately chosen layers of another type. The key is to create a band structure where the carriers are trapped in quantum wells (Figure 5.10). The band structure schematic in Figure 5.10 outlines the important features of such confined systems. $E_{e,n}$ and $E_{h,n}$ are the energies of the bound states (those where the carriers remain trapped in the well) for the electrons and holes, respectively, for $n = 1,2,3...$ Electronic wave functions are shown for the first two states in the well. The key point here is to understand the unique and counterintuitive features of quantum wells. If the electrons were trapped in a classical well in a semiconductor (e.g., a macroscopic piece of semiconductor where the well walls are the edges of the material), in their ground state, they would each have an energy right at the conduction band edge, and there would be a continuum of available energies above that. In a narrow quantum well, though, even the ground state electrons have an energy above the conduction band edge. Furthermore, there are a discrete number of energies within the well that the electrons can occupy (in Figure 5.10, they are $E_{e,1}$ and $E_{e,2}$). As wells become narrower, the energy gaps

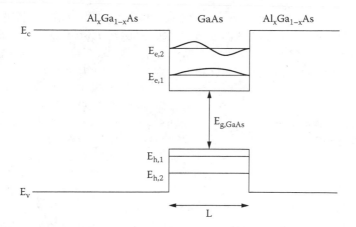

Figure 5.10 Band structure diagram of an $Al_xGa_{1-x}As/GaAs/Al_xGa_{1-x}As$ quantum well structure. L, the width of the quantum well, is typically 10 to 40 nm.

between the bound states become larger; thus, there are fewer bound states.

This quantization into individual levels of the carrier energies in a quantum well leads to fundamental differences in behavior between quantum wells and bulk semiconductors. For example, optical absorption in a semiconductor and optical emission of bulk semiconductor light-emitting diodes are governed by the band gap E_g. The wavelength emitted is

$$\lambda = \frac{h \cdot c}{E_g}$$

and the absorbed wavelengths are λ and shorter because electronic transitions into the continuum or states above the conduction band edge are possible. For quantum wells, though, those energies are modified. Typically for emission,

$$\lambda = \frac{h \cdot c}{E_{total}}$$

where $E_{total} = E_g + E_{e,1} + E_{e,2}$, for transitions between the electron and hole ground states. Absorption only happens strongly at discrete energies that match the gaps between allowed quantum well states. Those can be either from the conduction band to the valence band states or within the conduction or valence bands. Transitions within the conduction band are the basis of quantum well infrared photodetectors [84–88] and quantum cascade lasers [89–93]. Both are types of devices that rely on nanoscale layers and provide functionality that could not otherwise be achieved (i.e., conversion of infrared photons to current or current to infrared photons in a very narrow wavelength range). These are examples of devices where engineering the system at the nanoscale creates functionality that cannot exist in macroscopic devices.

In the above devices, the nanoscale layers are buried in a macroscopic semiconductor chip. However, the same arguments apply to semiconductor nanoparticles. In that case, the entire nanoparticle is a quantum well with the surface of the particle acting as the well wall. Such semiconductor nanoparticles can be fabricated in large quantities using simple wet chemistry [94–96]. This phenomenon allows the same material to be used as a fluorescent marker with a very large range of available wavelengths (Figure 5.11).

Semiconductor quantum dots have remarkable advantages when compared to traditional fluorescence techniques, which are commonly used in chemical and biological sensing and microscopy of biological processes [97–99], For years, these techniques have used fluorescent molecules to supply the fluorescence. Fluorescent molecules, however, only fluoresce for a relatively short period of time before they are photobleached. In photobleaching, the incoming photons eventually break down the molecule, at which point it stops fluorescing. Semiconductor quantum dots, on the other hand, do not suffer from this drawback. Because their fluorescence comes from electron-hole pair recombination in a semiconductor crystal, they do not photobleach. The short-wavelength light used to drive the fluorescence is not energetic enough to damage the semiconductor crystal.

The above phenomena in semiconductors can be understood using simple quantum mechanics. Particle in a box calculations suffice to predict the energy levels in the quantum wells, and therefore, the wavelengths of the fluorescence. In

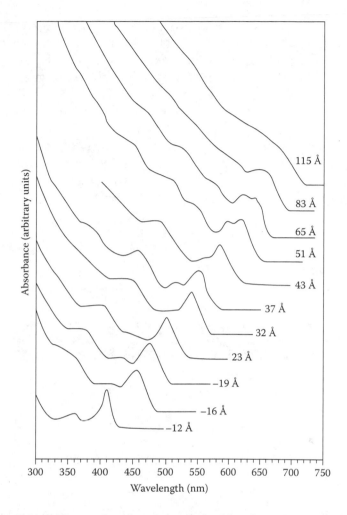

Figure 5.11 CdSe semiconductor nanoparticle absorption spectra with diameters from 1.2 to 11.5 nm. (From Murray, C.B., Norris, D.J., and Bawendi, M.G., *Journal of the American Chemical Society*, 115(19), 8706–8715, 1993. With permission.) The luminescence spectra of the quantum dots are also shown as a function of diameter.

metals, a more complicated phenomenon known as a surface plasmon resonance can create similar effects. In fact, without knowing it, such effects were harnessed centuries ago in the creation of stained glass [100]. Due to the surface plasmon absorption, the colors of metal nanoparticles can be very different from the colors of the bulk metals

[100–103]. Gold nanoparticles were used to make red glass and silver nanoparticles to make yellow glass. As with the semiconductor quantum dots described above, these metal nanoparticle-based dyes have lasted hundreds of years without fading, because this is an effect of the conduction electrons in a metal. This is in contrast to most paints or dyes where a few years of exposure to normal solar radiation will result in noticeable bleaching. As with many of the examples throughout this chapter, with both semiconductor and metal nanoparticles, there is a dramatic difference between the optical properties of the nanoparticles and their macroscopic counterparts. These differences arise only as the diameters of the particles drop below about 100 nm for metals and 20 nm for semiconductors. In contrast, throughout the many orders of magnitude of available size above those dimensions, the optical properties do not change significantly.

Conclusions

This chapter has served as a very brief introduction to a small fraction of the phenomena that can change dramatically with scale. A thorough listing and explanation of all such phenomena could clearly take up many volumes on its own. The purpose here has been to point out that there are two classes of effects that should be considered. One is those effects that vary continuously with scale, such as mass, volume, stiffness, and resonance frequency. Such phenomena also typically vary relatively gradually with scale. The other class is those phenomena that emerge relatively suddenly below a certain length scale and can lead to dramatic and unintuitive effects. These effects of scale are more difficult to design around in a typical systems engineering effort. These phenomena discussed here include purely laminar flow in microfluidic systems, molecular effects in nanofluidic systems, surface effects in MEMS devices, and quantum effects in many solids. The length scales at which the phenomena emerge vary greatly from hundreds of microns for fully laminar microfluidics to the micron-scale for surface stresses in MEMS devices to the nanoscale for Casimir forces and quantum confinement effects. For these and other such phenomena, an awareness of the length scales at which they become important is vital to any systems engineering effort that spans many orders of magnitude in size.

References

1. Fritz, J., et al., Translating biomolecular recognition into nano-mechanics. *Science*, 2000. 288(5464): 316–318.
2. Rebeiz, G.M., *RF MEMS: Theory, Design, and Technology.* 2003: Hoboken, NJ: John Wiley & Sons, 512.
3. Varadan, V.K., K.J. Vinoy, and K.A. Jose, *RF MEMS and Their Applications.* 2003, Chichester, England: Wiley & Sons.
4. Stowe, T.D., et al., Attonewton force detection using ultrathin silicon cantilevers. *Applied Physics Letters*, 1997. 71(2): 288–290.
5. Gimzewski, J.K., et al., Observation of a chemical reaction using a micromechanical sensor. *Chemical Physics Letters*, 1994. 217(5–6): 589–594.
6. Berger, R., et al., Surface stress in the self-assembly of alkane-thiols on gold. *Science*, 1997. 276(5321): 2021–2024.
7. Lang, H.P., et al., A chemical sensor based on a micromechanical cantilever array for the identification of gases and vapors. *Applied Physics A: Materials Science and Processing*, 1998. 66: S61–S64.
8. Datskos, P.G., and I. Sauers, Detection of 2-mercaptoethanol using gold-coated micromachined cantilevers. *Sensors and Actuators B: Chemical*, 1999. 61(1–3): 75–82.
9. Hu, Z., T. Thundat, and R.J. Warmack, Investigation of adsorption and absorption-induced stresses using microcantilever sensors. *Journal of Applied Physics*, 2001. 90(1): 427–431.
10. Hansen, K.M., and T. Thundat, Microcantilever biosensors. *Methods*, 2005. 37(1): 57–64.
11. Goeders, K.M., J.S. Colton, and L.A. Bottomley, Microcantilevers: Sensing chemical interactions via mechanical motion. *Chemical Reviews*, 2008. 108(2): 522–542.
12. Shu, W., E.D. Laue, and A.A. Seshia, Investigation of biotin-streptavidin binding interactions using microcantilever sensors. *Biosensors and Bioelectronics*, 2007. 22(9–10): 2003–2009.

13. Pinnaduwage, L.A., et al., Detection of 2,4-dinitrotoluene using microcantilever sensors. *Sensors and Actuators B: Chemical*, 2004. 99(2–3): 223–229.
14. Pinnaduwage, L.A., et al., Sensitive detection of plastic explosives with self-assembled monolayer-coated microcantilevers. *Applied Physics Letters*, 2003. 83(7): 1471–1473.
15. Casimir, H.B.G., On the attraction between two perfectly conducting plates. *Proc. Koninkl. Ned. Acad. Wetenschap.*, 1948. 51(7): 793.
16. Casimir, H.B.G., and D. Polder, The influence of retardation on the London-van der Waals forces. *Physical Review*, 1948. 73(4): 360.
17. Dzyaloshinskii, I.E., E.M. Lifshitz, and L.P. Pitaevskii, The general theory of van der Waals forces. *Advances in Physics*, 1961. 10(38): 165–209.
18. Genet, C., et al., Electromagnetic vacuum fluctuations, Casimir and Van der Waals forces. *Annales de la Fondation Louis de Broglie*, 2004. 29(1–2): 331–348.
19. Mohideen, U., and A. Roy, Precision measurement of the Casimir force from 0.1 to 0.9 µm. *Physical Review Letters*, 1998. 81(21): 4549–4552.
20. Harris, B.W., F. Chen, and U. Mohideen, Precision measurement of the Casimir force using gold surfaces. *Physical Review A*, 2000. 62(5): 052109.
21. Chan, H.B., et al., Quantum mechanical actuation of microelectromechanical systems by the Casimir force. *Science*, 2001. 291: 1941.
22. Chan, H.B., et al., Nonlinear micromechanical Casimir oscillator. *Physical Review B*, 2001. 81(21): 211801.
23. Munday, J.N., and F. Capasso, Precision measurement of the Casimir-Lifshitz force in a fluid. *Physical Review A*, 2007. 75(6): 060102(R).
24. Munday, J.N., F. Capasso, and V.A. Parsegian, Measured long-range repulsive Casimir-Lifshitz forces. *Nature*, 2009. 457(7226): 170–173.
25. Lee, S.-w. and W.M. Sigmund, Repulsive van der Waals forces for silica and alumina. *Journal of Colloid and Interface Science*, 2001. 243(2): 365–369.
26. Lamoreaux, S.K., Demonstration of the Casimir force in the 0.6 to 6 µm range. *Physical Review Letters*, 1997. 78(1): 5.
27. Serry, F.M., D. Walliser, and G.J. Maclay, The role of the Casimir effect in the static deflection and stiction of membrane strips in microelectromechanical systems (MEMS). *Journal of Applied Physics*, 1998. 84(5): 2501–2506.

28. Buks, E., and M.L. Roukes, Stiction, adhesion energy, and the Casimir effect in micromechanical systems. *Physical Review B*, 2001. 63(3): 033402.
29. Autumn, K., et al., Evidence for van der Waals adhesion in gecko setae. *Proceedings of the National Academy of Sciences of the U.S.A.*, 2002. 99(19): 12252–12256.
30. Autumn, K., et al., Adhesive force of a single gecko foot-hair. *Nature*, 2000. 405(6787): 681–685.
31. Autumn, K., and N. Gravish, Gecko adhesion: Evolutionary nanotechnology. *Philosophical Transactions of the Royal Society A—Mathematical Physical and Engineering Sciences*, 2008. 366(1870): 1575–1590.
32. Ruibal, R., and V. Ernst, The structure of the digital setae of lizards. *Journal of Morphology*, 1965. 117(3): 271–293.
33. Sethi, S., et al., Gecko-inspired carbon nanotube-based self-cleaning adhesives. *Nano Letters*, 2008. 8(3): 822–825.
34. Mahdavi, A., et al., A biodegradable and biocompatible gecko-inspired tissue adhesive. *Proceedings of the National Academy of Sciences*, 2008. 105(7): 2307–2312.
35. Parness, A., et al., A microfabricated wedge-shaped adhesive array displaying gecko-like dynamic adhesion, directionality and long lifetime. *Journal of the Royal Society Interface*, 2009. 6(41): 1223–1232.
36. Schubert, B., et al., Sliding-induced adhesion of stiff polymer microfibre arrays. II. Microscale behaviour. *Journal of the Royal Society Interface*, 2008. 5(25): 845–853.
37. Karch, J., R. Birringer, and H. Gleiter, Ceramics ductile at low temperature. *Nature*, 1987. 330(6148): 556–558.
38. Gleiter, H., Nanocrystalline materials. *Progress in Materials Science*, 1989. 33(4): 223–315.
39. Gleiter, H., Nanostructured materials: Basic concepts and microstructure. *Acta Materialia*, 2000. 48(1): 1–29.
40. Meyers, M.A., A. Mishra, and D.J. Benson, Mechanical properties of nanocrystalline materials. *Progress in Materials Science*, 2006. 51(4): 427–556.
41. Meyers, M.A., A. Mishra, and D.J. Benson, The deformation physics of nanocrystalline metals: Experiments, analysis, and computations. *Journal of Metals*, 2006. 58(4): 41–48.
42. Sanders, P.G., J.A. Eastman, and J.R. Weertman, Elastic and tensile behavior of nanocrystalline copper and palladium. *Acta Materialia*, 1997. 45(10): 4019–4025.
43. Kumar, K.S., et al., Deformation of electrodeposited nanocrystalline nickel. *Acta Materialia*, 2003. 51(2): 387–405.

44. Karimpoor, A.A., and U. Erb, Mechanical properties of nano-crystalline cobalt. *Physica Status Solidi A—Applications and Materials Science*, 2006. 203(6): 1265–1270.
45. Trelewicz, J.R., and C.A. Schuh, The Hall-Petch breakdown in nanocrystalline metals: A crossover to glass-like deformation. *Acta Materialia*, 2007. 55(17): 5948–5958.
46. Schiotz, J., F.D. Di Tolla, and K.W. Jacobsen, Softening of nano-crystalline metals at very small grain sizes. *Nature*, 1998. 391(6667): 561–563.
47. Schiotz, J., and K.W. Jacobsen, A maximum in the strength of nanocrystalline copper. *Science*, 2003. 301(5638): 1357–1359.
48. Schiotz, J., et al., Atomic-scale simulations of the mechanical deformation of nanocrystalline metals. *Physical Review B*, 1999. 60(17): 11971.
49. Brody, J.P., et al., Biotechnology at low Reynolds numbers. *Biophysical Journal*, 1996. 71(6): 3430–3441.
50. Brody, J.P., and P. Yager, Diffusion-based extraction in a micro-fabricated device. *Sensors and Actuators A—Physical*, 1997. 58(1): 13–18.
51. Stroock, A.D., et al., Chaotic mixer for microchannels. *Science*, 2002. 295(5555): 647–651.
52. Stroock, A.D., et al., Patterning flows using grooved surfaces. *Analytical Chemistry*, 2002. 74(20): 5306–5312.
53. Wang, H.Z., et al., Optimizing layout of obstacles for enhanced mixing m microchannels. *Smart Materials and Structures*, 2002. 11(5): 662–667.
54. Duffy, D.C., et al., Rapid prototyping of microfluidic systems in poly(dimethylsiloxane). *Analytical Chemistry*, 1998. 70(23): 4974–4984.
55. Sia, S.K., and G.M. Whitesides, Microfluidic devices fabricated in poly(dimethylsiloxane) for biological studies. *Electrophoresis*, 2003. 24(21): 3563–3576.
56. Unger, M.A., et al., Monolithic microfabricated valves and pumps by multilayer soft lithography. *Science*, 2000. 288(5463): 113–116.
57. Felton, M.J., The new generation of microvalves. *Analytical Chemistry*, 2003. 75(19): 429A–432A.
58. Man, P.F., et al., Microfabricated capillarity-driven stop valve and sample injector. *Micro Electro Mechanical Systems—IEEE Eleventh Annual International Workshop Proceedings*, 1998: 45–50.

59. Moroney, R.M., R. Amantea, and S.E. McBride, A passive fluid valve element for a high-density chemical synthesis machine. *1999 International Conference on Modeling and Simulation of Microsystems*, 1999: 526–529.

60. Duffy, D.C., et al., Microfabricated centrifugal microfluidic systems: Characterization and multiple enzymatic assays. *Analytical Chemistry*, 1999. 71(20): 4669–4678.

61. Burns, M.A., et al., An integrated nanoliter DNA analysis device. *Science*, 1998. 282(5388): 484–487.

62. Burns, M.A., et al., Microfabricated structures for integrated DNA analysis. *Proceedings of the National Academy of Sciences of the U.S.A.*, 1996. 93(11): 5556–5561.

63. Mastrangelo, C.H., et al., Microchips for DNA sequencing. *Proceedings of SPIE: Microfluidic Devices and Systems II*, 1999. 3877: 82–87.

64. Han, J., and H.G. Craighead, Separation of long DNA molecules in a microfabricated entropic trap array. *Science*, 2000. 288(5468): 1026–1029.

65. Han, J., S.W. Turner, and H.G. Craighead, Entropic trapping and escape of long DNA molecules at submicron size constriction. *Physical Review Letters*, 1999. 83(8): 1688–1691.

66. Hansen, C.L., et al., A robust and scalable microfluidic metering method that allows protein crystal growth by free interface diffusion. *Proceedings of the National Academy of Sciences of the U.S.A.*, 2002. 99(26): 16531–16536.

67. Li, L., et al., Nanoliter microfluidic hybrid method for simultaneous screening and optimization validated with crystallization of membrane proteins. *Proceedings of the National Academy of Sciences of the U.S.A.*, 2006. 103(51): 19243–19248.

68. Hansen, C., and S.R. Quake, Microfluidics in structural biology: Smaller, faster ... better. *Current Opinion in Structural Biology*, 2003. 13(5): 538–544.

69. Manz, A., et al., Planar chips technology for miniaturization of separation systems—A developing perspective in chemical monitoring. *Advances in Chromatography*, 1993. 33: 1–66.

70. Manz, A., et al., Miniaturization of separation techniques using planar chip technology. *HRC—Journal of High Resolution Chromatography*, 1993. 16(7): 433–436.

71. Lazar, I.M., P. Trisiripisal, and H.A. Sarvaiya, Microfluidic liquid chromatography system for proteomic applications and biomarker screening. *Analytical Chemistry*, 2006. 78(15): 5513–5524.

72. Lu, C.J., et al., First-generation hybrid MEMS gas chromatograph. *Lab on a Chip*, 2005. 5(10): 1123–1131.
73. Yager, P., et al., Design of microfluidic sample preconditioning systems for detection of biological agents in environmental samples. *Microfluidic Devices and Systems*, 1998. 3515: 252–259.
74. Zeng, J., et al., Design analyses of capillary burst valves in centrifugal microfluidics. *Micro Total Analysis Systems 2000, Proceedings*, 2000: 579–582.
75. Rice, C.L., and Whitehead, R., Electrokinetic flow in a narrow cylindrical capillary. *Journal of Physical Chemistry*, 1965. 69(11): 4017–4024.
76. Sweedler, J.V., et al., Interfacing biological measurements to microfluidic systems using nanocapillary array interconnects. *Abstracts of Papers of the American Chemical Society*, 2004. 227: U110.
77. Tulock, J.J., et al., Microfluidic separation and gateable fraction collection for mass-limited samples. *Analytical Chemistry*, 2004. 76(21): 6419–6425.
78. Kuo, T.C., et al., Gateable nanofluidic interconnects for multilayered microfluidic separation systems. *Analytical Chemistry*, 2003. 75(8): 1861–1867.
79. Noy, A., et al., Nanofluidics in carbon nanotubes. *Nano Today*, 2007. 2(6): 22–29.
80. Holt, J.K., et al., Carbon nanotube-based membranes: A platform for studying nanofluidics. *2004 Fourth IEEE Conference on Nanotechnology*, 2004: 110–112.
81. Sui, H.X., et al., Structural basis of water-specific transport through the AQP1 water channel. *Nature*, 2001. 414(6866): 872–878.
82. Roux, B., et al., Thermodynamic stability of water molecules in the bacteriorhodopsin proton channel: A molecular dynamics free energy perturbation study. *Biophysical Journal*, 1996. 71(2): 670–681.
83. Ando, T., A.B. Fowler, and F. Stern, Electronic properties of two-dimensional systems. *Reviews of Modern Physics*, 1982. 54(2): 437–672.
84. Choi, K.K., The physics of quantum well infrared photodetectors. 1997, *World Scientific*, River Edge, NJ.
85. Choi, K.K., QWIP technology: Advances and prospects. *Physics and Simulation of Optoelectronic Devices X*, 2002. 4646: 79–93.
86. Choi, K.K., et al., C-Qwip Focal Plane Arrays. *Infrared Physics and Technology*, 2009. 52(6): 364–370.

87. Henini, M., QWIPs enhance infrared detection. *III-Vs Review*, 1998. 11(3): 30–34.
88. Szweda, R., QWIPs-Multi-spectral mine clearance and medical. *III-Vs Review*, 2005. 18(2): 44–47.
89. Faist, J., et al., Narrowing of the intersubband electroluminescent spectrum in coupled quantum well heterostructures. *Applied Physics Letters*, 1994. 65: 94.
90. Faist, J., et al., Vertical transition quantum cascade laser with Bragg confined excited state. *Applied Physics Letters*, 1995. 66: 538.
91. Capasso, F., et al., Quantum cascade lasers: New resonant tunnelling light sources for the mid-infrared. *Philosophical Transactions: Mathematical, Physical and Engineering Sciences*, 1996: 2463–2467.
92. Gmachl, C., et al., Single-mode, tunable distributed-feedback and multiple-wavelength quantum cascade lasers. *IEEE Journal of Quantum Electronics*, 2002. 38: 569.
93. Gmachl, C., F. Capasso, and D.L. Sivco, Recent progress in quantum cascade lasers and applications. *Reports on Progress in Physics*, 2001. 64: 1533–1601.
94. Kortan, A.R., et al., Nucleation and growth of CdSe on ZnS quantum crystallite seeds, and vice versa, in inverse micelle media. *Journal of the American Chemical Society*, 1990. 112(4): 1327–1332.
95. Bawendi, M.G., M.L. Steigerwald, and L.E. Brus, The quantum-mechanics of larger semiconductor clusters (quantum dots). *Annual Review of Physical Chemistry*, 1990. 41: 477–496.
96. Murray, C.B., D.J. Norris, and M.G. Bawendi, Synthesis and characterization of nearly monodisperse CdE (E = S, Se, Te) semiconductor nanocrystallites. *Journal of the American Chemical Society*, 1993. 115(19): 8706–8715.
97. Gao, X.H., et al., In vivo cancer targeting and imaging with semiconductor quantum dots. *Nature Biotechnology*, 2004. 22(8): 969–976.
98. Sapsford, K.E., et al., Biosensing with luminescent semiconductor quantum dots. *Sensors*, 2006. 6(8): 925–953.
99. Somers, R.C., M.G. Bawendi, and D.G. Nocera, CdSe nanocrystal based chem-/bio-sensors. *Chemical Society Reviews*, 2007. 36(4): 579–591.
100. Kelly, K.L., et al., The optical properties of metal nanoparticles: The influence of size, shape, and dielectric environment. *Journal of Physical Chemistry B*, 2003. 107(3): 668–677.

101. Link, S., and M.A. El-Sayed, Shape and size dependence of radiative, non-radiative and photothermal properties of gold nanocrystals. *International Reviews in Physical Chemistry*, 2000. 19(3): 409–453.
102. Link, S., and M.A. El-Sayed, Size and temperature dependence of the plasmon absorption of colloidal gold nanoparticles. *Journal of Physical Chemistry B*, 1999. 103(21): 4212–4217.
103. Link, S., and M.A. El-Sayed, Spectral properties and relaxation dynamics of surface plasmon electronic oscillations in gold and silver nanodots and nanorods. *Journal of Physical Chemistry B*, 1999. 103(40): 8410–8426.

6

Micro Electro Mechanical Systems—Systems Engineering's Transition into the Nanoworld

Robert Osiander

First attempts as producing microchips
through top down methodology.

Contents

Introduction

The start of the endeavor into the micro- and nanoworld is probably considered by many to be Richard Feynman's "There's Plenty of Room at the Bottom" presentation during a meeting of the American Physical Society in 1959. He challenged the audience to design and build a tiny motor to fit into a cube 1/64th of an inch or to write the contents of a book at a scale 1/25,000 onto a small surface, and he offered prizes for his challenges. Using conventional tools, William McLellan was able to readily meet the first challenge. However, advancements in Micro Electro Mechanical Systems (MEMS) and nanotechnology were required to allow Tom Newman at Stanford, Palo Alto, California, to fulfill the second challenge.

Given the evolution of the field, and similar to the seemingly ubiquitous use of terms like *nanotechonology*, it is very difficult to define the categories of devices that can be considered

MEMS. Historically, MEMS has had its roots in microelectronics and microelectronics fabrication, and the *MEMS* term has been applied for almost everything smaller than an inch. While small devices fabricated with focused ion beams (FIBs) or microassembly could be considered MEMS, an important aspect is the use of batch photolithographic technology to produce multiple exact copies of the same device.

The first batch-fabricated MEMS device was a resonant gate transistor (RGT) produced by Harvey Nathanson from Westinghouse in 1964. This device joined a mechanical component with electronic elements. The fabrication was the first demonstration of what came to be called *surface micromachining* techniques, which is a process for patterning features onto a surface. In the late 1960s, H. A. Waggener introduced anisotropic etching in silicon, which was the start of bulk micromachining [1]. Kurt Petersen from IBM [2] was one of the pioneers in sensors to use resonators and pressure sensors and is considered one of the founding fathers of MEMS. His pressure sensors, which were used in blood monitoring devices, can be considered one of the earliest commercial MEMS successes and was probably the start of systems engineering in MEMS. Thermal inkjet technology based on MEMS nozzles was introduced by Hewlett Packard in 1979. In 1982, another MEMS fabrication process, X-ray Lithography *Galvanoforming* (German for electroplating) *und Abformung* (German for Molding) called LIGA was introduced, which allowed high aspect ratio microstructures to be fabricated either directly by electroplating or by compression molding using the LIGA structures as molds. Another technology, the deep-reactive ion etching (DRIE) was added to the MEMS tool-set by Bosch in 1994. It allows creation of steep-sided deep trenches in silicon wafers.

A breakthrough for MEMS came in 1993 when the Microelectronics Center of North Carolina (MCNC), with support from Defense Advanced Research Projects Agency (DARPA), created the Multi-User MEMS Process (MUMPs). While the Metal Oxide Semiconductor Implementation Service (MOSIS) foundry was the first one founded in 1981, this center was the first open MEMS foundry to allow different users (from nonsemiconductor industries or academic researchers) access to low-cost microsystem processing. Sandia offered their five-layer Sandia Ultra-planar Multi-level MEMS Technology 5 (SUMMiT IV™) MEMS process as a foundry service in 1998.

Continued commercial success of MEMS technology was seen in the late 1980s. In 1987, L. Hornbeck and W. E. Nelson at Texas Instruments invented the digital micromirror device (DMD), which is now the basis for many digital light projectors (DLPs) [3–5]. From 1987 it took years of engineering to introduce the first instrument using a DMD, the TI DMD2000 airline ticket printer in 1992. The first one-chip DLP subsystems were delivered to manufacturers in 1996 [6,7]. In 1993 Analog Devices introduced surface micromachined accelerometers in high volume and at a low cost, which allowed them to be used in air-bag sensors, making them available in all automobiles [8,9].

MEMS, and to some extent Nano Electro Mechanical Systems (NEMS), are, from a systems engineering point of view, the transition from macrosystems to nanosystems. Like semiconductor electronics, MEMS are integral parts in many macroscopic systems, from air-bag sensors in cars to gyroscopes in missiles and satellites. Their integration follows standard system engineering approaches for batch-produced components that are tested, packaged, and integrated.

Given the engineering design, manufacturing, and fabrication process, components involved, and inherent packaging, micro-electro-mechanical devices were considered systems from the very beginning. Unlike macroscopic devices, MEMS devices are not assembled from individually fabricated parts, but they are designed as a system. This is true for the most simple resonator or microphone, as well as Sandia National Laboratory's (SNL) motors, which consist of a number of springs, wheels, and gear trains [10]. Once designed, all the components of a MEMS device are batch fabricated in a series of processing steps to create the system, very much like the devices in an integrated circuit.

High integration of multiple lower-complexity MEMS devices is more comparable to a top-down nanofabrication approach. Examples of this are the mirrors for the Texas Instruments' digital micromirror device (DMD), which are used in digital light projector (DLP) technologies [11,12] or the shutters for the James Webb Space Telescope (JWST). In both cases, the devices consist of an array of thousands of movable mirrors or shutters, which are individually controlled. In the case of the DLP technology, every mirror was required to work, as no user would want even a single pixel of the display to fail. In this prime example of MEMS systems engineering, Texas Instruments invested billions of dollars into the DLP technology to ensure this level of device fidelity for each manufactured product. Given that it was a unique device and the funds were not available for development of a defect-free device, the system engineers for the JWST had to take a different view and somewhat redefine their requirements based on technology limitations. The approach is comparable to that taken for very-large-scale integration (VLSI) of integrated circuits where limitations in performance due to yield are overcome by hardware and software designs around it.

The packaging and integration of a completed MEMS component into a device can introduce a variety of design and engineering constraints. Independent of the instrument or device complexity, in many cases, the integration of a MEMS component into a device requires a systems engineering approach different from the approach that an electrical engineer would use to assemble an instrument from microelectronics components. While this is certainly true for large spacecraft instruments such as the MEMS shutters for the JWST instrument [13], even a small instrument with a MEMS sensor requires a system-level approach. The MEMS package is an integral part of the device system and is often built during the construction or fabrication process of the MEMS component [14]. For some MEMS systems, the packaging schemes incorporate a housing that can protect the MEMS component from a variety of external factors, such as particulates and chemical exposure, mechanical loading, and thermal input, which can cause disruption or damage. For an open MEMS device such as the JWST shutters, the design, assembly, and performance of the entire instrument need to take into account that even small particles can stop the MEMS device from working. Temperature-induced changes can also have a large impact on the MEMS device as resonance frequencies will shift and springs or structures can bend or distort if there is a thermal expansion mismatch between the MEMS device and the package or instrument.

Micro Electro Mechanical Systems (MEMS) Fabrication

There are two basic fabrication approaches for MEMS devices—additive and subtractive:

Additive processing is where features or material layers are patterned or blanket deposited onto a substrate and can include surface micromachining and molding techniques like the LIGA process. This substrate is usually a silicon wafer, which acts as a platform for the depositing and patterning of materials. For surface micromachining on a substrate, thin (0.2 to 3 μm) layers of material (typically polysilicon, silicon nitride, silicon dioxide, and metals) are deposited. Photoresist is spun on and patterned to expose the areas where material is to be removed using an applicable etching process. The same process is performed with the next material. In some cases, for example, SUMMiT IV, every deposition is followed by an oxide deposition and a planarization step for the wafer. Silicon dioxide (SiO_2) has a very special purpose, as it is typically used as a sacrificial material that is removed by an oxide or hydrofluoric acid (HF) etch as a final step in the MEMS process (called release or lift-off). This allows freestanding Si and metal structures, such as cantilevers and bridges, to be generated and features like-covered channels and fluidic systems to be patterned.

Subtractive processing is called bulk micromachining but can also include some selective etching processes mentioned above. This processing typically uses the substrate as the material in which to carve out the final device. Here photoresist is deposited and patterned on the substrate (silicon) as a protective layer exposing the bare areas of substrate that will be etched to form the final structure. This can be done using wet etching with potassium hydroxide (KOH), which selectively attacks the silicon along the edge of the crystalline structure to form bare planes, or *dry-etching*, which bombards any exposed surfaces with ions in a plasma field causing gradual material removal. For silicon, the DRIE process developed by Bosch allows vertical structures to be etched through the entire wafer thickness.

MEMS Packaging

In order to physically and functionally integrate a MEMS device into a system, special connectivity called *packaging* is required. As typically seen for standard microelectronics, the packaging requirements from one MEMS device to the next will vary a great deal. The package housing for a resonator can allow optical access, environmental exposure, or vacuum sealing. In addition to the housing of the device, packaging can include mechanical and electrical connectivity. This can include connection of moving parts or connection of electrical components via wire-bonded interconnects. The complexity and constraints involved in these types of housings and functional connections necessitate a systems engineering approach. In most cases, the MEMS package cannot be separated from the device design, and device and package need to be treated as a system.

MEMS Electronic Interface

One of the hardest problems in MEMS engineering is interfacing to the macroscale electronic world. Many of the sensor devices generate only a small signal or current, and considering the scale and capacitances involved, usually require immediate signal boosting by a nearby amplifier. Different approaches have been used to get around the different processing environments, which are destructive for the complementary metal-oxide semiconductor (CMOS) devices, from burying the electronics before MEMS processing or processing the electronics before the MEMS is released. More recently, chip-scale packaging allows the electronics wafer to be bonded and connected to the MEMS die, which is probably the most elegant way of hybrid packaging with two different dies. Wafer vias and other three-dimensional integration (3DIC) technologies are becoming more common in integrating MEMS and electronics. Interconnecting, driving, and reading the MEMS devices are still engineering challenges for which the solutions depend on the system in view.

MEMS Engineering

As described above, the techniques employed for the design and fabrication of MEMS systems come from the microelectronics/integrated circuit world. In order to understand the differences, let's take a look at systems engineering, focused on design and fabrication, in the microelectronics world first, and then see how this compares to MEMS fabrication.

Integrated Circuit Fabrication

The electronic components in an integrated circuit (IC), such as a CMOS transistor with a gate, source, and drain, are fabricated on an ultrapure silicon wafer. The wafers used in such processing are up to 300 mm in diameter and around 775 µm thick. For fabrication of a CMOS transistor, a defect-free silicon layer is deposited via epitaxy onto a prepared wafer followed by deposition of the gate dielectric (SiO_2). Using photolithography, the gate and the source and drain regions are patterned, and then dopants are implanted via ion beam implantation and annealing to achieve the desired electronic properties.

In order to form complete electrical circuits, the transistors (and other components) are interconnected with insulated metal conduction lines. For this processing, a thin film of SiO_2 is deposited, and trenches (for the interconnects) and holes (for the vias) are patterned and etched into the oxide. The trenches and holes are filled with a barrier layer and Cu. After each layer, chemical mechanical planarization (CMP) is used to ensure flat surfaces for the subsequent step.

An important aspect of this fabrication, which will continue into MEMS and nanotechnology, is parallel/batch fabrication. No longer are the components of the integrated circuit built separately as their own entities and assembled, but they are constructed at the same time using the deposition and patterning of successive layers over an entire substrate. The big advantage of this approach is that many completely identical devices can be built in each batch with only slight variations, which originate from defects in the masks and the wafers. One of the disadvantages in this type of construction method is the dependency of success on each serial process. Minor variations in the processing of one layer will compound in successive layers. Additionally, if a process at the end of the batch fabrication fails, the entire device and thus all previous steps are wasted. In order to mitigate these dependencies and reduce overall cost, the number of masks and, therefore, the number of layers and processes, needs to be kept to a minimum. This strategy will reduce the variety of the components across the wafer. For example, in a single process batch, the doping levels at each device will probably be kept the same, and it will be difficult to design a circuit, which requires different doping levels.

Integrated CMOS circuits are built in a semiconductor fabrication plant (fab) or a foundry (fab which only produces for outside designers). For each required process step, the IC designer sends a mask for each serial process in the design. In order to reduce cost for the designer, some foundries standardize their processes so multiple designs can be produced on a single wafer or potentially multiple wafers in a single piece of equipment. This can impose some design rules onto the designer, as well as a given order of processes.

Integrated Circuit Engineering

Depending on the complexity and hierarchy of the device, it is likely most integrated circuit or ASIC (application-specific integrated circuit) designers will ever see a photolithography mask layout or know of the different layers or processes in the device fabrication. As a means of abstraction, starting in the mid-1980s, *standard cells* were implemented where each ASIC manufacturer offered the use of functional blocks to the designer with known characteristics (propagation delay, capacitance, and inductance). These blocks, which separate design from fabrication, could be utilized in software programs that utilize hardware description language (HDL) to describe the functionality of the integrated circuit and facilitate efficient circuit layout. Logic synthesis tools now compile such HDL descriptions into a gate-level netlist, which enables the standard cell integrated circuit design. The different steps are shown below, which follow very well the systems engineering approach described in the beginning of this book:

Requirements analysis: Determine the required functions of the ASIC.

Register transfer level (RTL) Design: Describe the functions of the ASIC in HDL.

Functional verification: Software models and simulation are used to verify the RTL design.

Logic synthesis and layout: The RTL design is turned into a collection of standard cells and electrical connections between them, called a gate-level netlist. A placement tool places the standard cells on the ASIC, and a routing tool uses the netlist to create electrical connection between the physical placements of the standard cells. The output is a layout file that can be used to create the photomasks required to produce the integrated circuit in a foundry or fab.

Signoff analysis: From the final layout, parasitic resistances and capacitances are calculated to determine delay information and performance via timing analysis. A design rule checker determines if the designs follow the rules recommended by the foundry (e.g., with respect to width, pacing, and enclosure). After completion, a set of photomask designs is released to the foundry for chip fabrication.

Fabrication: Using photolithography, source and drain regions are patterned, dopants are implanted, and trenches and holes for vias and interconnects are patterned, etched, and filled.

Metrology: After each step, wafer test metrology verifies if the wafer has been damaged and needs to be scrapped.

Wafer-level device testing: Once fabricated, each die is tested at the wafer level. In many cases, test structures to allow and simplify these tests are included in the wafer design. The ratio of properly working dies on a wafer is referred to as *yield*.

Chip-scale packages: A modern package approach uses chip-scale packaging where the package (a silicon or glass cover die) is attached at the wafer level, and then a die-saw is used to dice the wafer into completely packaged devices.

Packaging: Each die is mounted into a ceramic or plastic package, the die bond pads are connected to the package pins via wire-bonding, and the die is capped or hermetically sealed if required.

Final retesting: This confirms performance of the integrated circuit.

As shown above, producing a microfabricated integrated circuit follows pretty much a standard systems engineering approach. The steps would be a little different if the IC design required fast design iterations for tweaking of speed, chip density, and so forth. For a custom device, given the front-end resources required, this standard cell approach can no longer be used. For custom designing, there is a much closer connection between the design functionality and the mask layout; therefore, knowledge of the physics and the fabrication process is required. If the custom design closely follows another with a previously established fabrication process, this cell-like approach may be employed. This is true only on a new development. Once a process has been developed and reduced into a standard cell-like approach, the approach sketched above applies again.

One approach, which revolutionized systems engineering approaches in semiconductor design, was the introduction of VLSI. In order to integrate these large electronic systems like a microprocessor onto a single die, die size or device density or both needed to increase. This requirement obviously increases fabrication complexity and negatively impacts the batch yield. Additionally, because modularity is decreased, it requires dies with less and smaller defect density and photolithography patterning with higher resolution and improved alignment accuracy. When considering the potential sources of variation in materials and processing and the decreasing tolerances brought about by small features and higher feature densities, it might actually be impossible to create a single die without any defect. Therefore, to compensate, new approaches were taken during the design process allowing for a certain number of defects in the die by expecting these in the design of the die. Once produced, a die can be tested, and the defective areas can be identified and assessed. Depending on the defect density and functional location of defects, this information is stored and can be used to determine the potential operating range in the application of the die. For example, even with a defect that impacts speed, the die could still be used with a speed lower than that for which it was originally designed.

MEMS Fabrication

Given the evolution of the semiconductor field and the technologies available for patterning microscale structures, it is no surprise that the fabrication steps for MEMS devices are similar to those for integrated circuits. MEMS devices can consist of mechanical components, such as membranes, cantilevers, springs, actuators, and so forth, or electrical components, such as conduction lines or sensors that detect changes in specific phenomena (impedance, electrical fields, temperature, etc.). MEMS devices can consist of a single functional or structural component (e.g., a MEMS microphone, a resonator) or a number of integrated components such as springs, proof-masses, and rotors.

As is the case for integrated circuits, batch fabrication involving the series deposition and patterning of blanket layers of material is required to construct each device. Despite the similarities in processing methodology of CMOS and MEMS fabrication, there are differences in the processing (i.e., temperature), structural geometries, and materials used depending on the mechanical requirements of the final device. Whereas the design of CMOS devices focuses on the electrical resistivity and conduction and the thermal issues, MEMS devices need to account for mechanical factors like material elasticity and durability. For example, an extreme elevation in a processing parameter like chamber temperature might be required to achieve superior mechanical performance. Additionally, whereas an IC device will be separated from the wafer and batch by dicing the substrate around it, many nonplatform-based MEMS devices need to be physically detached or "lifted-off" from the substrate following fabrication.

MEMS Engineering

The MEMS devices are typically fabricated in foundries such as MCNC's MUMPs process (now MEMS CAP), or SNL's SUMMiT IV process. As described before, each foundry has a predefined series of processes for the successive deposition and patterning of material layers and thicknesses to allow multiple designs to be placed on the same wafer. Custom MEMS devices with special materials and thicknesses require unique processing; therefore, they are usually more expensive to manufacture. Typically, these special processing runs are used only for special mass products like Analog Devices Accelerometers, Texas Instruments' DMDs, or certain high-impact applications requiring specific performance. The different steps in a standard MEMS design and fabrication process are shown below and follow a similar path as the steps for microelectronics:

Requirements analysis: Determine the required functions of the MEMS device.

Functional model: This is often a multiphysics model where electrical behavior is coupled to mechanical behavior. In some applications, fluidic behavior (gas or liquid) also needs to be included, as does thermal behavior, particularly for thermally sensitive devices or thermally actuated devices.

Functional verification: Simulation software is used to verify the performance of the MEMS model.

Layout: The mechanical model is translated into a MEMS design built of different layers. Different foundries have libraries of such devices, which take to required parameters (frequency, capacity, etc.) as input and create a multilayer design. For custom designs, software is available (Coventor, L-Edit, Sandia's MEMS design tools with AutoCAD, etc.), which helps to translate the model into layers and follow the design rules; however, there is still significant manual layout involved. Placement and routing are typically done manually, because the device densities are small enough. The output of the layout software is a layout file, which can be used to create the photomasks required to build the devices at a fab.

Signoff analysis: For some standard MEMS designs, the performance can be estimated using certain software modules, which will "model" the fabrication and generate a 3D model of the end result. This process still requires processing expertise and, typically, a design review. Once at the foundry, the design rules will be checked before the photomask designs are released for the fabrication.

Fabrication: Using photolithography, each layer is patterned and deposited and possibly planarized, etched, or annealed.

Metrology: For more complex MEMS processes, metrology verifies if the wafer has been processed correctly or needs to be scrapped. A common approach for process development is to start a large number of wafers in the process and only run a partial lot to have some backup wafers in case the processes did not work. Test structures are typically included by the foundry to measure the performance.

Wafer-level device testing: Usually, the device's functionality cannot be assessed before the devices are released, but some of the electric properties (the E in MEMS) can be tested.

Dicing and release: This process is unique to MEMS devices. Using etchants, a sacrificial layer is removed which generates freestanding MEMS devices. Depending on the device, these can be very sensitive, and often the release process is performed after the wafer is packaged and diced.

Wafer-level packaging: Wafer-level packaging (WLP), where a silicon or glass cover die is attached at the wafer level, was introduced for MEMS devices initially and has become more important for MEMS because it allows release processes at the wafer level before die-saw. The wafer level package keeps the particles generated from the die-saw away from the sensitive MEMS components. This also allows sensitive mechanical structures to be packaged in vacuum or sealed in atmosphere and from then can be used like an electronic die in a plastic package.

Packaging: Many MEMS devices are packaged into ceramic or plastic packages, but often the package is part of the engineering design, and special requirements such as environmental access or damping restrictions require custom packaging approaches.

Final testing of the packaged device confirms performance of the MEMS device. Even though this seems to be a straightforward engineering approach, the geometric and material variability and processing parameter ranges can create processing interdependencies that complicate the overall systems design. For example, etching selectivity can preclude the use of certain materials or result in additional processing to pattern protective layers. The breadth of materials expands the application *ad infinitum* as much as if one would include all semiconductors into IC processing.

MEMS Systems

From the large number of MEMS devices that have been integrated into systems, few have had commercial impact, and those that have are still in an engineering spiral for performance enhancement or miniaturization. Other devices have been specifically developed for unique applications such as a spacecraft mission. Depending on the production level, application, and device complexity, the systems engineering approach for MEMS devices is quite different.

Only a few MEMS systems are finalized at the component level with little device integration or packaging needed. Examples for those are resonators or MEMS switches. These devices can be treated like electronics components requiring just electrical connects and simple packaging to maintain a preset pressure whose precision controls the resonator quality factor.

Most MEMS devices are designed to interact mechanically with the environment (e.g., directing optical beams, detecting sound waves, or directing liquid flow). In this case, the devices are integrated into a larger system, and package and microscale-to-macroscale interface is the important aspect for the systems engineer. The complexity of this packaging can range from simple electrical connectivity to high precision physical alignment of moving mechanical parts. A simple example for packaging is the integration of a MEMS microphone into a circuit board using standard electronic assembly processes. More complex examples are the integration and environmental considerations for MEMS micromirrors from Texas Instruments or the shutters for the James Webb Space Telescope. Requiring even more development are medical and biomedical devices, which require not only biological compatibility but also ensured benefit for each patient.

One large advantage that MEMS components provide is miniaturization. Although this can provide tremendous benefits for integration into space- and weight-limited devices, it can require special considerations during the system engineering approach for packaging. A good example of the benefit provided for MEMS miniaturization was the development of stand-alone accelerometers and gyroscopes, both components designed to be inertial measurement units (IMUs). The incorporation of these components into car air-bag deployment systems was a big breakthrough for the MEMS field. This miniaturization benefit propagated the development of technologies to amplify the very small charges generated in the reduced-impedance capacitive readout. Soon after, gyroscopes with six degrees of freedom on a single die were successfully developed. Even though the MEMS device did not change much, with exception to some evolutionary performance improvement, the device as a whole, the IMU, suddenly became the complete microsystem. These motion-detecting systems, which are presently in use in commercial entertainment systems, eventually contained both the electronics for translating MEMS outputs into coordinates and the microprocessor capability to translate velocity data into motion signals such as shake, up, down, and so forth. The MEMS device becomes part of a more integrated microsystem, which utilizes MEMS technology to integrate multiple components into a small package.

The following will discuss some commercial and spacecraft applications that have been developed from MEMS devices and now are integrated MEMS systems.

MEMS Microphone

Given that the first MEMS devices were pressure sensors using small freestanding membranes, taking the next step to build a MEMS microphone is not all that exciting by itself, and there has not been a commercial or technical driver for a long time. At their inception, MEMS microphones could not compete with electret microphones, which were already very inexpensive, more sensitive, and nearly as small. This, however, changed with the introduction of a newly popular mass-production process called lead-free solder reflow assembly. Instead of point-soldering each stand-alone electrical component for integration into a circuit board, this new process heated the entire board to reflow all patterned solder at once. Although MEMS devices could withstand the processing temperatures required during the assembly cycles, electret microphones needed to still be placed and soldered individually. Given its allowable use in this mass-production process, the development of the Knowles SiSonic™ MEMS microphone was targeted toward high-volume consumer electronic products where cost is a key factor [15].

For a complete auditory system (in addition to the MEMS microphone sensor), accessory circuit components are required to boost and process the MEMS microphone signal. Established and inexpensive CMOS amplifiers are well suited to provide this amplification and are small enough to be electrically packaged to the microphone and allow the sensor to be open to the environment.

The microphone is a fully clamped round polysilicon membrane about 0.5 mm in diameter and 1 μm thick, which is micromachined on a standard silicon wafer. During fabrication, the silicon below the membrane is removed using deep reactive ion etch (DRIE). The total die size is 1.65 × 1.65 mm, and the electrical packaging for the microphone contains a series of separately constructed CMOS amplifiers to boost the signal from the membrane. Given the limited capacity between this membrane and the substrate, these amplifiers are placed very close to the MEMS device to reduce stray capacities and prevent the introduction of electromagnetic interference (EMI).

Without a commercial foundry to offer joint MEMS and integrated CMOS electronics fabrication, the dies that contain the individual fabricated microphone and amplifiers are subsequently packaged together by using standard semiconductor packaging equipment at an extremely low cost. The packaging sufficiently exposes the microphone sensor to the environment yet still protects the MEMS and CMOS die from both physical damage and unwanted electrical noise.

MEMS Inertial Measurement Systems

MEMS inertial systems such as accelerometers and gyroscopes can probably be considered to be the MEMS devices with the largest commercial impact. Driven by the needs of air-bag technology and entrance into a potentially large commercial market, Analog Devices introduced the ADXL50 in 1996 as the world's first integrated MEMS accelerometer [13,14].

For the development of the ADXL50 accelerometer, there are multiple systems engineering considerations. A more holistic view looks at the inertial navigation system (INS) and its spiral development into a microsystem. An inertial measurement unit typically consists of three-axis accelerometers and three-axis gyroscopes and the read-out electronics that determine direction and velocities. An INS also requires a processor, such as Systron Donner's MMQ50 INS system [16], to perform *Kalman* filtering and provide continuous position information. The development of such a system requires an initial plan for integration of these individual components followed by the design and construction of each component. This system design and integration has led to the development of INS devices like InvenSense's MPU-6000 three-axis Gyroscope, three-axis Accelerometer, and nine-axis sensor fusion [17], which can incorporate motion-processing libraries to allow use in smartphone and gaming devices.

Looking at the development of inertial MEMS devices from a historical view reveals the same spiral steps that have to be taken when designing an integrated MEMS INS starting with the MEMS accelerometers.

Analog Devices' success with the ADXL50 accelerometer was based on the integration of analog electronics with the MEMS structure using Analog Device's *i*MEMS® process [18]. Their process allowed the electronics and the MEMS devices to be built on the same die, which was a critical development. This close integration allowed parasitic capacities to remain small, increasing the overall measuring resolution of the accelerometer [19,20].

Another important functional requirement for a complete IMU is the measurement of orientation using a gyroscope. Most MEMS gyroscopes measure the effect of the Coriolis force onto a vibrating mass. This measurement uses an object in a rotating reference system where an applied pseudo-force proportional to the rotation rate acts perpendicular to the direction of motion. The difference in this accelerometer design is that the mass is driven in one direction, and the deflection is measured in the other direction. The ADXRS150 and the ADXRS300 with full-scale ranges of 150°/s and 300°/s from Analog Devices were the first commercially available surface-micromachined angular rate sensors with integrated electronics [21].

Although combining two axes onto a die is straightforward, the integration of a third axis requires a different design for either accelerometers or gyroscopes. In 2006 Analog Devices introduced the ADXL330, which provided three sense axes in a 4 mm × 4 mm Lead Frame Chip Scale Package (LFCSP) by combining both MEMS structures and ASIC circuitry on a single die [21–23]. Just recently, Analog Devices introduced the ADXL345, a triple-axis accelerometer that uses separate MEMS and ASIC dies wire-bonded together in a single package. While this separates the fabrication processes, the increased modularity allows for higher complexity in the ASIC circuitry. A similar approach is used by InvenSense for their MPU 6000 using the ASIC die as the cap for the MEMS die [24].

MEMS DMD

One of the biggest success stories in Micro Opto Electro Mechanical Systems (MOEMS) is the Texas Instruments Digital Mirror Display (DMD). First invented in 1987 and introduced in 2000, the technology now powers more than 1,500,000 projectors [25,26]. DMD is an array of mirror-switches each 16 µm by 16 µm in size that reflect light in and out of the optical path.

Texas Instruments' development of the DMD for digital light projection is an excellent example for systems engineering involving MEMS. It started with the work of Larry Hornbeck in 1977 at Texas Instruments to develop a device to modulate light for the Department of Defense. The first prototype was a deformable mirror based on a metalized polymer membrane, which was controlled by a charge-coupled device (CCD) array.

Further development in the 1980s led to an analog system that incorporated a reflective metal cantilever controlled by n-channel metal-oxide-semiconductor (nMOS) field-effect transistor circuitry. At this point, Texas Instruments decided to investigate the incorporation of mirror display technology into their printer systems. However, the analog mirror did not meet the requirements for the printer or display applications. As a result, focus was placed on the utilization of digital switches that could turn the mirror in or out and modulate the intensity via pulse-width modulation. The first DMD, a linear 512 pixel array, was built in 1987, and in 1992 Texas Instruments utilized a 840 pixel array in their high-speed airline ticket printers.

In 1991 Texas Instruments formed a corporate-level venture project to develop digital video applications using the DMD, which then became digital light processing (DLP).

Texas Instruments built a dedicated wafer fabrication facility for the development of this concept, which replaced the pixels with arrays of close-packed micromirrors on torsional flexures hidden under the mirror. This new design helped to remove scatter from the projector light. They then developed a pulse-width modulation scheme based on electromechanical latching of the mirror to the electrodes displaying the optical signal while memory is being reloaded. One of the impressive breakthroughs for this DMD device is that the MEMS components are integrated directly with control electronics on the same chip. This provides advantages for both simplified connectivity and control of each mirror and just requires careful selection of materials to ensure fabrication process compatibility for both the MEMS and CMOS components. Given the number of mirrors on each device, this direct connectivity is necessary because it would be virtually impossible to wire-bond sufficient connections from the mirrors to the chip to pass all switching information. The aluminum alloy used to construct mirrors allows low-temperature processing compatible with the silicon electronics on the chip. With this process and device integration, the first DLP projection system was introduced in 1997, and it captured a 20% share of the market very quickly.

Since the release of the initial projection system, DMD technology has been expanded to digital cinema applications, volumetric displays, and spectral processors. As of 2008, DMD chips were available in sizes up to 2048 × 1080 mirrors, each mirror moving up to 3,000 times per second or 10^{12} cycles per 100,000 hours.

In the case of the DMD technology, the engineering development began at the manufacturing process. The DMD array is a monolithic device with the aluminum DMD superstructure built onto a CMOS address circuit on a silicon wafer. The DMD structure composes three layers of an aluminum alloy developed for the requirements of the system with two layers of sacrificial photoresist for release. After release, a self-assembled monolayer is deposited onto the device during packaging to act as a lubricant to prevent sticking. With this design, the aluminum alloy for the torsion fixture allows reliability to more than 20,000 hours. The packaging requirements for this system include optical access through a high-quality window, image-limiting aperture, package headspace for reliable operation (hermeticity and contaminant removal via getters), and temperature control.

The DMD is one of the most complex machines ever built with 1 million moving parts working reliably over trillions of cycles. The challenge for the device is that each pixel needs to work reliably as a single failure is readily identifiable. Due to the requirement for very high fidelity, a Failure Modes and Effects Analysis (FMEA) was used to ensure reliability, and a test-to-failure implementation was used to probe the limits of the devices.

Space Technology 5 Satellites— MEMS Louvers

In addition to the large-scale manufacturing for commercial products, MEMS devices can be uniquely designed for many noncommercial and limited-production systems. Although development for these systems still requires iterative design, fabrication, packaging, and assembly, only limited numbers of working devices are constructed and used for the prescribed and dedicated functionalities. Many examples of these devices originate from development for military and space applications, and include the shutters used for the Space Technology 5 (ST5) satellites' MEMS radiator [27,28] and the James Webb Space Telescope [29].

Given the scope, unique deliverables, and the defined budgets for these types of projects, different requirements exist for device development. A typically limited budget does not allow the same investment as Texan Instruments undertook to build a complete MEMS product platform. This is especially true when considering the limited application space and production requirements of these unique devices. For devices used in space, special environmental considerations can impose engineering constraints and boundary conditions. In all, the following unique conditions and requirements expected for a device launched into space can complicate the design: low-power requirements, resistance to humidity and severe mechanical inputs while on the launch pad, viability in space vacuum and in extreme temperatures ranging from −45°C to 65°C, limited outgasing, resistance to micrometeorites in space, and the need for backup systems and redundancy to prevent complete failure. Given the cost of launching devices into space, extensive testing is also needed to ensure performance.

One of the first space devices comparable in complexity to the DMD was the MEMS shutter array for the National Aeronautics and Space Administration's (NASA) New Millennium Mission ST5, launched in 2006. The development of this device resulted from a collaboration between NASA's Goddard Space Flight Center, the Johns Hopkins Applied Physics Laboratory, and Sandia National Laboratories. This device functioned with minimal energy to open and close a 10 × 10 cm radiator in order to control the emitted radiation, and therefore the temperature, of the satellite. As a forerunner for small micro- and nanosatellites, the successful demonstration of this MEMS device to provide active thermal control validated the applicability of MEMS in space.

The initial shutter design used arrays of large louvers, 1 × 0.4 mm, that were opened and closed by MEMS motors (see Figure 6.1). Although this design allowed good coverage and greater material selection for the high emissivity surface (the exposed material), these larger active areas appeared to be too delicate for space flight. Additionally, the initial fabrication process that was used did not allow for high-efficiency electrostatic motors or reliable shutter hinges. Thermal expansion mismatches between the gold coating and the silicon substrate of these large louvers also led to material strains large enough to induce unintended bending and curling. The subsequent design compartmentalized the large louvers into arrays that contained shutters that were each 6 μm wide, 150 μm long, and 6 μm apart. These arrays, which were 3 × 6 mm in size, could be opened and closed by teams of six integrated electrostatic motors powered with voltages up to 30 V. In the first prototype, nine such arrays were on each die, as shown in Figure 6.2. These prototypes were fabricated with Sandia's Summit V process, gold coated, and then subjected to electrical, shock, and temperature tests. The design utilized "mushroom" bondpads, which were designed to allow

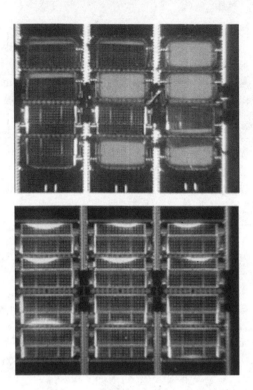

Figure 6.1 Micro Electro Mechanical Systems (MEMS) louvers, partially open (top) and all closed (bottom). (Courtesy of JHU/APL.)

the entire array to be gold coated without creating shorts at the bondpad–silicon interface. For the same reason, the design included buried interconnects between the bondpads and the motors. In this design, the movement of the shutters was sensitive to any type of debris, either mechanically disabling the shutter action or inducing a short in the drive circuit as only one in five devices had no defects and complete functionality.

As a compromise between higher yield, which can be dependent on device size and total coverage area, the next design used arrays of 300 × 300 shutters, and a die size of 12.65 × 13.03 mm. Instead of independent control, all shutter arrays were driven by a single control supply. In order to ensure that an electrical short in any shutter would not short the entire array, each shutter was connected to this drive power with a MEMS fuse. In the case of a short, this fuse would blow and disconnect the broken shutter from the rest of the shutter array.

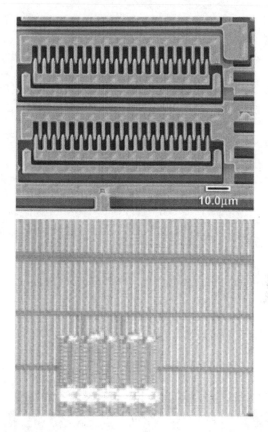

Figure 6.2 Close-up of electrostatic motor and shutters (top) and shutter array (bottom). (Courtesy of JHU/APL.)

Given the number of moving devices, fidelity remained a main challenge. Each die was tested semiautomatically, and after initial power-up, nonworking shutters would blow their individual fuses. By quantifying these broken fuses, the yield of working shutters could be better quantified. For this specific application in space, the goal was to achieve roughly 95% of the shutters working.

The sensitivity to particulate contamination created unique packaging challenges for protecting the shutter elements during launch. This device was mounted to the exterior of the spacecraft; therefore, grounding for the spacecraft electrical surface needed to be considered as well. The final package covering for the radiator utilized a window made from 3 μm CP1 polymer that was coated with indium-tin oxide (ITO), which is a transparent conductor, to meet the conductivity requirements. To manage the thermal expansion mismatch between aluminum and silicon for the survival temperature range (−45°C to 65°C), an intermediate carrier made from aluminum nitrite was used.

For assembly, sets of six die with wire-bonds connecting all the common inputs were attached to an aluminum nitride substrate with conductive epoxy. Six of these die sets were then themselves attached to the aluminum radiator with epoxy. The overall shutter control was closed-loop, utilizing a thermistor temperature sensor colocated on the underside of the radiator chassis with a heater to allow both heating and cooling control of the radiator temperature. A top-view picture of the radiator is shown in Figure 6.3.

Prior to launch, the package went through a full space qualification including thermal vacuum, vibration, shock, and acoustic testing at NASA Goddard Space Flight Center. The presence of MEMS devices open to the environment required some changes in the testing work-flow. The radiator was protected by a "remove before launch" cover, which, while protective, was

Figure 6.3 Space Technology 5 (ST5) radiator, six substrates with six Micro Electro Mechanical Systems (MEMS) shutter arrays each. (Courtesy of JHU/APL.)

not kept under vacuum. In air, condensation could potentially cause shorts and stiction, and any voltage on the devices would blow the respective fuses disconnecting these specific shutters and reducing device performance. Therefore, testing was only performed while under vacuum. This ST5 radiator device and its MEMS components finished its mission successfully in 2006 working for the full 3 months of the mission life.

James Webb Space Telescope (JWST) MEMS Shutters

The MEMS shutters for the James Webb Space Telescope (JWST) provide an example of a high-complexity device that was designed around the MEMS component [29,30]. These shutters selectively expose an infrared spectrometer, the Near-Infrared Spectrograph (NIRSpec), to different sections of an image to perform simultaneous spectroscopy on up to 100 very dim targets. To further illustrate the complexity, one of the original NASA requirements was an open fill-factor of more than 65%. Two potential solutions existed: small micromirrors (similar to the Texas Instruments' DMD), which would reflect undesired objects out of the optical path, or shutters, which would block unwanted parts of the image.

The final implemented solution was an array of silicon shutters with a magnetic coating, which were controllable by a proximal magnetic field moving across the array. The shutters could be independently opened by an electrostatic electrode allowing for complete control of the image that passed through the array. Figure 6.4 shows an earlier version of the shutters with one of them being opened mechanically. Each of the four 41.2 × 44.6 mm microshutter arrays contains 171 × 365 shutters, and each shutter is 204 μm long and 105 μm wide. In addition to achieving space qualification (thermal vacuum, shock, vibration, and acoustic testing), these shutters had to work at cryogenic temperatures (35°K) and have the ability to successfully open and close more than 100,000 times. Unlike the ST5 shutters that are all on a common signal line, this shutter device needed to demonstrate independent shutter control. Using a strategy employing physical electrical conduction lines would require a large number of connections between the shutter die and a control board. Because these

Figure 6.4 The James Webb Space Telescope (JWST) microshutter array. (Courtesy of the National Aeronautics and Space Administration.)

Figure 6.5 The James Webb Space Telescope (JWST) shutter array mounted onto the substrate. (Courtesy of the National Aeronautics and Space Administration.)

connections also needed to survive the transition to cryogenic temperatures, NASA used indium bump bondpads to connect the shutter array to the silicon substrate. Figure 6.5 shows such a shutter array mounted on the substrate. To reduce the possibility of failure, approaches were taken to mitigate risk and ensure array performance. For example, failed shutters were sealed, and any shorts in the metal traces were removed using laser ablation. Figure 6.6 shows all four shutter arrays mounted in the NIRSpec instrument.

Figure 6.6 The Near-Infrared Spectrograph (NIRSpec) microshutter assembly. (Courtesy of the National Aeronautics and Space Administration.)

From Micro- to Nanosystems

The goal for this chapter is to define the unique requirements and conditions for MEMS system engineering as compared to standard system engineering principles and techniques and to reveal some past MEMS successes and their unique systems approach. Background on a variety of devices was provided to demonstrate where MEMS have been integrated into engineering systems. The MEMS louvers that were built for spacecraft heat dissipation are a good example with which to examine the systems engineering approaches as analogous devices exist at the macroscopic, microscopic, and potentially at submicron/nanolevels.

The comparison of engineering approaches for macro- and microscale devices provides some insight for the development of nanoscale devices. A macroscopic system, which includes an assembly of modular components, is always built top-down as the development of each subsystem is carried out independent from each other. For example, a macroscopic louver consists of a holding frame, the louvers, the driving motor, and the electronics control module. The final device will utilize one modular radiator design, which is potentially constructed and incorporated multiple times. Each of the shutter and motor subcomponents will be constructed in parallel and then subsequently assembled. In addition to reducing the reliance on construction yield for device success, modular design also allows for independent testing prior to assembly.

As illustrated previously, for a MEMS system, both the design approach and construction processes have significant differences compared to macroscale systems. MEMS systems cannot readily utilize subsystem components, because the requirement for assembly and packaging at the microscale needs to be minimized. Where modularity and reuse of subsystem components would be implemented in a macroscale system, the MEMS solution would instead reuse a particular design or feature multiple times, which would be batch fabricated at once. Even though mass production is possible using batch fabrication, it carries with it a higher reliance on batch yield, because the serial replacement of any nonfunctioning features in a MEMS component is either not possible or is extremely resource intensive. When considering the development of the ST5 shutters, the bondpads had to be designed in a way that the final gold coat of the entire array, which was not part of the SUMMiT V process, would not short to the substrate. The removal of these shorts required individual laser ablations, which did not repair the device but simply prevented its complete failure. Given that features of the MEMS component are fabricated together, thorough knowledge of each fabrication process parameter and its effect on those features is important from the initial design and forward. These considerations range from processes like wet-etching silicon, where the etching angle depends on the orientation of the silicon crystal lattice, to blanket etches using deep-reactive ion etching (DRIE), where the etch rate depends on the feature dimensions.

Once the system is designed, all components are built in parallel on the wafer, layer by layer, and again, the main reliance for device success is the yield. Yield is a seminal issue for semiconductor manufacturing, and given the reliance on similar construction techniques, it extends to MEMS. As described previously for the ST5 shutters, the batch yield decreases with increasing size according to a power law. Increasing complexity of the MEMS component that increases the number of fabrication processes can decrease yield as well. This has to be taken into account in the very beginning of a project, in the requirements definition phase, and has to be integrated into the system design.

Two approaches for reliance on yield have been discussed: the refinement of the fabrication process to increase the yield and the introduction of compensatory design features to allow device performance despite nonperfect feature yield. Texas Instruments invested to develop processes and approaches that, after the infusion of millions of dollars, delivered error-free devices with a reasonable yield. The ST5 MEMS shutters program chose the second approach by incorporating fuses to prevent an electrical short from causing the failure of the entire device.

Assuming we could scale the MEMS structures down to the submicron and nanoscale to produce Nano Electro Mechanical Systems (NEMS), it is likely that the design approach would closely follow the top-down approach that is used for semiconductor devices and MEMS. NEMS devices share similar design challenges with MEMS and have promising benefits for continued miniaturization and reduced power requirements, but nanoscale material and scaling-based limitations ultimately limit the overall application space. For example, in a nanoscale shutter system similar in concept to the ST5, once the shutter becomes smaller than the wavelength of the light, it will no longer transmit all wavelengths. In addition, the metrology and testing techniques available to MEMS, even something as simple as visualization, would not scale down appropriately or provide insufficient resolution for NEMS. Characterization and validation of the device performance would be difficult or even impossible. Given these complexities and limitations, the top-down systems approach of a nanoscale device is still prohibitively challenging at this point.

There are, however, examples of nanosystems that have been developed using a bottom-up system approach. The integration of carbon nanotubes (CNTs) has been successful in a variety of technologies, because they can provide unique structural, geometric, thermal, and electrical properties. They can be used to increase the conductive surface area and, therefore, reflection when spread on top of a substrate (e.g., by an electric field), or, given their high emissivity when standing up, to radiate heat. This could be used to switch the emissivity of a CNT-coated substrate via an applied electric field. This emissivity change can only be observed for a forest of CNTs, because a single CNT does not change its emissivity whether it is flat on the substrate or straight up. This new property is called an emergent property, existing due to the assembly of the CNTs only. In contrast, conductive polymers will change emissivity within an applied electrical field independent of the dimensions of the polymer. This differentiates a nanosystem from a material such as the polymer.

Like the deposition of thin-film SiO_2 onto a silicon wafer or the careful etching of nickel to produce nanosized islands, the process of growing CNTs is an established MEMS fabrication technique. Process refinement for the construction of semiconductor and MEMS components will increase yield and repeatability, but the process conditions for growing an individual carbon nanotube cannot be controlled precisely enough to manipulate the shape or permit exact reproduction of a specific carbon nanotube design. Given the scale and the method of CNT growth, the environmental conditions (chemical availability, thermodynamics, etc.) seen at the chemical bond level contain too much variability, and increased reproducibility requires the development of microscale technologies dedicated to enhancing this control.

Although controlled growth of individual nanotube elements is currently limited, a variety of design parameters for the development of CNT systems, such as nanotube density, length, thickness, and potentially in the near future, chirality, can be controlled. Given enough research and resources, the processes can be set to achieve these design parameters, but within the distribution set by environmental conditions such as processing temperature. Despite the dimensional considerations, a NEMS designer can use an understanding of phenomena like van der Waal forces and CNT behavior in an electrical field to manipulate, among other things, the direction of CNT growth. Fine-tuning the density, length, and so forth, of the nanotubes, which affects overall emissivity, can be employed as a means to control electrical functionality. However, many uncertainties in the development of this relatively well-understood nanotechnology area still exist. The performance of this and other nanosystems cannot be reliably controlled or designed, just optimized within certain limits. For nanotechnology in general and the development of NMES, fundamental discovery is required to overcome these extensive challenges.

Concluding Remarks

Within the last two decades, the field of MEMS has achieved enough maturity to be integrated into a wide breadth of engineered systems from gaming devices, to projectors, to specialized space applications. For MEMS development, the systems engineering approaches developed for macrosystems cannot only be adapted well, but are required, to couple the environment, the MEMS device, and the electronic and macroscopic world. Given how close in size MEMS are to the threshold of the nanoworld, the systems engineering approaches discussed here are in many ways applicable to systems engineering for nanotechnology and nanoscale devices.

Acknowledgments

The author would like to thank Greg Nielsen, Sandia National Laboratories, and Brock Wester, the Johns Hopkins University Applied Physics Laboratory, for helpful discussion and their thorough review of the manuscript.

References

1. Waggener, H.A., "Electrochemically controlled thinning of silicon," *Bell Syst. Tech. J.*, vol. 49, 473–475 (1970).
2. Petersen, K.E., "Silicon as a Mechanical Material," *Proceedings of the IEEE,* vol. 70, 420–457 (1982).

3. Hornbeck, L.J. and W.E. Nelson, "Bistable deformable mirror device," *OSA Technical Digest Series, Vol. 8, Spatial Light Modulators and Applications*, 107 (1988).
4. Nelson, W.E. and L.J. Hornbeck, "Micromechanical spatial light modulator for electrophotographic printers," *SPSE Fourth International Congress on Advances in Non-Impact Printing Technologies*, 427 (1988).
5. Hornbeck, L.J., "Deformable-mirror spatial light modulators (invited paper)," Spatial Light Modulators and Applications III, *SPIE Critical Reviews*, vol. 1150, 86–102 (1989).
6. Sampsell, J.B., "An overview of Texas Instruments digital micromirror device (DMD) and its application to projection displays," *Society for Information Display International Symposium Digest of Tech. Papers*, vol. 24, 1012–1015 (1993).
7. Hornbeck, L.J., "Current status of the digital micromirror device (DMD) for projection television applications (invited paper)," *International Electron Devices Technical Digest*, 381–384 (1993).
8. Harney, K.P., "Standard semiconductor packaging for high reliability low cost MEMS applications," *Proceedings of SPIE*, vol. 571, *Reliability, Packaging, Testing and Characterization of MEMS / Moems IV*, 1–8 (2005).
9. Goodenough, F., "Airbag boom when IC accelerometer sees 50G," *Electronic Design* August 8 (1991).
10. See Sandia National Laboratories, MEMS at www.mems.sandia.gov/index.html.
11. See, for example, the ASME Historic Mechanical Engineering Landmark at www.asme.org/Communities/History/Landmarks/53_Digital_Micromirror_Device.cfm.
12. The Digital Micromirror Device, A Historic Mechanical Engineering Landmark, ASME, at http://files.asme.org/asmeorg/Communities/History/Landmarks/14607.pdf.
13. See NASA's JWST at http://jwst.gsfc.nasa.gov/microshutters.html.
14. Darrin, A.M.G. and R. Osiander, "MEMS packaging materials," in *MEMS Materials and Processes Handbook*, R. Ghodssi and P. Lin, Eds., in *MEMS Reference Shelf*, Series Editors S. Senturia, R.T. Howe, and A.J. Ricco., Springer, NY (2011).
15. Loeppert, P.V. and S.B. Lee, "SiSonic™—The first commercialized MEMS microphone," *Technical Digest of the 2008 Solid-State Sensors, Actuators, and Microsystems Workshop*, pp. 27–30 (2008).

16. See www.systron.com/products/mmq50.
17. See http://invensense.com/mems/gyro/mpu6000.html.
18. Lewis, S., S. Alie, T. Brosnihan, C. Core, T. Core, R. Howe, J. Geen, D. Hollocher, M. Judy, J. Memishian, K. Nunan, R. Paine, S. Sherman, B. Tsang, and B. Wachtmann, "Integrated sensor and electronics processing for >10↑8 'iMEMS' inertial measurement unit components," IEEE International Electron Devices Meeting 2003, 39.1.1–4 (2003).
19. Sherman, S.J., W.K. Tsang, T.A.I. Core, and D.E. Quinn, "A low cost monolithic accelerometer," 1992 Symposium on LSI Circuits. Digest of Technical Papers, 34–35 (1992).
20. Geen, J.A., S.J. Sherman, J.F. Chang, and S.R. Lewis, "Single-chip surface-micromachined integrated gyroscope with 50°/hour root Allan variance," 2002 IEEE International Solid-State Circuits Conference. Digest of Technical Papers vol. 1, 426–427 (2002).
21. Hollocher, D., X. Zhang, A. Sparks et al. "A very low cost, 3-axis, MEMS accelerometer for consumer applications," Proceedings of IEEE Sensors 2009, 953–957 (2009).
22. Dixon-Warren, St.J., "Inside analog devices' new MEMS technology," Chipworks, at www.memsindustrygroup.org/i4a/pages/index.cfm?pageid=3804.
23. Felton, L.E., N. Hablutzel, W.A. Webster, and K.P. Harney, "Chip scale packaging of a MEMS accelerometer," 2004 Proceedings of the 54th Electronic Components and Technology Conference vol. 1, 869–873 (2004).
24. Seeger, J., M. Lim, and S. Nasiri, "Development of high-performance, high volume consumer MEMS gyroscopes," Technical Digest of the 2008 Solid-State Sensors, Actuators, and Microsystems Workshop (2010).
25. Van Kessel, P.F., L.J. Hornbeck, R.E. Meier, and M.R. Douglass, "MEMS-based projection display," Proceedings of the IEEE 86(8), 1687–1704 (1998).
26. Douglass, M.R., "DMD reliability: A MEMS success story," Proceedings of SPIE 4980, 1–11 (2003).
27. Osiander, R., J.L. Champion, and M.A. Darrin, "MEMS in aerospace applications—Thermal control shutters," Proceedings of SPIE 4587, 7–13 (2001).
28. Osiander, R., S.L. Firebaugh, J.L. Champion, D. Farrar, and M.A. Darrin, "Microelectromechanical devices for satellite thermal control," IEEE Sensors Journal, vol. 4(4), 525–531 (2004).

29. Li, M.J., I.S. Aslam, A. Ewin, R.K. Fettig, D. Franz, A.S. Kutyrev, S.H. Mosley, C. Monroy, D.B. Mott, and Y. Zheng, "Fabrication of microshutter arrays for space application," Proceedings of SPIE 4407, 295–303 (2001).
30. See www.jwst.nasa.gov/nirspec.html.

7

Introduction to Nanotechnology

Jennifer L. Sample

"Would you like the Bottom up
or Top down section?"

Contents

Introduction

Nanotechnology is the creation, manipulation, and exploitation of materials with at least one dimension ≤100 nm. For only a few decades now, thanks to advances in microscopy, manipulation, and materials synthesis and deposition, scientists and engineers have been able to purposely create, control, and characterize materials at this scale. These scientific and engineering advances have led to the realization of some of the phenomenological explorations highlighted by Richard Feynman in his famous lecture, "There's Plenty of Room at the Bottom."[1]

Feynman posed a challenge to the scientific community to explore the rich physics at this scale. Although the technology was not yet available to manipulate nanostructures individually, it was known that fundamental physical length scales, such as the penetration depth in superconductors, the Bohr exciton radius in semiconductors, and the magnetic coercivity critical thickness had nanoscale dimensions. Feynman correctly predicted that the creation and study of nanomaterials would be an exciting field of research.

Several techniques critical to nanomaterial fabrication have been available for decades, including chemical synthesis and milling or grinding (bulk material degradation). A nanoparticle consists of on the order of 1,000 to 10,000 atoms and can be created either by assembling the particle from atoms or by processing bulk, macroscopic materials down to the nanoscale. Starting with atoms to create nanomaterials is generally referred to as *bottom-up* assembly or manufacturing, whereas starting from bulk materials is referred to as *top-down* manufacturing or processing. An example of bottom-up manufacturing is wet chemical synthesis of nanoparticles by reduction of metal salts or electrodeposition of metal nanowires. Bottom-up assembly leads to flexibility in materials design, for example, ligands, metal deposition, alignment, and so forth. Top-down fabrication leads to precisions not currently achievable with bottom-up assembly.

Molecular beam epitaxy (MBE), invented in the 1960s, is a top-down fabrication technique allowing deposition of nearly perfect layers of atoms onto substrates. MBE, combined with photolithography and electron microscopy, provided a foundation for creating microscale structures from the top down. Generally, top-down lithography is currently used in nanotechnology to create nanostructures, to characterize nanomaterials, and to make devices. And, of course, it is also used for integrated circuit fabrication.

Atomic imaging was invented in 1981 by researchers at IBM who created the scanning tunneling microscope, a microscope that exploits the exponentially distance-dependent tunneling current between two metal surfaces to characterize individual atoms of a surface.[2] This invention paved the way for significant advances in nanotechnology by allowing researchers to understand the arrangement of atoms in nanomaterials such as carbon nanotubes.

In some sense, nanomaterial fabrication began centuries ago. The colors of stained glass and glazed pottery can be attributed to dispersions of nanoparticles or colloidal nanoparticles. However, reproducible creation of nanomaterials with well-defined, controllable size and shape came much later. Semiconductor nanoparticles often called *quantum dots* were first synthesized by chemists in 1993.[3] These particles were found to exhibit a size-dependent bandgap or color and were one of the first nanomaterials to be commercialized. They are now available for biomedical imaging applications and as light-emitting diodes (LEDs).[4]

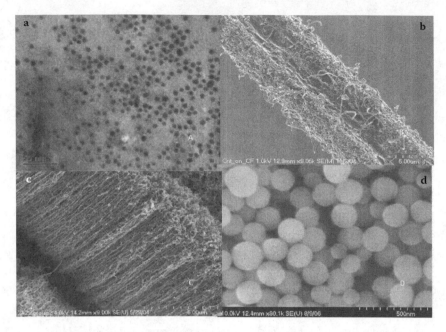

Figure 7.1 (a) CdSe quantum dots in polyphenylenevinylene polymer. (b) Carbon nanotubes (CNTs) grown onto a carbon fiber. (c) CNT array film. (d) Silica nanospheres.

Carbon nanotubes (CNTs), single sheets of graphene with remarkable mechanical, thermal, and electrical properties, were discovered in 1991.[5] CNTs have made their way into devices and commercial technology and will be explored later in this chapter as a case study in the "generations of nanotechnology." Figure 7.1 shows electron microscope images of nanomaterials including carbon nanotubes grown as a film and onto a carbon fiber to increase the surface area of the fiber and nanoparticles alone and grown into a conductive polymer for bulk heterojunction, flexible solar cell applications.

Nanotechnology is frequently based on exploitation of a property found at the nanoscale but not in bulk, such as increased wear resistance of a nanograined ceramic[6] due to the Hall-Petch effect. Another example is nanowire sensors.[7] One reason that nanotechnology is interesting scientifically is that new phenomena exist at the nanoscale, including some that do not exist in bulk. For example, one photon can excite two or more excitons (electron-hole pairs) in semiconductor nanoparticles, a property that is extremely useful for solar cell applications.[8] This effect arises because of quantum confinement, which increases the interaction between the high-energy single exciton state and the first-excited multiexciton state.

More often, properties are just different at the nanoscale below a certain critical length-scale. Electrical, optical, physical, magnetic, surface properties, and reactivity can all be different at the nanoscale.

Emergent Nanoscale Properties

Emergent properties are essentially functions of a system that are not attributable to its individual parts. Thus, a difference exists between *nanosystems* based on simple exploitation of a nanoscale phenomenon and emergent behavior in a nanomaterial or nanoscale system. This distinction may be relevant to the application of systems engineering principles to nanoscale systems.

As an example of nonemergent behavior, Thaxton et al. (Figure 7.2) demonstrated detection of prostate serum antigen (PSA) in postprostatectomy patients.[9] Detection of this antigen in these patients is difficult due to the extremely low levels in blood after prostate removal, but it is important in predicting cancer recurrence. The researchers used an elegant nanostructure-based amplification technique to detect extremely small (femtogram) quantities of antigen exploiting surface area and binding between DNA- and antibody-coated nanostructures. This sensor relies on multiple nanoscale interactions, and its performance is determined by those interactions in a known way (specificity and number of binding interactions).

By contrast, numerous nanotechnology-based systems exhibit emergent properties. Perhaps the most widely known example is the photonic crystal. Like an atomic crystal lattice, a photonic crystal diffracts electromagnetic radiation of wavelength of order of the structure's periodicity. Photonic crystals are created using periodic arrays of nanostructures and have been assembled from silica nano- and microparticles

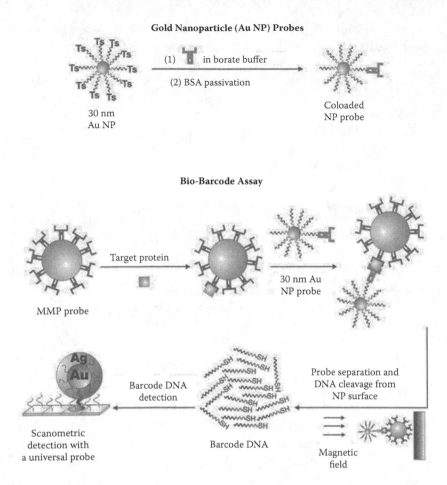

Figure 7.2 Prostate specific antigen (PSA) detection scheme showing nanoparticle probes (upper) and bio-barcode assay (lower). This assay detects PSA at 330 fg/mL (see C.S. Thaxton, R. Elghanian, A.D. Thomas, S.I. Stoeva, J-S. Lee, N.D. Smith, A.J. Schaeffer, H. Klocker, W. Horninger, G. Bartsch, C.A. Mirkin, *PNAS*, 106(44), 18437–18442, 2009). In this figure, NP refers to nanoparticle, Au and Ag are the elements, BSA is a protein, and the MMP is a magnetic bead.

(Figure 7.3) of various sizes,[10] and other materials, notably including polystyrene microspheres. These crystals can also be inverted (filled with another material, such as a metal) with the template material removed afterward in order to achieve a higher degree of dielectric constant contrast within the structure.

Figure *7.3* Photonic crystal made from self-assembled polystyrene microspheres.

The photonic bandgap is an emergent phenomenon arising specifically in these photonic crystal arrays. This gap essentially corresponds to wavelengths that are disallowed in the structure due to the diffraction-like conditions set up by the periodic dielectric constant. These structures have been explored for waveguide applications because certain wavelengths cannot pass through the structure. The ordered array of nanoparticles rather than the nanoparticles are manipulating light; hence, this represents an emergent phenomenon.

Other nanomaterial systems with emergent properties include shear thickening fluids, magnetorestrictive, magnetorheological, and ferrofluids, dye-sensitized solar cells, quantum-dot-based quantum cellular automata (QCAs), single electron transistors, and magnetic nanoparticle-based artificial cilia structures.[11]

It is interesting to note that the concept of emergent properties also defines a system. For example, the definition of a system from the International Council on Systems Engineering (INCOSE) is a "homogeneous entity that exhibits predefined behavior in the real world and is composed of heterogeneous parts that do not individually exhibit that behavior and an integrated configuration of components and/or subsystems."[12]

At a very basic level, the ability of nanomaterials to self-assemble (bottom-up versus engineered top-down assembly) may represent emergent behavior. For example, when placed onto the surface of water, a biomodally sized population of silver nanoparticles will spontaneously form arrays and shapes such as wires and disks according to their diameter, diameter distribution, and concentration.[13] The ability of nanoparticles to self-assemble is a direct result of interparticle interactions; therefore, not emergent, the interactions of groups of assembled particles leading to egg-shaped arrays with large nanoparticle "yolks" and small particle "whites" arises from more complex interactions including interparticle attractions and repulsions, intercluster attractions and repulsions, and interactions of the clustered particles with the water subphase.

The dynamics of nanoscale systems are such that the scale of attractive and repulsive forces between nanoparticles relative to thermal energy dominates the interactions of these materials. Thus, they are configurable and dynamically reconfigurable. The transition between nanoscale and bulk (coupled behavior of nanoparticles) properties in nanomaterial systems has been extensively studied for phenomena including metal-to-insulator[14] and superconductor-to-insulator transitions, which in some cases are reversible due to the dynamic nature of the forces governing their interactions. In other cases, the transition is irreversible or reversible once due to the nature of the probing experiment. For example, the superconductor-to-insulator transition has been probed discretely and, thus, irreversibly by varying the coupling interactions via ligand length between lead nanoparticles.[15] The metal-to-insulator transition has been visualized and observed via second harmonic generation at the nanoparticle/water interface in a system similar to the one described above but with highly ordered nanoparticles. The Teflon barriers of the Langmuir-Blodgett trough are used to physically apply pressure to the nanoparticle array until the interparticle spacing decreases to essentially zero and the array becomes metallic. This transition is observed with an optical change from red (particle film appears red due to surface plasmon resonance of separate nanoparticles) to silver.

At any scale, both naturally occurring and engineered complex systems exist. Engineered systems are typically designed to have specific emergent properties, whereas natural systems simply exhibit these properties. Engineered systems are often delicate with respect to their natural counterparts, which are often more robust, flexible, and adaptable. These differences may be due to the top-down or hierarchical design approach often used in human engineered systems. Nature designs systems from the bottom-up from existing or modified components and assembles without performance constraints or functional goals beyond survival according to the principles of natural selection. Engineered systems instead are designed with the end in mind, often with the guidance of rigorous systems engineering principles.[16]

Nanotechnology Integration Challenges

Nanomaterials have remarkable properties often not available with conventional materials. Thus, they have been considered as candidates for very demanding materials applications including for various systems, including those exposed to extreme environments such as outer space. However, it can be difficult to create functional devices or technology from nanomaterials because of various material property-related challenges, such as strong interparticle van der Waals interactions, tendency to phase segregate, insolubility in appropriate solvents, incompatibility with cellular environments, toxicity, and so forth. These challenges are illustrated by the example of carbon nanotubes.

Carbon nanotubes (CNTs) are hollow tubes consisting of one sheet of carbon atoms rolled into a tube. They were discovered in 1991, and it took approximately eight years for them to become commercially available due to bulk fabrication challenges. It took another several years for researchers to develop techniques to disaggregate them so that their individual properties could be studied. Once isolated, their chirality-dependent properties were investigated. The electrical properties of CNTs depend on the structural arrangement of their carbon atoms. Possible arrangements are shown in Figure 7.4.[17]

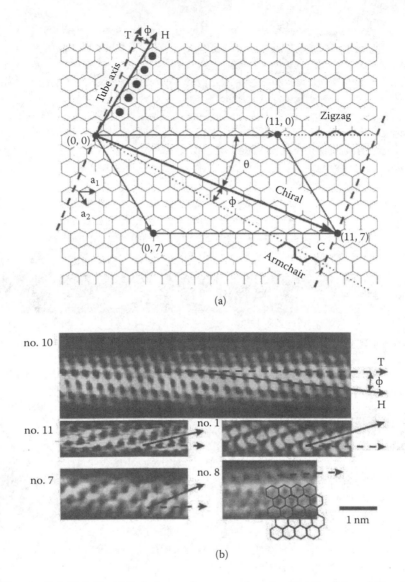

Figure 7.4 Relationship between hexagonal carbon lattice (a) and the chirality of carbon nanotubes. (b) Atomically resolved STM images of carbon nanotubes of varying chirality.

CNT-based transistors were investigated for electronics applications by many groups in the early 2000s.[18] Their very small diameter and ballistic electron transport over macroscopic length scales captured the imagination of scientists,

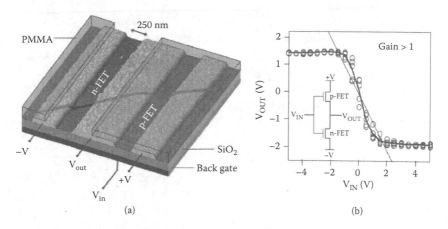

Figure 7.5 Carbon nanotube (CNT) transistor (a) and performance data (b).

making them seem ideal to replace silicon as a transistor channel material. Many such devices were made and tested despite significant initial device fabrication integration challenges, including isolation and making ohmic contact to a single nanotube. These fabrication challenges were overcome using dilute dispersions of carbon nanotubes spun onto a silicon wafer with alignment markers. This resulted in isolated nanotubes randomly distributed on the surface. After the nanotubes were deposited, they were imaged by scanning electron microscopy, their locations were noted, and electrodes were patterned on top of them. Alternatively, CNTs were deposited on top of prepatterned electrodes, and those appropriate for making measurements were found by imaging and electrical measurements. Figure 7.5 is a photo of a CNT transistor and a plot of its device performance that shows gain, an essential property of transistors.[19]

An interesting aspect of CNT transistor technology, however, is the historical perspective on the promise of this technology versus the outcome. As is often the case with research into new functional materials, especially nanomaterials, realizing the promise of the materials can be difficult. A mix of semiconducting and metallic nanotubes exists in most samples because of the way they are grown. Much research is underway to either separate out the nanotubes based on their electronic properties or to specifically grow nanotubes with desired electronic properties, but this technology also has yet to be realized.[20,21] The fabrication requirements for CNT growth also present a challenge with device integration. CNTs hold promise as a channel material and may ultimately find use as part of hybrid devices, which hybridize conventional complementary metal-oxide semiconductor (CMOS) technology with nanomaterials or as a silicon-CMOS replacement. Conventional CMOS is based on silicon devices whose performance is substantially degraded when exposed to the elevated temperatures required to grow CNTs. Thus, it is difficult to grow CNTs on devices and difficult to deposit them onto devices after they are grown due to their varying electronic properties. CNTs have not yet achieved their promise as the next generation of transistors because of these fabrication challenges.

Examples of Applied Nanotechnology

Nanomaterials have also found widespread application in biomedical (diagnostic and therapeutic) and personal care consumer product applications. Nanomaterials can be found in consumer products such as cosmetics and sports equipment, mostly as nanoparticle-based pigments, emulsions, or nanofiber/nanocomposite materials. Commercial products as diverse as socks and ski wax are reported to contain nanotechnology. Nanoparticles are commercially sold for biological imaging applications (quantum dots) and therapeutics (e.g., nanoshells[22]).

Various active, functional nanodevices are in development for sensing and electronics applications. Silicon nanowire sensor systems have been developed for highly sensitive detection of biomolecules.[23] The conductance through very small wires, compared to large wires, is more sensitive to binding events and, specifically, to the binding of charged molecules to the wire surface. This is because in bulk, macroscopic, or micron-sized wires, the percentage of atoms at the surface relative to interior atoms is small, and current through the wire is relatively unaffected by what is happening on the wire surface. However, with nanowires, the surface is a much larger fraction of the total wire, and a change such as the binding of a charged molecule will affect the conductance through the wire by field effect (the field is set up by the bound charged molecules).[24] An array of sensors operating on this principle and data demonstrating the nanowire conductance response to various concentrations of bound analyte are shown in Figure 7.6.

The marriage of nanotechnology and biology is a fascinating area of nanotechnology applications. Significant advances have been made in nanomedicine, including the use of nanoscale drug delivery platforms for more effective and targeted drug delivery, nanoscale vectors for nonviral gene therapy, and novel therapeutics using nanomaterials, such as gold/silica nanoshell nanoparticles, which aggregate in a tumor and then absorb light and heat up, thereby shrinking and killing the tumor.[25]

A more esoteric and imaginative merging of nanotechnology and biology is shown in Figure 7.7: the DNA walker. While it remains questionable what the ultimate utility of DNA-based robots will be, it is fascinating to consider the extent to which DNA can be manipulated simply as a self-assembling material regardless of its biological function.

It is possible to specifically engineer DNA to have desired shapes (such as smiley faces, Figure 7.8)[26] and assembly properties, because DNA readily self-assembles to its complementary strand in a highly controllable way. The DNA walker is an advanced example of this concept. Free DNA and a "track" of DNA are engineered such that as the DNA hybridizes and dehybridizes to its complementary strands on the track, it will walk along the track. This motion can be observed by incorporating into the moving part fluorescent reporter dyes that indicate the position of the DNA "bot."

Maturation of Nanotechnology

The four generations of commercialized nanotechnology were outlined by Roco in a 2004 National Nanotechnology Initiative (NNI) report.[27] The generations begin with the synthesis of nanomaterials (first generation, see Table 7.1) and progress

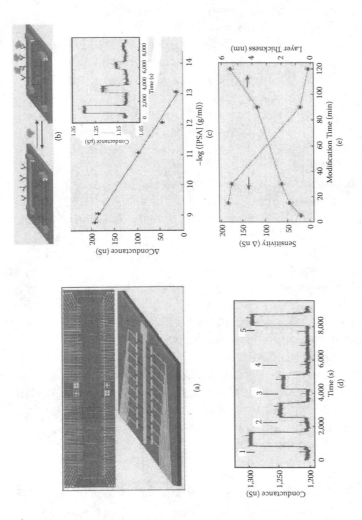

Figure 7.6 Nanowire sensor arrays and detection data. (a) Optical image of a nanowire device array, with white lines corresponding to metal electrodes connecting individual nanowire devices. (b) Two nanowire devices with nanowires modified to detect two different analytes. (c) Change in conductance versus concentration of prostate specific antigen (PSA) analyte, with buffer added in between PSA concentrations. (d) Conductance versus time for additional PSA solutions. (e) Thickness dependence of "gate" (analyte) layer on device properties.

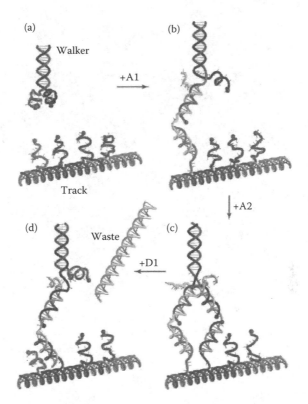

Figure 7.7 DNA walker locomotion mechanism. Dyes are used to detect the movement of the walker. The diagram depicts (a) unbound walker, (b) walker attached to branch 1, (c) walker attached to branches 1 and 2, and (d) walker released from branch 1, yielding waste DNA strand. (See Shin, J.-S, and Pierce, N.A., *Journal of the American Chemical Society*, 126, 10834, 2004.)

by generations as nanomaterials are applied and matured to complex, all-nanotechnology-based nanosystems. These generations categorize technological development and maturity from academic laboratories to commercialized products.

The current state of nanotechnology overlaps the first, second, and third generations. Incremental improvements in maturity in some technologies and major advances in others will be required to progress from the first to fourth generations. For example, the presentation "Productive Nanosystems" by Drexler[28] envisions a manufacturing-style assembly of complex systems from molecules and atoms. This is a compelling

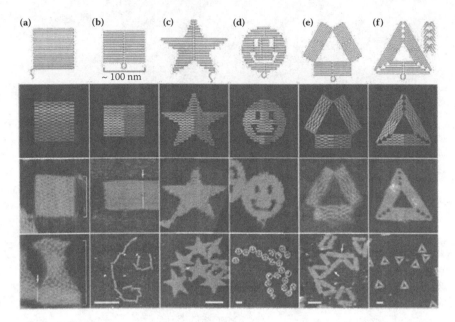

Figure 7.8 Shapes formed using "DNA origami." Structure-programmed folding paths are shown in the top row, followed by a row showing diagrams of the structures, followed by scanning electron microscope (SEM) images (third-row panels are all the same size), scale bars for the bottom five right-most images are (b) 1 mm and (c–f) 100 nm.

vision for highly controlled materials and device synthesis, yet it may require significant technological advancements to be made practically useful and cost effective for many applications. For example, the precision required to place individual atoms into precise arrangements is in contrast with what is achievable in the presence of thermal energy (room temperature). Also, chemistry and local control of chemistry are often temperature dependent. Thus, very low temperatures or precise temperature control may be required for this type of manufacturing, making it impractical at this time for mass production of devices such as laptops.

New properties of nanomaterials are still being discovered. For example, carbon nanotubes were recently discovered to act as fuses and burn very quickly producing energy in the process.[29] This type of energy source has not been studied, but new advances in energy technology are always needed. Several

TABLE 7.1

Four Generations of Commercialized Nanotechnology

First generation	Passive nanostructures: nanomaterials fabrication and commercialization including carbon nanotubes, quantum dots, nanomaterial dispersions, required isolation and functionalization, characterization of novel properties, and standardization
Second generation	Active nanostructures: devices exploiting nanoscale phenomena, nanowire field effect transistors, quantum well infrared photodetectors, stimulus-responsive nanomaterials,[a] therapeutic nanomaterials[b]
Third generation	Three-dimensional (3D) nanosystems and systems of nanosystems: carbon nanotube (CNT) memory devices, MRAM devices, quantum dot polymer solar cells
Fourth generation	Heterogeneous molecular nanosystems: systems in which each molecule in the nanosystem has a specific function, molecular manufacturing[e]

Note: See Rothemund, P.W.K., *Nature*, 440, 297–302, 2006.

[a] See Technology Roadmap for Productive Nanosystems, www.foresight.org/roadmaps/.

[b] See Heath, J.R., Kuekes, P.J., Snider, G.S., and Williams, R.S., *Science*, 280, 1716–1721, 1998.

barriers to implementing this technology as a battery replacement exist, for example, providing long-lasting energy rather than rapid burning, and most notably, efficient conversion of heat into electricity. However, in this case, the marriage of engineering with nanotechnology may yield a productive solution to this problem, and work in this area is certainly currently underway.

In order for any nanomaterial to find widespread application, it must first be available to scientists and engineers. Initial fabrication of nanomaterials occurred in academic laboratories, and to work with the materials it was necessary to collaborate or somehow otherwise procure the material. The first commercial nanotechnology was simply nanomaterials for purchase from companies specializing in their fabrication. Numerous examples of commercially available nanomaterials exist, including carbon nanotubes, quantum dots, and magnetic, metallic, silica, and polymer nanoparticles. Table 7.2 lists the year of first discovery and commercialization of several nanomaterials.

TABLE 7.2
Examples of Nanomaterial Discovery and
Commercial Availability

Material	First Reports	Year Commercially Available
Carbon nanotubes (CNTs)	1991	1999
Quantum dots (QDs)	1993	2002
Nano Electro Mechanical Systems (NEMS)	1995	2005

The fourth generation of commercialized nanotechnology is interesting to consider, and it is particularly interesting to predict whether or not this generation will actually be realized or rather redefined. For heterogeneous molecular nanosystems such as those envisioned in "molecular manufacturing"[35] to be realized, significant control over local chemical environments of atoms and molecules will be required. Whether or not it will be necessary, practical, or possible to create nanosystems under these conditions remains to be seen.

Self-assembly, or making functional structures from the bottom up, does hold promise for creating heterogeneous molecular nanosystems. Self-assembly typically exploits reversible interactions including van der Waals interactions, electrostatic interactions, and hydrophobic/hydrophilic interactions. These forces, depending on the specific materials and the interactions involved, can be comparable to thermal energy; therefore, imperfect materials may be produced ultimately. Defect-tolerant architectures have been proposed to accommodate defect densities arising from self-assembling components.[36] The impact of these defects ultimately depends upon the role of the nanomaterials in the system.

The possibility of achieving the fourth generation of commercialized nanotechnology is interesting from the perspective of systems engineering. If autonomous nanosystems whose functions exploited nanoscale effects not achievable at the macroscale, or emergent phenomena existed, it is interesting to consider whether or not the traditional systems engineering spiral would be sufficient to address risk associated with the nanocomponents.

Systems Engineering Applied to Nanotechnology: Issues and Conclusions

Configuration management (i.e., maintaining consistency in a system's performance), is a primary goal and of great importance in systems engineering. To achieve the required level of performance consistency in nanotechnology, further standardization of nanomaterials and related device performance characterization may be necessary. Historically, much progress has been made toward this goal; however, many challenges remain.

It was certainly true near the beginning of nanoscience as a field (1990s) that nanomaterials were highly variable from batch to batch as produced. In the late 1990s, researchers were still learning how to create and characterize nanostructures and variables such as temperature, chemical composition of reactants and catalysts, and other external factors had yet unknown effects on products. In 2000, Richard Smalley, one of the pioneers of nanotechnology, started a company that sold carbon nanotubes, Carbon Nanotechnologies, Inc. (CNI).[37] Variability in the purchased nanotubes from batch to batch was common, such that experiments optimized from one batch had to be reoptimized when that batch ran out. In 2003, the National Institute of Standards and Technology (NIST) began to tackle the challenge of defining metrics by which nanotubes could be standardized. Today, NIST standard reference nanoparticle and nanotube materials are available.

Still, some of the published research and experimental results derived from nanomaterials-based technologies remain difficult to reproduce, largely because the art or skill or system familiarity required to generate the reported data is very difficult to attain. Similar challenges exist with transitioning basic research results into system design regardless of the type of technology.

Additional issues arise when systems engineers face using nanotechnology as a system component. Many of these are addressed in other chapters of this book:

- System design tools—new tools may be required
- Material suppliers and their consistency—standardization of nanomaterials
- Reliability—smaller scaling may lead to higher failure rates
- Architecture—redundancy may counter effects of smaller scaling
- Circuit manufacturing technology
- Circuit architectures
- Cost
- Profit margin
- Toxicity
- Public fear/marketing
- Unforeseen risks

Toxicity and health effects of many nanomaterials are still unknown. Health effects of CNTs are under study particularly because of their resemblance to asbestos fibers in diameter (when bundled) and aspect ratio. A correlation has been found between length, dispersity, and inflammation. Bundled and aggregated nanotubes appear to be relatively inert to the body.[38]

Public fear of nanotechnology is a concern. Health effects associated with nanomaterials and nanotechnology-based consumer products are certainly a cause for public concern. However, science fiction also shapes the image of nanotechnology. Michael Crichton's fictional account of nanotechnology gone wrong in *Prey*[39] raises ethical and unanticipated consequence issues. Although fictional, such scenarios could be alarming to readers without specific knowledge about the actual technology and its capabilities and limitations. Of course, even those working in the field may be surprised by as yet unanticipated risks.

Summary

Recent issues of journals from most disciplines contain research involving nanotechnology or nanoscale phenomena and effects. Nanotechnology has found wide application, and due to new phenomenology, it is still relevant because there remain a lot of unexplored phenomena at the nanoscale. It remains to be seen when and how nanotechnology will be integrated into multiscaled functional systems designed by systems engineers; however, with increased technological maturity and attention to risk mitigation, it should be possible.

References

1. Feynman, R. "There's Plenty of Room at the Bottom," 1959.
2. Hansma, P. K. and Tersoff, J. *J. Appl. Phys.* 61, R1–R23, 1987.
3. Murray, C.B., Norris, D.J., and Bawendi, M.G. "Synthesis and characerization of nearly monodisperse CdE (E=S, Se, Te) semiconductor nanocrystallites." *J. Am. Chem. Soc.* 115, 8706–8715, 1993.
4. Evident Technologies, http://www.evidenttech.com/news, 12/27/10
5. Iijima, S. (7 November 1991). "Helical microtubules of graphitic carbon." *Nature* 354, 56–58.
6. Schiøtz, J. and Jacobsen, K.W., *Science,* 301 (5638), 1357–1359, 2003.
7. Zheng, G., Gao, X., and Lieber, C.M., "Frequency domain detection of biomolecules using silicon nanowire biosensors," *Nano Lett.* 10, 3179–3183, 2010.
8. Ellingson, R.J., Beard, M.C., Johnson, J.C., Yu, P., Micic, O.I., Nozik, A.J., Shabaev, A., and Efros, A.L., "Highly efficient multiple exciton generation in colloidal PbSe and PbS quantum dots," *Nano Letters,* 5, 865–871, 2005.
9. Thaxton, C.S., Elghanian, R., Thomas, A.D., Stoeva, S.I., Lee, J.-S., Smith, N.D., Schaeffer, A.J., Klocker, H., Horninger, W., Bartsch, G., and Mirkin, C.A., Nanoparticle-based bio-barcode assay redefines "undetectable" PSA and biochemical recurrence after radical prostatectomy, *PNAS,* 2009, vol. 106, no. 44, 18437–18442.
10. Colvin, V.L. "From opals to optics: Colloidal photonic crystals," *MRS Bulletin,* 26, 637–641, 2001.
11. Benkoski, J.J., Deacon, R., Land, B.H., Baird, L.M., Breidenich, J.L., Srinivasan, R., Clatterbaugh, G.V., Keng, P., and Pyun, J., "Preparation of Dipolar Nanoparticles and Magnetic Filaments," *Soft Matter* 2010, 6, 602–609.
12. *INCOSE Systems Engineering Handbook,* v. 3.1.

13. Sear, R.P., Chung, S.-W., Markovich, G., Gelbart, W.M., and Heath, J.R., "Spontaneous patterning of quantum dots at the air-water interface," *Phys. Rev. E,* 59, R6255-6258, 1999.

14. Collier, C.P., Henrichs, S., Shiang, J.J., Saykally, R.J., and Heath, J.R., "Reversible tunning of silver quantum dot monolayers through the metal-insulator transition, *Science,* 277, 1978–80, 1997.

15. Weitz, I., Sample, J., Ries, R., Spain, E., and Heath, J.R., "Josephson coupled quantum dot artificial solids," *J. Phys. Chem. B Letter,* 104(18), 4288 (2000).

16. Sample, J.L. and Benkoski, J.J., "Systems engineering and nanotechnology," *INCOSE Insight,* 13, 18, 2010.

17. Wildoer, J.W.G., Venema, L.C., Rinzler, A.G., Smalley, R.E., and Dekker, C., "Electronic structure of atomically resolved carbon nanotubes," *Nature,* 39, 59, 1998.

18. Derycke, V., Martel, R., Appenzeller, J., and Avouris, Ph., Carbon nanotube inter- and intramolecular logic gates *Nano Lett.,* 2001, 1 , pp. 453–456

19. Derycke, V., Martel, R., Appenzeller, J., and Avouris, Ph., Carbon nanotube inter- and intramolecular logic gates, *Nano Letters,* 2001, 1 (9), 453–456.

20. Huh, J.Y., Walker, A.R.H., Ro, H.W., Obrzut, J., Mansfield, E., Geiss, R., and Fagan, J.A., "Separation and characteriation of double-wall carbon nanotube subpopulations," *J. Phys. Chem. C,* 2010, 114 (26), 11343–11351.

21. Harutyunyan, A.R., Chen, G., Paronyan, T.M., Pigos, E.M., Kuznetsov, O.A., Hweaparakrama, K., Kim, S.M., Zakharov, D., Stach, E.A., and Sumansekera, G.U., "Preferential growth of single-walled carbon nanotubes with metallic conductivity," *Science,* 326, 116-120, 2009.

22. http://www.nanospectra.com/.

23. He, B., Morrow, T.J., and Keating, C.C., "Nanowire sensors for multiplexed detection of biomolecules," *Curr. Opin. Chem. Biol.,* 12, 522–528, 2008.

24. Zheng et al., "Multiplexed electrical detection of cancer markers with nanowire sensor arrays," *Nature Biotechnology,* 23, 1294, 2005.

25. Gobin, A.M., Moon, J.J., and West, J.L., "EphrinAl-targeted nanoshells for photothermal ablation of prostate cancer cells," *Int. J. Nanomed.,* 2008, 3 351–358.

26. Rothemund, P.W.K., "Folding DNA to create nanoscale shapes and patterns," *Nature,* 440, 297-302, 2006.

27. Roco, M., Nanoscale Science and Engineering: Unifying and Transforming Tools, *AIChE Journal,* 50 (5), May 2004, 890.
28. http://www.foresight.org/roadmaps/.
29. Choi, W., Hong, S., Abrahamson, J.T., Han, J.H., Song, C., Nair, N., Baik, S., and Strano, M.S. "Chemically driven carbon-nanotube-guided thermopower waves," *Nature Mat.* 9 (2010), 423–429.
30. Technology Roadmap for Productive Nanosystems, http://www.foresight.org/roadmaps/.
31. Heath, J.R., Kuekes, P.J., Snider, G.S., and Williams, R.S., "A defect-tolerant computer architecture: Opportunities for nanotechnology," *Science,* 280, 1716-1721, 1998.
32. This company has since merged with Unidym, Inc, http://www.nanotech-now.com/news.cgi?story_id=21369, accessed 8/30/10.
33. Poland, C.A. et al. "Carbon nanotubes introduced into the abdominal cavity of mice show asbestos-like pathogenicity in a pilot study," *Nature Nanotech.*, 3, 2008, 423.
34. Prey, M.C., New York: HarperCollins, 2002.
35. Benkoski, J.J.. R.M. Deacon, H.B. Land, L.M. Baird, J.L. and Breidenich, R. Srinivasan, G. V. Clatterbaugh, P.-Y. Keng, and J. Pyun, "Dipolar assembly of ferromagnetic nanoparticles into magnetically driven artificial cilia," *Soft Matter,* 6 (2010) 602–609.
36. Cho, Y., Shi, R., Borgens, R., and Ivanisevic, A., "Repairing the damaged spinal cord and brain with nanomedicine," *Small,* 4, 1676–1681, 2008.
37. http://www.nantero.com/.
38. Allwood, D.A., Xiong, G., Faulkner, C.C., Atkinson, D. Petit, D., and Cowburn, R.P., "Magnetic domain-wall logic," *Science,* 309, 1688–1692, 2005.
39. Huynh, W.U., Dittmer, J.J., and Alivasatos, A.P., "Hybrid nano-rod-polymer solar cells," *Science,* 295, 2425–2427, 2002.
40. Shin, J.-S. and Pierce, N.A., "A synthetic DNA walker for molecular transport," *J. Am. Chem. Soc.,* 126, 10834, 2004.

8

Nanoscale Systems—
Top-Down Assembly

Jeffrey P. Maranchi

When top down meets
bottom up construction techniques.

Contents

Introduction

In the world of nanoscale systems, there are three routes to take to assemble a nanoscale material or component into a useful system that meets a well-defined requirement as defined by the systems engineer. One route is to use a top-down assembly approach and guide the nanoscale components or materials into a desired configuration using techniques borrowed from traditional system fabrication processes such as printing, depositing, or etching. Another route is to "let nature take its course" and to rely on bottom-up assembly techniques such as self-assembly, which rely on the minimization of free energy to assemble nanoscale materials and components into useful structures, devices, and ultimately systems. The third method is a hybrid method that when necessary in system processing steps, utilizes techniques that are both top-down and bottom-up in nature to leverage the best parts of both assembly styles at the right time in the fabrication. The top-down assembly route to create nanoscale systems will be discussed in detail in this chapter, while bottom-up assembly will be examined closely in Chapter 9. The hybrid method will not be discussed further, as it is a variable method that will draw from the techniques described in the top-down and bottom-up chapters of this book.

There are several key nanoscale system manufacturing "grand challenges" from both the cost and performance perspective. In particular for nanoscale electronic or optoelectronic systems, the challenges include the following:

- The prohibitive cost of state-of-the-art manufacturing systems for "standard" photonic-based lithographic tools (e.g., extreme-ultraviolet (UV) lithography is greater than $25 million per tool) and the low throughput speed of techniques such as electron-beam lithography must be overcome.
- Today's techniques such as nanoimprinting and soft-lithography are not suitable for developing complex three-dimensional (3D) nanoscale systems.
- The reliability of self-assembled nanoscale systems needs to be improved to reduce defect densities.

In particular, it is interesting to explore the world where nanoscale systems for advanced electronic devices could potentially break us free from Moore's law. For example, Figures 8.1 and 8.2 illustrate Moore's law for feature size and device complexity. Although the development of true nanoscale systems using more sophisticated tools and approaches compared to conventional tools and fabrication techniques could further decrease the feature size and concomitantly increase complexity (e.g., transistors per die), Figure 8.3 shows the cost of advanced tools as a function of time and illustrates the need for more cost-effective, scalable approaches to nanoscale systems engineering and development as a whole, using both top-down and bottom-up assembly processes.

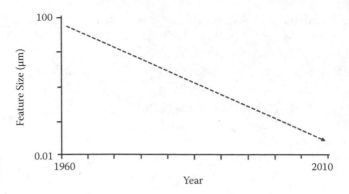

Figure 8.1 The decrease in feature size over time.

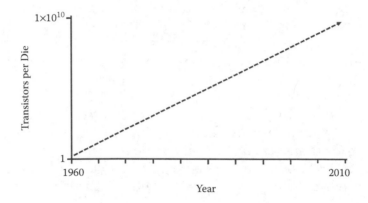

Figure 8.2 The increase in complexity over time.

It is important to take a forward look at methodologies to cost-effectively produce nanoscale systems in the remaining sections of this chapter, but we should pause for a moment to examine our starting point. In particular, we will briefly introduce or refresh the reader on the current and past top-down methods to produce *micro*scale systems. We can only briefly examine the top-down fabrication of microscale systems in this section, so the reader is strongly encouraged to refer to Campbell's excellent treatment of the subject in his book entitled *The Science and Engineering of Microelectronic Fabrication.*[1] The subject of top-down microelectronic fabrication methods can be broken down into the following categories: raw materials for integrated circuits, patterning methods,

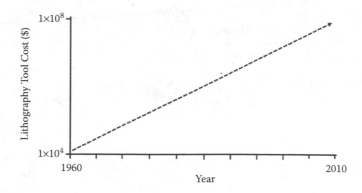

Figure 8.3 The exponential rise in lithography tool cost (per tool) over time.

thin film deposition techniques, etching, and interconnect technologies. When one thinks of microelectronics, one thinks of resistors, capacitors, transistors, diodes, and many other electrical circuit elements. In order to illustrate a simple case of microelectronic circuit fabrication, one can consider how one may produce a rectangular thin film capacitor on a silicon wafer. One possible top-down fabrication method for the capacitor would involve the steps shown in Figure 8.4.

The process of forming a thin film capacitor using semiconductor processing techniques as described above is a good illustration of many of the common top-down unit processes used today in the semiconductor device industry and which have been developed and refined over the past ~40 years to yield very sophisticated high-yield devices and components. The categories of unit processes used in semiconductor device fabrication include the following:

1. Single crystal growth and wafer formation—Common methods of wafer growth include techniques such as Czochralski, Bridgeman, and Float Zone growth methods. The boules (large single crystals) grown using these methods are cut into rough wafers, ground, polished, and lapped to yield very flat, low-defect density semiconductor wafers that can be over a foot in diameter.

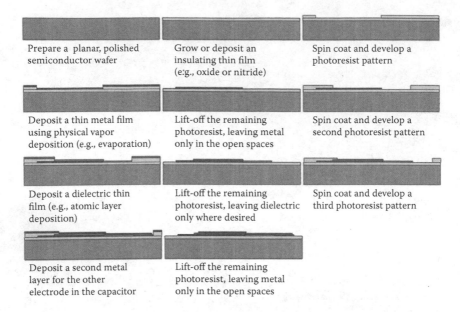

Prepare a planar, polished semiconductor wafer	Grow or deposit an insulating thin film (e.g., oxide or nitride)	Spin coat and develop a photoresist pattern
Deposit a thin metal film using physical vapor deposition (e.g., evaporation)	Lift-off the remaining photoresist, leaving metal only in the open spaces	Spin coat and develop a second photoresist pattern
Deposit a dielectric thin film (e.g., atomic layer deposition)	Lift-off the remaining photoresist, leaving dielectric only where desired	Spin coat and develop a third photoresist pattern
Deposit a second metal layer for the other electrode in the capacitor	Lift-off the remaining photoresist, leaving metal only in the open spaces	

Figure 8.4 The process steps necessary for one possible method of formation of a thin film capacitor on a semiconductor wafer.

2. Insulator growth/deposition—Silicon-based devices are so abundant because silicon forms a wonderful native oxide naturally when it is thermally oxidized. The silicon oxide can be formed using thermal anneals in an oxidizing atmosphere at elevated temperature and can differ based on the amount of moisture present during the anneal (wet/dry oxide). Silicon oxide and silicon nitride can also be deposited using chemical vapor deposition (CVD) from various precursors such as tetraethylorthosilicate (TEOS) and ammonia. Furthermore, the stress in the oxide films can be tailored based on the deposition/growth parameters.

3. Patterning—Patterning of films is typically performed by spin coating a photosensitive polymer onto a substrate and exposing ultraviolet (UV) light through a photomask pattern (e.g., chrome thin film pattern on a glass plate) and then washing away the exposed/unexposed areas depending on whether or not one used a positive/negative photoresist, respectively. The pattern of photoresist serves to protect underlying films/

components such that etching or deposition can be performed on exposed regions.

4. Etching—Etching is performed using wet chemicals such as acids or bases (e.g., hydrochloric acid (HCl) to etch indium gallium arsenide (InGaAs) or hot potassium hydroxide (KOH) to etch silicon) or reactive ion etching (RIE) in which a plasma is used to activate a gas molecule making it more reactive to etch away the desired material. Some etching methods are directional (anisotropic) following crystallographic planes or normal to the plane of an electrode in a plasma system, while other etching methods are more isotropic in nature.

5. Deposition—Deposition methods include physical vapor deposition methods (e.g., thermal evaporation, electron beam evaporation, sputtering), CVD methods (e.g., plasma-enhanced CVD, metallorganic CVD [aka organometallic vapor phase epitaxy]), molecular beam epitaxy, atomic layer deposition, pulsed laser deposition, and other methods.

Top-Down Assembly of Nanoscale Electronic and Opto-Electronic Systems

There are several emerging top-down assembly technologies that have significant promise to tackle the nanoscale system manufacturing "grand challenges" outlined above for nanoscale electronic and nanoscale opto-electronic systems. Top-down assembly manufacturing processes will be discussed such as plasmonic imaging lithography,[2] nano-imprint lithography,[3] dip pen nanolithography, and heterogeneous 3D electronic devices made by using printed nanomaterials.[4] Particular emphasis will be placed on the materials, reliability of the processes, scalability potential for volume production, and integration with preexisting manufacturing unit processes for each of the top-down assembly processes.

Plasmonic Imaging Lithography

Plasmonic imaging lithography (PIL) is a relatively recent photolithography exposure technique that allows researchers to surpass the diffraction limit of light using the special properties of surface waves or plasmons. In particular, researchers have recently shown that patterned thin metallic films that facilitate the formation of surface plasmon polaritons (SPPs) can be used to focus and guide UV light that can subsequently be used to form patterns at length scales less than the wavelength of the incident light.[5] Therefore, one can think of patterning larger focusing structures (e.g., circles or ellipses) on a mask and illuminating the patterned mask with UV light. The focused surface plasmons from the patterned structures could be focused in the near field to pattern photosensitive polymers with nanometer-scale resolution. Rather than using a relatively large array of two-dimensional (2D) patterned structures to generate the focused plasmons, one could leverage the disk drive infrastructure and technology and pattern a single structure or a linear array of structures on a flying head (i.e., plasmonic flying head) that could be used with a rotating substrate to generate the desired subwavelength features as shown in Figure 8.5.[6] A prototype plasmonic imaging lithography system as shown in Figure 8.5 has been used successfully to pattern the features shown in Figure 8.6 at length scales in the 80 nm range. A favorable operating energy analysis comparison of PIL has been made compared to conventional optical projection lithography (OPL) and electron beam lithography.[2] In fact, the operational energy analysis has shown that PIL really shows an advantage in situations with low numbers of prototypes or cases where 11 or fewer wafers are used per design change. The advantage lies in the energy cost to generate the OPL photomask, which is fairly high compared to the energy required to implement the PIL system.

Nano-Imprint Lithography

Nano-imprint lithography is an excellent example of a relatively new, scalable, technique in which a nanostructured master template (typically made by electron beam lithography) is utilized to create a working template (e.g., polymer such as polyvinylchloride [PVC] or polydimethyl siloxane [PDMS]) that can then be used to pattern a thermoplastic at elevated temperature or a UV curable resin when illuminated with UV light. For example, conventional antireflective (AR) coatings

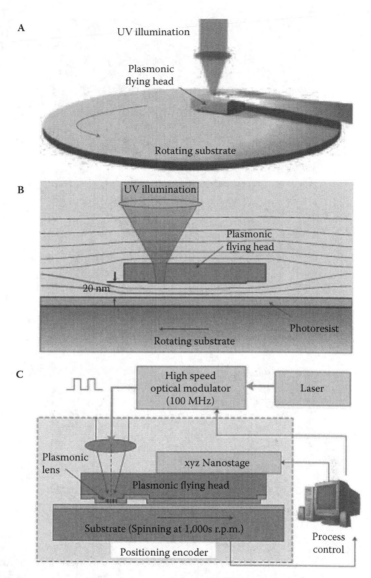

Figure 8.5 One method of implementing plasmonic imaging lithography. (A) The lens array focusing ultraviolet (365 nm) laser pulses onto the rotating substrate to concentrate surface plasmons into sub-100 nm spots. However, sub-100 nm spots are only produced in the near field of the lens, so a process control system is needed to maintain the gap between the lens and the substrate at 20 nm. (B) Cross-sectional schematic of the plasmonic head flying 20 nm above the rotating substrate that is covered with photoresist. (C) A process control system. (Reprinted by permission from Macmillan Publishers Ltd: *Nature Nanotechnology*, Vol. 3, p. 733–737, Copyright 2008.)

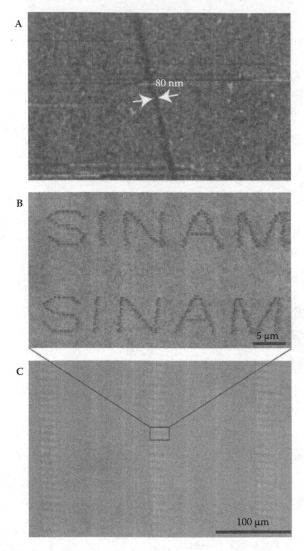

Figure 8.6 Maskless lithography by flying plasmonic lenses at the near field. (A) Atomic force microscope (AFM) image of a pattern with 80 nm linewidth on the TeOx-based thermal photoresist. (B) AFM image of arbitrary writing of "SINAM" with 145 nm linewidth. (C) Optical micrograph of patterning of the large arrays of "SINAM." (Reprinted by permission from Macmillan Publishers Ltd: *Nature Nanotechnology,* Vol. 3, p. 733–737, Copyright 2008.)

consist of low and high refractive index multilayer dielectric films. Even though the conventional film-based AR coatings are highly effective at reducing reflections for a particular

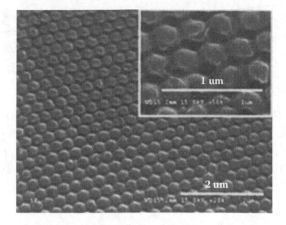

Figure 8.7 Scanning electron microscope (SEM) image of the moth's eye pattern imprinted using ultraviolet-nano-imprint lithography (UV-NIL) in a polymer on a glass substrate.

wavelength and angle of incidence, they are not effective over broad wavelength regions and angles of incidence. Engineers have taken a cue from nature and are attempting to mimic the nano- and microstructure of the moth's eye to make a better AR surface. A moth's eye is composed of 200 to 300 nm tall pillars that enable them to see well in the dark and also minimize reflections off of their eyes, camouflaging them from predators. This same nanostructure has been fabricated in master templates, transferred to PVC working stamps, and used to pattern a UV curable resin with the moth's eye structure.[7] The nanostructured moth's eye serves as a continously graded refractive index coating that grades from the refractive index of air (or other surrounding medium) to the index of the imprinted polymer. The continuously changing refractive index mitigates Fresnel reflections. A scanning electron microscope (SEM) image of the imprinted pattern is shown in Figure 8.7. The normal incidence transmission for a double-side moth's eye treated glass slide was improved by ~4 – 6% over the entire visual spectrum while the same measurement with a 30 degree angle of incidence showed a similar improvement but over a slightly reduced wavelength range of ~ 450 nm to 700 nm. The effect pattern pitch and dual-side patterning of the moth's eye AR coating created via nano-imprint lithography (NIL) was also systematically studied by Choi et

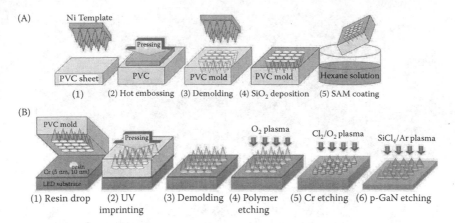

Figure 8.8 (A, B) Process of top-down nanoscale light-emitting diode (LED) system with nanotextured surface for antireflective properties. (Reprinted from Materials Science & Engineering B, Vol. 163, Hong et al., Fabrication of moth-eye structure on p-GaN layer of GaN-based LEDs for improvement of light extraction, pp. 170–173, Copyright 2009, with permission from Elsevier.)

al.[8] Furthermore, a practical implementation of a moth's eye AR coating on a GaN light-emitting diode (LED) was recently described by Hong et al.[9] The process for treating the LED surface with the nanotextured pattern is shown in Figure 8.8. The transmittance of the AR-coated LED was increased by 1.5 to 2.5 times compared to the untreated LED. Furthermore, the photoluminescence from the treated LED was improved by five to seven times over the untreated LED. Soft NIL has also been applied to a sensing application that relies on surface-enhanced raman spectroscopy (SERS) to identify target molecules on a surface. The technique described by Baca et al. is important because it is a method for fabricating and utilizing large-area, planar SERS substrates that in a cost-effective manner with uniform amplification laterally across the substrate can be engineered for multiwavelength operation.[10] The NIL processing to produce the SERS substrates used soft lithography in negative photoresist (SU-8) to pattern the SU-8 with nanocavity arrays with periodicities ranging from 0.49 to 1.75 mm and diameters ranging from 0.17 to 1.12 mm and cavity depths of ~360 nm. Figure 8.9 shows the cross section of the plasmonic cavity array, SEM views of the array and individual cavities, and a photograph of an entire SERS substrate.

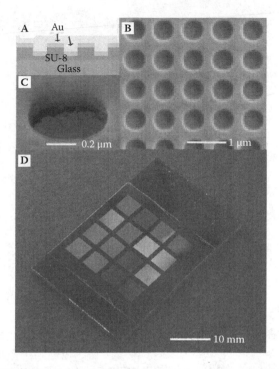

Figure 8.9 (A) Cross section of the nanocavity surface-enhanced raman spectroscopy (SERS) substrate architecture. (B) Scanning electon micros-copy (SEM) of array of nanocavities. (C) Zoom-in SEM of individual nano-cavity. (D) The complete SERS substrate. (Reprinted with permission from Applied Physics Letters, Vol. 94, pp. 243109-1–243109-3, Copyright 2009, American Institute of Physics.)

Baca et al.[10] and Piner et al.[11] did an excellent job at develop-ing a scalable, simple approach to fabricate a highly uniform nanoscale system that delivers reproducible responses.

Dip Pen Nanolithography

Dip pen nanolithography (DPN) is an excellent method devel-oped circa 2000 by Mirkin et al., which is important because it addresses individual nanowire or nanomaterial assembly at the very low nanometer scale (e.g., 1 to 10 nm dimension con-stituent length scale components).[11] DPN is also advantageous over conventional lithography and other processes because it eliminates processing steps (e.g., multiple protection/deprotec-tion steps) saving time and money during the manufacturing process. In short, DPN uses an atomic force microscope (AFM) to deliver an ink to a substrate in a user-defined, arbitrary configuration. The ink can be composed of a volatile solvent with a nanoparticle dispersion, a polymer dissolved in a solution, or other liquid media carrier dispersions. The size limitation (minimum feature size) of DPN is driven by the probe shape, the free surface energy of the liquid media and the substrate, and the chemical nature of the ink. One group has successfully used DPN to deposit luminescent conductive polymer nanowires (MEH-PPV) in an array on a substrate.[12] Upon deposition using the DPN's AFM probe, the nanowires were subsequently imaged using a confocal microscope to determine their fluorescence intensity. This work led the authors to discover that another important parameter is probe translation speed. Slower probe translation speeds across the substrate led to an increase in the polymer nanowire diameter. While the optical properties (fluorescence) of DPN deposited nanowires were of interest in the work of Noy et al.,[12] the technique of DPN for producing electrical contacts to nanowires (i.e., single-walled carbon nanotubes) developed by Wang et al. expanded the realm of single nanowire electronic device applications.[13] The process that Wang et al. developed is described in Figure 8.10. The first step in the process is to deposit a thin film of gold in desired locations. The second step involves DPN of an etch resist layer (16-mercaptohexadecanoic acid, MHA) in a user-defined geometry. Finally, a gold wet etchant is used to remove all gold not patterned with MHA via DPN. The MHA geometry is selected such that it produces contacts to the two ends of a predeposited carbon nanotube on a substrate. It should be noted that recent work is pointing toward the evolution of new system-level approaches to make DPN a high-throughput process. In particular, Haaheim et al. pioneered a new Micro Electro Mechanical Systems (MEMS)-enabled DPN process for Directed Nanoscale Deposition and commercialized it as a system called NanoInk's 2D nano PrintArray™. They used 55,000 tip-cantilevers across a 1 cm² chip in a manner described schematically in Figure 8.11.[14] The manufacturability advantages of massively paralleled DPN heads in terms of resolution, flexibility, and life-cycle cost are shown in Table 8.1.

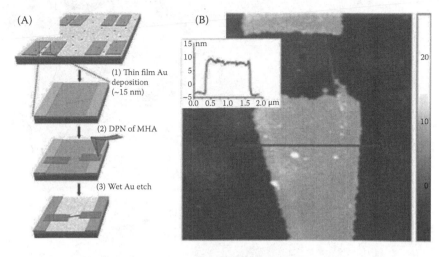

Figure 8.10 (A) Dip pen nanolithography (DPN) process to make contacts to carbon nanotubes on a rigid substrate. (B) Atomic force microscopy (AFM) topographic images of the mercaptohexadecanoic acid (MHA)-masked gold contacts and carbon nanotube. (Reprinted in part with permission from ACS Nano, Vol. 3(11), pp. 3543–3551. Copyright 2009, American Chemical Society.)

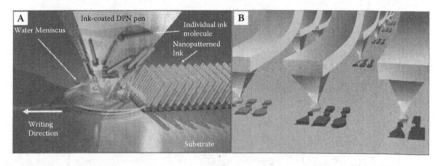

Figure 8.11 (A) Dip pen nanolithography (DPN) concept illustration. (Reprinted by permission from Macmillan Publishers Ltd: Nature Nanotechnology, Vol. 2, pp. 145–155, Copyright 2007.) (B) Concept of two-dimensional (2D) nano-PrintArray™ lithographic process. (Reprinted by permission from John Wiley & Sons Inc.: Scanning, Vol. 30, pp. 137–150, Copyright 2008.)

Printing Nanomaterials in 3D to Make Devices

When one thinks of the ultimate in systems engineering to make a scalable nanomaterial-based electronic or opto-electronic device, one can envision a notional process that includes

TABLE 8.1
Comparing Nanopatterning Techniques—Dip Pen Lithography's (DPN) Competitive Advantages

Approach	Nanopatterning Technique	Serial/ Parallel	Material Flexibility	Direct-Write Placement of Nanoscale Features	Litho Resolution	Litho Speed	Registration Accuracy	Cycle Time	Cost Purchase	Cost Operation
Top-down	Photolithography	Parallel	No	No	~35 nm	Very fast	High	Weeks	>$10M	High— masks
	E-Beam Lithography	Serial	No	No	~15 nm	Medium	High	Days	>$1M	High
	Nanoimprint Lithography (NIL)	Parallel	No	No	~10 nm	Fast	High	Days–week	>$500k	Moderate— molds
Enables both	Dip Pen Nanolithography (DPN)	Serial or parallel pens	Yes	Yes	14 nm	Highly scalable	Extremely high	Hours— change on the fly	<$250k	Low
Bottom-up	Microcontact Printing (μCP)	Parallel	Yes	No	~100 nm	Fast	Low	Days–week	~$200k	Moderate— masks
	Scanning Tunneling Microscopy (STM)	Serial	Limited	No	Atomic	Very slow	Extremely high	Days	>$250k	Low

Source: Reprinted by permission from John Wiley & Sons Inc.: Scanning, Vol. 30, p. 137–150, Copyright 2008.

a scalable method of producing high-quality constituent nano-materials, a process of assembling them using a top-down process (which may mean that humans control the kinetics of a process rather than nature, as may be the case in self-assembled bottom-up processes), and a method of directing the placement of those assembled nanomaterial constituents into useful systems. In fact, the process described above has recently been realized by the Rogers group at the University of Illinois at Urbana-Champaign.

The method of stamp transfer printing of nanomaterials was shown by Ahn et al. to yield high-performance, heterogeneous 3D nanomaterial-based circuits.[15,16] In that work, the authors followed a procedure illustrated in Figure 8.12. First, high-quality semiconducting nanomaterials such as single-wall carbon nanotubes (SWNTs), GaN ribbons, Si ribbons, or GaAs nanowires were prepared on a source "donor" wafer via a CVD process in the case of the SWNTs or a photolithography and etching procedure for the other materials. An elastomeric stamp composed of a polymer such as polydimethylsiloxane (PDMS) was used to gently pick up the nanowire array and transfer it to the destination substrate. The destination substrate is coated with a thermal or UV curing thin liquid film such as polyimide. The "inked" PDMS stamp is contacted to the coated substrate while the thin liquid cures. After curing, subsequent peeling of the PDMS leaves behind the embedded nanomaterials. Iterative stamping and curing processes can be used to build up multi-layer electronic circuits composed of nanomaterial components. Furthermore, all of the aforementioned processes can be done with excellent registration between deposited layers.

Top-Down Assembly of Nanoscale Composite Materials for Structural, Thermal, and Energy Harvesting Systems

A second focus of this chapter will be on the top-down assembly of nanoscale composite materials that are not electronic or opto-electronic in nature, but rather fulfill other functionalities in larger engineered systems (e.g., structural, thermal,

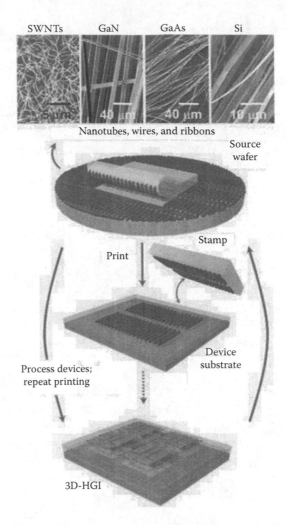

Figure 8.12 A scalable manufacturing method for printing heterogeneous three-dimensional (3D) electronics using nanomaterial constituents. (From Ahn et al., Heterogeneous Three-Dimensional Electronics by Use of Printed Semiconductor Nanomaterials, Science, Vol. 314, pp. 1754–1757. Reprinted with permission from AAAS.)

or energy harvesting functionalities). For example, top-down assembly practices such as mechanical stretching, spin-casting, wet spinning, melt fiber spinning, and electrospinning have all been used by researchers to produce preferentially oriented nanomaterial-based polymer composites.

Top-Down Assembly of Nanoscale Structural Material Systems

Interestingly, materials such as carbon nanotubes and nanoscale diameter cellulose fibers exhibit mechanical strengths and moduli that are extraordinary compared to traditional composite reinforcement materials such as glass or carbon fibers. However, translating those impressive individual fiber mechanical properties to macroscopic, practical structural materials proves challenging. One reason is that, unlike micron scale glass fibers, the nanomaterial fibers are difficult to create in long straight runs. Therefore, one obtains, at best, individual segments of nanofiber that are in the ~100s of μm to several mm in length. Making composites with those short segment nanomaterials tends to reduce the overall composite's mechanical properties below the ultimate possible in the nanomaterial reinforcements constituents. All that being said, recent progress in academia has suggested that naturally occurring, bacteria-derived nanocellulose fiber mats may yield composite materials with excellent mechanical properties.[17–22] Recently, efforts have been initiated at the Johns Hopkins University's Applied Physics Laboratory (JHUAPL) to refine the work done in academia and make it amenable for producing larger-scale composites. The overall concept for a top-down manufacturing approach to generate large-scale nanocellulose mat–based structural composites with interesting optical

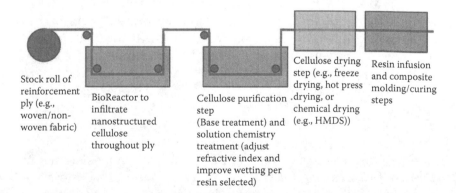

Stock roll of reinforcement ply (e.g., woven/non-woven fabric)

BioReactor to infiltrate nanostructured cellulose throughout ply

Cellulose purification step (Base treatment) and solution chemistry treatment (adjust refractive index and improve wetting per resin selected)

Cellulose drying step (e.g., freeze drying, hot press drying, or chemical drying (e.g., HMDS))

Resin infusion and composite molding/curing steps

Figure 8.13 A scalable manufacturing concept for making nanocellulose structural materials.

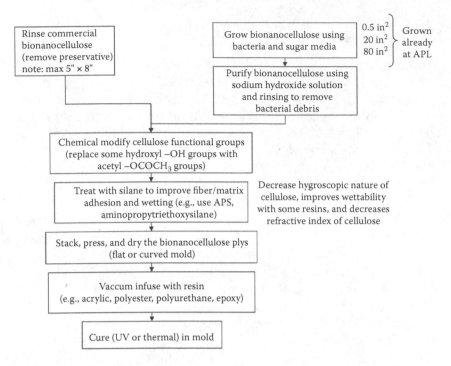

Figure 8.14 Processing details for making nanocellulose structural materials using top-down methods.

properties is shown in Figure 8.13. More details on specific processing steps with rationale are provided in Figure 8.14.

To date, emphasis has been placed on scaling the thickness of the nanocomposite materials from the ~60 um thickness reported in academia, retaining good mechanical reinforcing capability (high modulus and high strength) and maintaining high transmittance in the visible region of the electromagnetic spectrum. The bacterial cellulose nanofibers are shown in the SEM images of Figure 8.15. The bacteria ingest sugar and secrete six pure cellulose chains per pore from an array of ~50 pores on its backside. The resultant fibers aggregate to form ~14 nm × 50 nm ribbon cross-sectional fibers that are hundreds of microns in length. The images shown in Figure 8.15A through 8.15C are images from a drying experiment in which the effect of drying technique on result fiber mat morphology was examined. From the images, it can be seen that hexamethyldisilazane (HMDS) chemical drying and

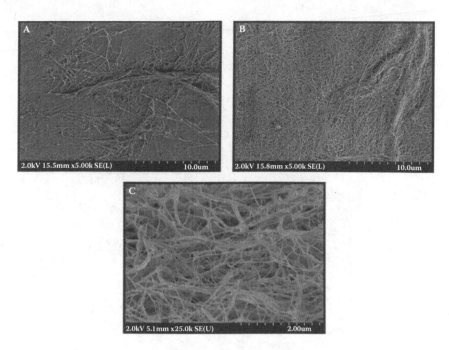

Figure 8.15 Scanning electron microscopy (SEM) images of (A) ethanol dried bacterial cellulose mat, (B) hexamethyldisilazane (HMDS) dried bacterial cellulose mat, and (C) supercritical CO_2 dried bacterial cellulose mat.

supercritical CO_2 drying are both very effective in maintaining discrete cellulose fibrils, while ethanol dried cellulose tends to agglomerate and form very dense, rather impenetrable mats. The open discrete fibril structure is conducive for the next step, which is resin infiltration. Typically, resins, such as Norland Optical Adhesive resins, were vacuum infused into the cellulose mats over a period of hours. After complete wetting, the infused cellulose mat was transferred to a flat, two-part mold with either 110 µm or 250 µm spacers. The top mold was typically a UV transparent quartz plate. Then, a UV curing lamp was used to cure the composite. Following careful demolding, the composite was aged overnight at 60°C and subsequently characterized for its optical and mechanical properties. A typical bionanocellulose-reinforced polyester sample is shown in Figure 8.16. The mechanical properties of a 0.25 mm thick sample were measured and compared to the native, unreinforced resin properties. The stress–strain curve is shown in

Figure 8.16 Transparent cured bionanocellulose composite in front of a flower at the Johns Hopkins University's Applied Physics Laboratory (JHU/APL). Sample thickness was measured to be 110 μm.

Figure 8.17. The calculated ultimate tensile strength was ~60 MPa and the modulus was 9 GPa. This result is significant because the modulus of the neat resin was improved by an order of magnitude and the tensile strength was improved by four times. Furthermore, as shown in Figure 8.18, the transmittance across the visible range is still very high. At ~550 nm, the wavelength of highest photopic sensitivity for humans, the total transmittance of the composite is as high as 87% with ~85% direct transmittance, which implies high optical clarity (very little "fuzziness") in images seen through the sample even when the sample to background object/observer is significant. If the sample measured in Figure 8.18 had been treated on the front and rear surfaces with a perfect antireflective coating, one could expect to achieve the top blue curve in Figure 8.18, which has an ~95% total transmittance at 550 nm.

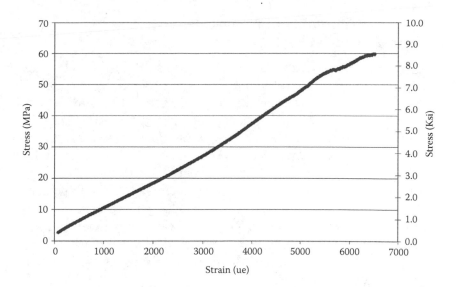

Figure 8.17 Stress–strain curve of a 0.25 mm thick bionanocellulose composite.

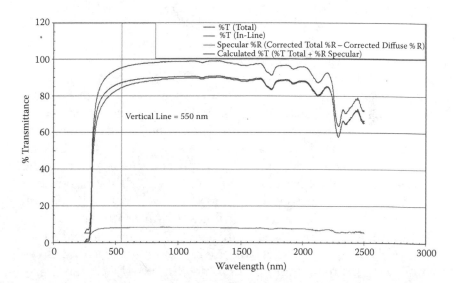

Figure 8.18 Ultraviolet-visible near-infrared (UV-VIS-NIR) transmittance curves (measured and calculated) for total transmittance, direct transmittance, and notional total transmittance with perfect antireflective front–back surface coatings.

Top-Down Assembly of Nanoscale Thermal Material Systems

Top-down assembly can also be used to create nano-, micro-, or macroscopic material systems with desired thermally conductive characteristics. For example, desired characteristics may be patterned heat spreader applications, but more interesting applications for which nanomaterials are particularly suited are those that require flexibility in the end product. Making thermally conductive materials flexible and stretchable can be a challenge with conventional materials in the materials scientists' toolbelt. However, researchers have recently had separate successes in using top-down assembly to create flexible, stretchable, high thermal conductivity material substrates and composites that incorporate ultrananocrystalline diamond (UNCD) and carbon nanotube thermal materials.

Diamond is a well-known material with extremely high thermal conductivity on the order of 1000 to 2000 W/(m · K). In comparison, traditional engineering polymers typically exhibit thermal conductivities on the order of 0.25 to 1 W/(m · K). The same group at the University of Illinois at Urbana-Champaign who developed the stamp transfer technique for making heterogeneous 3D circuits from nanomaterials has also applied the technique to make UNCD heat spreader layers for flexible devices. Kim et al. showed that while UNCD films could not be grown successfully on flexible polymer substrates due to the high deposition temperature, the UNCD films could be grown on a rigid Si/SiO_2 source wafer, patterned and etched into UNCD microstructures (long platelets), and then transferred onto a flexible substrate to enable improved thermal spreading on the flexible substrate.[23] Figure 8.19 is an illustrated flowchart of the process to create the UNCD microstructures and then stamp transfer them to a flexible substrate. In order to test the heat spreading capabilities of the stamp transferred/printed UNCD microstructures, the authors fabricated three Au/Ti serpentine heaters on flexible 75 μm polyethylene terephthalate (PET) substrates. The first heater was uncoated. The second heater was coated with a thin film of poly(methyl methacrylate) (PMMA) polymer. The third heater was coated with an array of UNCD stamp transfer printed platelets that

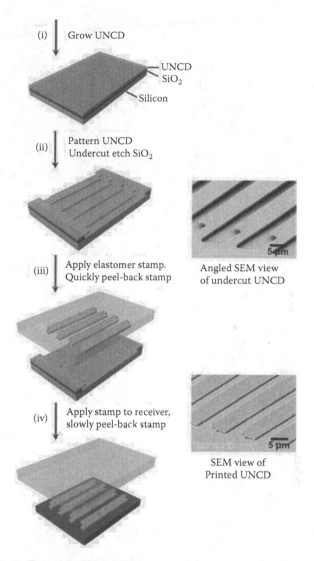

Figure 8.19 Process (i)–(iv) to create ultrananocrystalline diamond (UNCD) microstructure patterns on a flexible substrate using polydimethyl siloxane (PDMS)-based stamp transfer techniques. (Kim et al., Ultranano-crystalline diamond with applications in thermal management. Advanced Materials. 2008. Vol. 20. pp. 2171–2176. Copyright Wiley-VCH Verlag GmbH & Co. KGaA. Reproduced with permission.)

were 800 μm × 800 μm × 400 nm thick. Figure 8.20 shows the results for each heater when heated at three different applied power levels of 53 mW, 97 mW, and 160 mW. One can see that

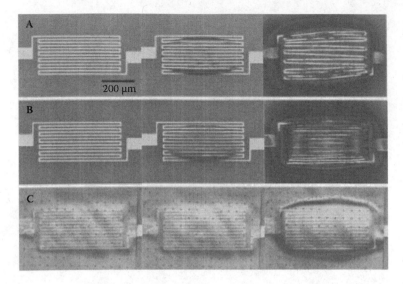

Figure 8.20 Heat spreader functionality shown with (A) uncoated serpentine heater, (B) poly(methyl methacrylate) (PMMA) polymer coated serpentine heater, and (C) ultrananocrystalline diamond (UNCD) platelet array coated serpentine heater. Images in each column are at three different applied power levels of 53 mW, 97 mW, and 160 mW, respectively, from left to right. (Kim et al., Ultrananocrystalline diamond with applications in thermal management. Advanced Materials. 2008. Vol. 20. pp. 2171–2176. Copyright Wiley-VCH Verlag GmbH & Co. KGaA. Reproduced with permission.)

at the highest power level, the substrate is damaged in all of the images, and the heater survived for only the UNCD coated sample.

In the case of carbon nanotube thermal materials, they have very good promise in applications such as printed wiring boards, thermal interface materials, and heat sinks. However, large jumps in thermal conductivity of materials have not been seen by forming composites with low volume fraction of carbon nanotubes. Although the thermal conductivity of an individual nanotube is high, the thermal conductivity of the resultant carbon nanotube composites appears to be limited by the extremely high thermal resistance at the carbon nanotube to resin interfaces. Common top-down assembly methods used for carbon nanotube composite fabrication, such as solution blending, melt blending, and in situ polymerization have not led to substantial improvements in thermal conductivity over the neat resins. However, chemical vapor deposited carbon nanotube arrays have shown significant promise for thermal interface material applications when they were synthesized with injection molded silicone matrices.

Top-Down Assembly of Nanoscale Energy Harvesting Systems

Energy harvesting from natural and man-made sources in the environment has the potential to significantly increase the efficacy of sensors and devices by increasing their endurance or operational characteristics (e.g., pulse repetition interval). It is believed that there are opportunities to significantly advance the state-of-the-art of harvesting kinetic energy using the piezoelectric effect by leveraging recent work in two respective fields: stamp transfer–based assembly of nanomaterial-based electronic devices and new nanostructured piezoelectric materials. It is believed that with proper engineering, a mechanically flexible energy harvesting material that can harvest energy 10 times more effectively than polyvinylidiene fluoride (PVDF) from vibrations (e.g., cantilever or membrane geometry), flowing water (e.g., bluff body/eel geometry), or wind (e.g., bluff body/fluttering flag geometry) can be demonstrated by developing an innovative, scalable, low-temperature manufacturing process to fabricate a piezoelectric energy harvesting composite that includes nanowire piezoelectric energy harvesting materials and nanoscale electrode materials.

In earlier work, others used piezoelectric thin crystals or film-based bimorph cantilevers made of materials such as lead zirconium titanate (PZT) to harvest vibration energy and zinc oxide (ZnO) piezoelectric nanowires attached radially to a polymer fiber to harvest frictional energy. Specifically relevant to the aforementioned concept, PVDF artificial "eels" have demonstrated the ability to harvest energy from the currents of flowing water as a succession of rotational vortices behind a bluff body strain the piezoelectric elements. In a previously funded Defense Advanced Research Projects Agency (DARPA) effort, the piezoelectric eel, roughly the dimensions of a scarf, was able to harvest 10 mW in flowing water.

As shown in Table 8.2, other materials such as lead zirconia niobate–lead titanate (PZN-PT) exhibit much higher

TABLE 8.2
Strain Coefficients (d_{33}) of Various Bulk Piezoelectric Materials

	Polyvinylidiene Fluoride (PVDF)	Zinc Oxide (ZnO)	Lead Zirconium Titanate (PZT)	Lead Zirconia Niobate– Lead Titanate (PZN-PT)
d_{33} (pC/N)	30	10	300–650	2,000

mechanical to electrical coupling properties and proportionally higher harvestable power compared to PVDF.

The concept for novel nanomaterial-based nano-piezogenerator (NPG) composite devices for energy harvesting synthesis is described below. One would synthesize high-quality piezoelectric nanowires of PZN-PT or PZT, assemble them using a polydimethylsiloxane (PDMS) stamp transfer technique, and align them using a electrophoretic alignment process. Using these aligned nanowires, one would construct a piezoelectric/polymer composite that is flexible and thin with electrode layers (Ag nanoparticles) immediately above and below the edges of the piezoelectric nanowires. It is believed that the nanowire composite approach and the selected harvester architecture have significant advantages over PVDF, such as having a higher electromechanical coupling coefficient that results in more charge production for a given stress; being more damage tolerant, flexible, and having better piezoelectric material quality than the PZT bimorph thin crystals or films (discrete nanowires lead to more flexibility, and the small dimensions of nanowires may have few or single crystals and less losses in the piezoceramic during operation); and having better performance than the semiconducting ZnO direct current radial nanowire-on-polymer fiber approach (friction will lead to failure at some cycle number, and the architecture had electrode contact at only discrete portions of the nanowire). Also, while macrofiber composites and microfiber composites exist that exhibit piezoelectric properties, the nanowire proposed devices should be superior in terms of ultimate bend strain to failure (fewer defects in the nanowires compared to bulk microfibers of piezoelectric ceramics), and also because of the small nature of the nanowires, they may be single crystals or few grains, leading to less electrical losses (heating in the

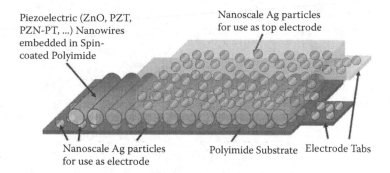

Piezoelectric (ZnO, PZT, PZN-PT, ...) Nanowires embedded in Spin-coated Polyimide

Nanoscale Ag particles for use as top electrode

Nanoscale Ag particles for use as electrode

Polyimide Substrate   Electrode Tabs

Figure 8.21 Three-dimensional (3D) cross section of a single layer of the proposed nano-piezo-generator. The multilayered ABCABCABC stacking sequence (nano-Ag electrode layer/piezoelectric nanowire layer/nano-Ag electrode layer...) is easily realizable.

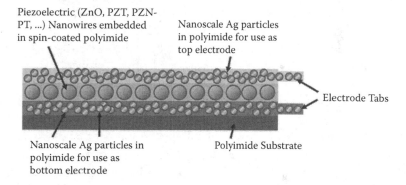

Piezoelectric (ZnO, PZT, PZN-PT, ...) Nanowires embedded in spin-coated polyimide

Nanoscale Ag particles in polyimide for use as top electrode

Electrode Tabs

Nanoscale Ag particles in polyimide for use as bottom electrode

Polyimide Substrate

Figure 8.22 Cross section of a single layer of the novel nano-piezo-generator.

material) as the material is cycled in an alternating bending stress.

The technical challenge to make a highly efficient energy harvesting flexible material will be to synthesize high-quality piezoelectric nanowires (such as PZN-PT) and assemble them into a piezoelectric/polymer composite as shown in Figures 8.21 and 8.22. The resultant composite should be flexible and thin with electrode layers immediately above and below the edges of the piezoelectric nanowires. The PZN-PT nanowire synthesis method can be described as a solvothermal synthesis technique in which one loads an autoclave with lead, zirconium, niobium, and titanium metal oxide and alkoxide precursors and solvents in appropriate ratios, seals the autoclave, and raises the temperature and pressure to nucleate and grow the PZN-PT nanowires. Another potential synthesis method suitable for PZN-PT nanowire fabrication includes growing the nanowires in a template, such as a nano-porous anodic alumina template, followed by dissolution of the alumina template. It should be noted that these would be truly novel materials, as there are currently no published articles describing the synthesis of PZN-PT nanowires.

To create the piezoelectric nanowire-based composite, one could employ a polydimethylsiloxane (PDMS) stamp transfer technique. A thermo- or photo-curing polymer (e.g., polyimide) could be spun coat on a substrate as a base layer, and a patterned PDMS stamp would be used to transfer a patterned electrode (e.g., Ag nanoparticles) layer (layer A) into the wet thin polymer layer using a specified contact pressure. Heat or UV light would be used to cure the polymer with the embedded Ag nanoparticles. The process would be repeated by spin coating another layer of thermo- or photo-polymer and then, using a different PDMS patterned stamp, transferring a single layer of piezoelectric nanowires (layer B). (Note that in some embodiments, the layer of piezoelectric nanowires may be aligned between interdigitated electrodes using dielectrophoresis before being stamp transferred. Dielectrophoresis is the polarization and manipulation of neutral dielectric objects in a nonuniform electric field, which can be used to manipulate the nanowires at the nanoscale between interdigitated electrodes on a substrate.) Again, heat or UV light will be used to cure the polymer with the embedded piezoelectric nanowires.

The process would be repeated with spin coating more polymer and transferring another electrode layer using the electrode patterned PDMS stamp (layer C). Curing the polymer in this layer creates a single-layer composite. However, the layers can be repeated with any desired permutation of A, B, and C (or other patterned stamp layers) until a suitably thick multilayer composite piezoelectric device has been developed. This will fabricate an ABCABCABC stacking sequence, where A and C are interleaved nano-Ag electrode layers and B is a sandwiched piezoelectric nanowire layer. Finally, a diamond knife and conductive epoxy can be used

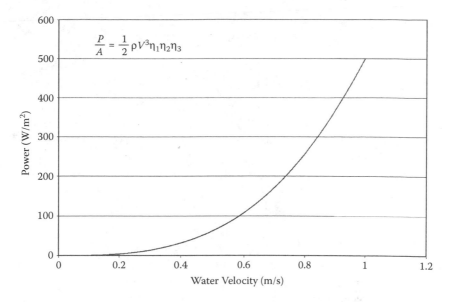

$$\frac{P}{A} = \frac{1}{2}\rho V^3 \eta_1 \eta_2 \eta_3$$

Figure 8.23 Model of the power per area produced by flowing water as a function of water velocity. At 1 m/s, the theoretical power output is 500 W/m^2.

to make electrical contact to all of the electrode layers in any desired configuration.

Sensing in oceanic, river, and stream environments is critical for commercial and defense applications. Current systems may have limited lifetime due to battery life or rotating parts (turbine-based systems) that limit endurance of the sensing activity. Continuous energy harvesting to trickle charge a battery that powers the sensor will enable extended duration sensing activities. As a result of this and knowledge of typical flow rates, one could optimize the development of a nano-piezo-generator for an underwater application with flowing water currents of 0.5 to 1 m/s.

As is shown in the model in Figure 8.23, the theoretical power available from water flowing at 1 m/s is approximately 500 W/m^2. However, for piezoelectric devices, this number is normalized using three efficiency parameters, η_1, η_2, and η_3. The first parameter, η_1, is the hydrodynamic efficiency that takes into account the efficiency of converting water energy into mechanical energy. This value is denoted by the Betz number (C_b) that computes the energy extracted by slowing flowing water. The equation for the Betz number is

$$C_b = \frac{1}{2}\left(1 - \frac{v_2^2}{v_1^2}\right)\left(1 + \frac{v_2}{v_1}\right)$$

where v_2 and v_1 are the downstream and upstream flow velocities, respectively.

With a velocity relation (v_1: v_2) of 3:1, the Betz number reaches a maximum value of 0.592. However, practical values of the number are in the range of 0.35 to 0.45, and one can select the conservative value of 0.35 for this efficiency. The second parameter, η_2, is the efficiency of the electronic circuitry. For the notional nano-piezo-generator devices, one can envision using a switched resonant power conversion circuit. For the frequencies of interest in the low Hz range, this type of circuit has calculated potential efficiencies of 37%. For our calculations, we are using a conservative value of ~25%. The last parameter, η_3, is the efficiency of a specific piezoelectric material to convert mechanical energy into electrical energy. For PZN-PT, this number has been reported as high as 88%, but again we will use a conservative value of 25%. For PVDF, this efficiency is 1%. Using these efficiencies, the theoretical power output of a PVDF device in 1 m/s flowing water is 0.4 W/m². Our power output estimate of a PZN-PT device in the same water is 10 W/m², due to the higher mechanical to electrical coupling.

The experimental approach for the energy harvesting, as schematically shown in Figure 8.24, incorporates a von Karman vortex street/fluttering flag geometry to harvest energy from flowing water. The Strouhal equation, which describes oscillating flow mechanisms, is

$$f_S = \frac{0.2 \cdot V}{d}$$

where f_S is the Strouhal frequency, V is the velocity of the fluid, d is the diameter of the bluff body in which the piezoelectric

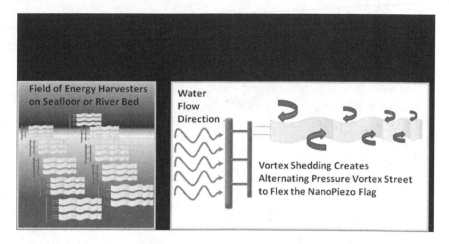

Figure 8.24 The nano-piezo-generator in a fluttering flag geometry for an underwater energy harvesting application.

composites are attached, and 0.2 is the Strouhal number for fluids with low Reynolds numbers. From this equation, at a water velocity of 1 m/s and a bluff body diameter of 1.5 cm, the flags will flutter at a frequency of 13 Hz. By adjusting the diameter of the bluff body, the frequency of oscillations can be optimized for the composite to maximize strain and resulting power output.

From the power equation above, we conservatively predict, using the kinematic viscosity of water, a flow velocity of 1 m/s, the efficiencies described above, and an additional 10% efficiency value, the fabricated device will produce 215 mW/m², a five times increase in the power produced by the PVDF electric eel (39 mW/m²).

The described composite nano-piezo-generator device can be considered a multifunctional apparatus. As mentioned above, one potential application area is in the area of energy harvesting from a flowing water source such as an ocean current. The piezoelectric composite device would be placed on an inactive flexible support sheet and arranged in a flag-type configuration behind a bluff body (e.g., something like a flagpole that would disrupt the normal flow of water to create a von Karman vortex street of alternating vortices at some frequency) as shown in Figure 8.24. An array of such "flags" could be used to harvest the necessary amount of power for a given application. Although so far in the discussion we have focused the nano-piezo-generator concept of operations on an underwater energy harvesting system, the final device's utility could be applicable to a wide variety of environments. For instance, the conformal devices could be used to cover large surface areas (truck beds, vehicle hoods, entire exteriors of buildings, etc.) to harvest vibrational energy, even at low frequencies. Furthermore, the piezoelectric device fluttering flag geometry could be used for wind harvesting in the same form. However, due to the power conversion equation governing piezoelectrics (power is proportional to density), the much lower density of air compared to water will severely limit the harvestable energy. Additionally, by changing the approach from a piezoelectric nanowire design to a thermoelectric nanomaterial (nanowire/nanoparticle) design, the described nanodevice could harvest energy from temperature differentials on vehicles, woven into fabrics to be worn by soldiers to recharge batteries, etc. One can also envision a hybrid piezoelectric:thermoelectric multimodal energy harvesting device assembled using nanomaterials, nanowires, and the stamp transfer technique.

Top-Down Assembly of Nano Electro Mechanical Systems (NEMS)

Finally, the chapter concludes with a brief examination of the top-down assembly processes and potential scalability associated with Nano Electro Mechanical systems (NEMS). For example, electrofluidic "tools" have been shown to be very effective in the assembly of NEMS.[24] Furthermore, a mass spectrometer on a chip-NEMS-based system to create a high-performance instrument is an excellent example of nanoscale top-down assembly.[25]

Electrofluidic Top-Down Assembly of a NEMS

NEMS are typically fabricated by spatially patterning a single crystal mechanical layer (Si, GaAs, etc.) followed by an etch release of an underlying sacrificial layer. Such NEMS mechanical components are useful for ultrasensitive sensing applications and low-power radio frequency (RF)-range nanomechanical signal processors.[24] However, traditional top-down nanomachining approaches have damaged the surfaces of the mechanical resonators limiting their quality factors and ultimately their sensitivity. Recently, Evoy et al. demonstrated a promising new method: electrofluidic top-down assembly of a NEMS.[25] In their approach, the researchers combined electrodeposition of a noble high-conductivity metal (Rhodium) into anodized alumina membrane nanopores to synthesize Rhodium nanorods. The nanorods were then assembled on an electrode Si/SiO_2 wafer using a dielectrophoresis electrofluidic assembly technique. Electrodes were deposited on one or both ends of the Rhodium rods. Finally, the underlying SiO_2 was etched away to produce suspended Rh rods with attached electrodes, a NEMS device.

Top-Down Assembly of a NEMS-Based Mass Spectrometer on a Chip

Researchers recently announced the first successful fabrication and testing of a NEMS-based mass spectrometer. The system is shown schematically in Figure 8.26. The system uses some traditional mass specification components, such as electrospray ionization to ionize the species of interest and magnetic fields to guide the species, but the heart of the system is the NEMS device. As the individual molecules and nanoparticles arrive at the NEMS sensor, the increased mass due to their adsorption causes increases of mass of the sensor beam and concomitant decreases in the resonance frequency. Their direct observation was reported for the first time in 2009 in the NEMS-MS system.[25] The numerous potential benefits of NEMS-MS systems equipped with arrays of NEMS sensors and additional technologies are very promising for the future. The top-down assembly of the NEMS sensor has made this attractive single molecule/nanoparticle detection method possible.

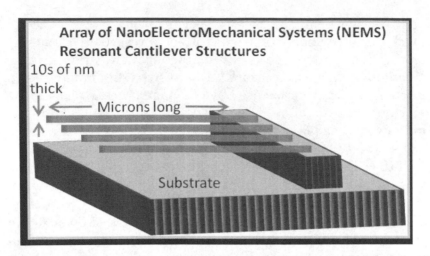

Figure 8.25 Conceptual diagram of an electrofluidic assembled Rh Nano Electro Mechanical Systems (NEMS).

Conclusion

The science and engineering related to the top-down assembly of nanoscale systems is still in the early stages. The research in this area is growing but is still sparse. Many researchers are still studying nanomaterials at the individual nanolevel. Significant, targeted government funding in the area of top-down assembly of scalable, cost-effective nanosystems is warranted to increase the number of researchers and quality of research in this important area. However, even though the existing body of literature is small in this area, this chapter has highlighted numerous success stories where engineers have been able to translate the properties of a single nanoparticle or nanomaterial into a larger system to create unique functionalities. Those functionalities have spanned multiple domains, including nano-electrical, nano-optical, nanostructural, nano-thermal, nano-energy, and nano-electro-mechanical. Now that proof-of-concept systems have been demonstrated, new challenges await the research engineers as they move toward the daunting task of making their systems cost-effective, scalable, and integration worthy into larger, more complex systems with heretofore unseen functionality. Systems engineers may begin

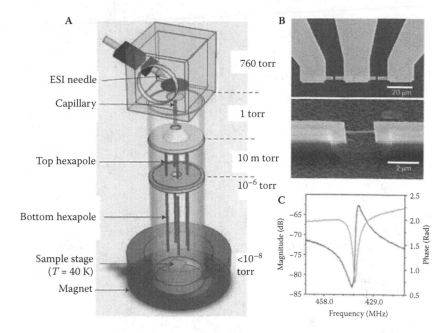

Figure 8.26 First-generation nano-electro-mechanical–mass spectrometer system. A. Simplified schematic of the experimental configuration (not to scale), showing the fluid-phase electrospray ionization (ESI) and injection, the system's three-stage differential pumping, and its two-stage ion optics. B. Scanning electron micrographs showing one of the doubly clamped beam NEMS devices used in these experiments. It is embedded in a nanofabricated three-terminal UHF bridge circuit. C. Magnitude and phase of the UHF NEMS resonator's response displaying a prominent fundamental-mode resonance near 428 MHz. (Reprinted by permission from Macmillan Publishers Ltd: Nature Nanotechnology, Vol. 4, pp. 445–450, Copyright 2009.)

to take note of the new, exciting capabilities of some of the newly demonstrated top-down assembled nanosystems and hopefully build previously unthinkable requirements into projects.

References

1. Campbell, S.A., *The Science and Engineering of Microelectronic Fabrication*, 2nd ed., Oxford University Press, New York, 2001.

2. Zhang, T.W. et al., Operational energy analysis of plasmonic imaging lithography, *Proceedings of the 2007 IEEE International Symposium on Electronics and the Environment*, May 7–10, 97–101, 2007.
3. Guo, L.J., Nanoimprint lithography: Methods and material requirements, *Advanced Materials*, 19(4), 495–513, 2007.
4. Ahn, J.H. et al., Heterogeneous three dimensional electronics by use of printed semiconductor nanomaterials, *Science*, 314, 1754, 2006.
5. Liu, Z. et al., Focusing surface plasmons with a plasmonic lens, *Nano Letters*, 5(9), 1726–1729, 2005.
6. Srituravanich, W. et al., Flying plasmonic lens in the near field for high-speed nanolithography, *Nature Nanotechnology*, 3, 733, 2008.
7. Bae, B.J. et al., Fabrication of moth-eye structure on glass by ultraviolet imprinting process with polymer template, *Japanese Journal of Applied Physics*, 48, 010207, 2009.
8. Choi, D.G. et al., Effects of pattern size, dual side patterning, and imprint materials in the fabrication of antireflective structure using nanoimprint, nanoengineering: Fabrication, properties, optics, and devices V, edited by E.A. Dobisz, L.A. Eldada, *Proceedings of SPIE*, 7039, 70391E, 0277-786X/08, 2008.
9. Hong, E.-J. et al., Fabrication of moth-eye structure on p-GaN layer of GaN-based LEDs for improvement of light extraction, *Materials Science and Engineering B*, 63, 170–173, 2009.
10. Baca, A.J. et al., Molded plasmonic crystals for detecting and spatially imaging surface bound species by surface-enhanced Raman scattering, *Applied Physics Letters*, 94, 243109, 2009.
11. Piner, R.D.; Jin, Z.; Feng, X.; Seunghun, H.; and Mirkin, C.A. "Dip-pen" nanolithography, *Science*, 283, 661–663, 1999.
12. Noy, A. et al., Fabrication of luminescent nanostructures and polymer nanowires using dip-pen nanolithography, *Nano Letters*, 2(2), 109–112, 2002.
13. Wang, W.M. et al., Dip-pen nanolithography of electrical contacts to single-walled carbon nanotubes, *ACS Nano*, 3(11), 3543–3555, 2009.
14. Haaheim, J.R. et al., Microfluidics, BioMEMS, and Medical Microsystems VII, edited by W. Wang, *Proc. of SPIE*, 7207, 720706-1, 2009.
15. Ahn, J.H. et al., Heterogeneous three-dimensional electronics by use of printed semiconductor nanomaterials, *Science*, 314, 1754, 2006.

16. Sun, Y. et al., Printed arrays of aligned GaAs wires for flexible transistors, diodes, and circuits on plastic substrates, *Small*, 2(11), 1330–1334, 2006.
17. Yano, H. et al., Optically transparent composites reinforced with networks of bacterial nanofibers, *Advanced Materials*, 17(2), 153, 2005.
18. Nogi, M. et al., Optically transparent bionanofiber composites with low sensitivity to refractive index of the polymer matrix, *Applied Physics Letters*, 87, 243110-1, 2005.
19. Nogi, M. et al., Property enhancement of optically transparent bionanofiber composites by acetylation, *Applied Physics Letters*, 89, 233123-1, 2006.
20. Nishino, T. et al., All-cellulose composite prepared by selective dissolving of fiber surface, *Biomacromolecules,* 8, 2712–2716, 2007.
21. Gindl, W. et al., Drawing of self-reinforced cellulose films, *Journal of Applied Polymer Science*, 103, 2703–2708, 2007.
22. Hsieh, Y.-C. et al., An estimation of the Young's modulus of bacterial cellulose filaments, *Cellulose* 15, 507–513, 2008.
23. Kim, T.-H. et al., Printable, flexible, and stretchable forms of ultrananocrystalline diamond with applications in thermal management, *Advanced Materials*, 20, 2171–2176, 2008.
24. Evoy, S. et al., Electrofluidic assembly of nanoelectromechanical systems, *MRS Fall Conference,* Symposium B, 2001.
25. Naik, A.K. et al., Towards single-molecule nanomechanical mass spectrometry, *Nature Nanotechnology*, 4, 445–450, 2009.

9

Nanoscale Systems— Bottom-Up Assembly

Jason Benkoski

Nano child's play.

Contents

Nano-Intuition

Because forces do not scale linearly with length, our physical intuition can falter at the nanoscale. The textbook example is gravity. Weight is a central design consideration for everything from transportation to personal electronics. At the nanoscale, gravity and inertia are practically nonexistent. In contrast, adhesion can be difficult to achieve between macroscopic objects, but it frequently dominates all other forces at the nanoscale. It is not unusual for self-assembling nanosystems to be held together by nothing more than the surface tension of water.

Nanotechnology would be counterintuitive enough if only the relative magnitudes of forces were different. Even less intuitive is the fact that thermal fluctuations are large enough to overcome the forces holding nanoscale systems together.[1] At times a challenge and an opportunity at others, the importance of random thermal vibrations means that the state of each nanocomponent is probabilistic. It makes more sense to speak in terms of a most probable state rather than a precise condition. Each nanocomponent may spend, for example, 80% of its time in the "on" state and 20% in the "off" state, switching thousands of times each second. Alternatively, one can take a snapshot of thousands of identical nanocomponents to reveal that 80% of them are "on" at any given instant.

Another way to conceptualize thermal fluctuations is to consider that the signal-to-noise ratio is extremely low. The total energy holding a system together might be only 3 or 10 times larger than the kinetic energy of an atom vibrating at room temperature. Many nanoscale systems, therefore, lie just on the edge of complete randomization. A vivid illustration is that ordinary physical properties do not have well-defined values. For instance, the adhesive strength between a pair of nanoparticles is never the same twice. It has a probabilistic range of values depending on how fast they are pulled apart.[2,3] Two particles pulled slowly will break apart at a much lower force than those pulled quickly. At infinite time, there is no force required. The reason is that if you wait long enough, a random thermal vibration will eventually come along to kick the two particles apart. If they are pulled quickly, less time is available for a helpful push.

No discussion of self-assembly is complete without mentioning the importance of entropy. Intuitively, one thinks about energy as the capacity to do work, such as the potential energy of a bowling ball placed at the top of a hill. However, energy can also be stored by increasing the order of a system. This energy of disorder is observed whenever salt dissolves in water. The system starts out ordered (pure salt and pure water), but then the drive for greater disorder causes mixing between the two pure compounds. Energy that depends on the degree of disorder is said to be entropic in origin. Entropy receives less attention at the macroscale because it is usually small enough to ignore.

The potential energy of the bowling ball in the above example is equal to the mass of the bowling ball times the acceleration of gravity times the height of the hill. Entropy is not large enough to affect the position of the bowling ball, but it will actually overcome gravity to roll a nanoparticle uphill. Defying gravity is only one of the counterintuitive consequences of entropy that will be explored later in this chapter.

Macroscale Manufacturing

These same inconsistencies carry over into nanoscale manufacturing. Macroscopic manufacturing is characterized by accuracy, precision, repeatability, and six-sigma quality management. Nanoscale manufacturing, by contrast, is inherently variable. Rather than producing exactly 1,000 identical computers, a nanoscale manufacturing process might create an unknown number of particles with a distribution of diameters, variable aspect ratio, and a fixed fraction of thermodynamically unavoidable defects. The final product may appear to be little more than a pile of powder,[4] a vial of solution,[5] or an invisible surface coating.[6] The manufactured "part" in this case can have greater variability than a collection of snowflakes and yet still outperform the most precisely constructed macroscale object.

The basic principles of manufacturing at the macroscale place the design paradigms of nanoscale self-assembly into context. As shown in Figure 9.1, manufacturing can be categorized into three types: transformative, subtractive, and additive. Transformative refers to any process that causes a shape change with no change in volume. Examples include injection molding, metal working, and pottery. The idea is to begin with a pliable material and then deform it into its final shape. The final step may or may not include a hardening step such as freezing, sintering, vitrifying, or cross-linking.

The second type of manufacturing is called subtractive, because one starts with a large, monolithic block of material and then removes material to achieve the desired shape. The shape change results from a loss of material. At its basic level, subtractive manufacturing is synonymous with sculpting.

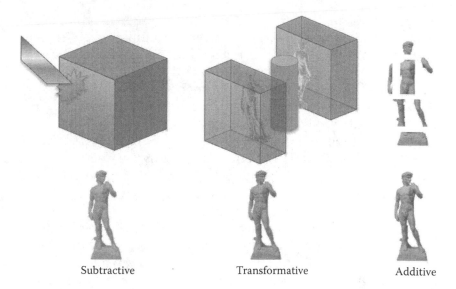

| | | |
| Subtractive | Transformative | Additive |

Figure 9.1 The three primary categories of manufacturing are (1) subtractive, which involves a shape change with a loss in volume, (2) transformative, which involves a shape change with no change in volume, and (3) additive, which involves a shape change with an increase in volume.

Industrial examples include etching and computer numerical control (CNC) machining. Unlike transformative processes, subtractive manufacturing necessarily begins with a solid material. This method is frequently employed precisely because the starting material is hard and brittle (e.g., marble).

The final category of manufacturing is additive processing. Additive processing most frequently refers to the assembly of smaller solid objects into a larger object. Additive processing may also occur through the addition of pliable materials, however, as with welding and three-dimensional (3D) printing. Therefore, the broadest sense of additive manufacturing is a shape change achieved through an increase in volume.

Nanoscale Manufacturing

These definitions set the stage for describing the two major classes of nanoscale manufacturing: top-down and bottom-up.

As described in the previous chapter, top-down nanomanufacturing is a subset of subtractive manufacturing. It was
originally developed by the semiconductor industry for the production of microchips. Broadly speaking, it is an iterative process involving the deposition of a monolithic film followed by
patterning with a template and then etching. Deposition processes include physical vapor deposition (PVD) and chemical
vapor deposition (CVD). Patterning may be achieved through
photolithography, nano-imprint lithography, or electron beam
lithography. Finally, wet etching, plasma etching, or reactive
ion etching carves out a permanent impression of the lithographic pattern on the starting material.

Whereas top-down nanomanufacturing resembles microchip fabrication, bottom-up nanofabrication resembles chemistry. Many of the building blocks for bottom-up assembly are
synthesized through a wet chemistry process. Polymer synthesis, for example, occurs through the successive addition of
organic repeat units into long chains. Another common bottom-
up process is the nucleation and growth of solid nanoparticles
from a liquid salt solution. As one example, gold nanoparticles
originate from reduction of auric acid ($HAuCl_4$) in water.[7] In
fact, few details of nanoparticle synthesis differ from routine
precipitation save for the ability to limit crystal growth and
minimize polydispersity. The subsequent process of nanoparticle assembly is also performed in a liquid medium. Liquid
media are generally crucial to this process, for they mediate
the interactions between the particles and prevent uncontrolled aggregation.

The differences between top-down and bottom-up manufacturing extend far beyond the methods of construction.
Moreover, the resulting structures, properties, and functions
generally fall along different lines as well. From a morphologi-
cal standpoint, top-down structures are characterized by geometric primitives, repeating patterns, and long-range order. In
many respects, they look similar to macroscale machines and
electronics. Bottom-up structures, on the other hand, appear
much more familiar to a microbiologist. As such, they contain
fractal geometries, polydisperse size distributions, and mostly
short-range order. The similarity to biological systems is not
coincidental. Bottom-up assembly shares many of the same

physical principles employed by cells during protein synthesis, protein assembly, and biomaterial synthesis.

Any technology that is manufactured by a process that resembles wet chemistry results in structures that resemble biology and obey quantum physics is bound to hold surprises for even the most seasoned engineer. The following sections will explain the physical principles that govern nanoscale assembly and then describe how they can be exploited to generate the desired structures and properties.

Relationship of Bottom-Up Fabrication to Chemistry

Chemistry is a natural starting point for nanotechnology. Indeed, nanotechnology traces back to the Middle Ages when alchemists made red pigments for stained glass from gold nanoparticles.[8] Most molecules are inherently nanoscopic. A routine chemical reaction typically produces about 10^{23} identical molecules, each of which measures a fraction of a nanometer across. In fact, one could say that nanoscale objects are large by chemistry standards. One might therefore mistakenly assume that nanomaterial synthesis would be relatively easy.

Although reducing the size is the greatest challenge for top-down nanomanufacturing, the reverse is true for bottom-up assembly. Consider how most molecules consist of fewer than 50 atoms. A 10 nm nanoparticle may consist of 1,000,000 atoms. The larger number of atoms leads to a larger opportunity for error.

The concepts of large molecules and polydispersity first became widespread with the onset of polymer chemistry. Polymers are built up from the addition of identical chemical repeat units, or monomers. A typical polymer chain consists of about 1,000 monomers linked end-to-end in a linear chain. Because initiation and termination of polymerization are random processes for each molecule, the number of monomers differs from one chain to the next. It would not be unusual for the largest polymer chain to be a million times larger than the smallest chain in the same reaction.

Nanoparticle synthesis follows similar behavior. The polydispersity is governed by the classical nucleation/diffusional growth model.[9-11] In the initial stages of the reaction, atoms or molecules nucleate into a fixed number of seeds, and particles then continue to grow by diffusion-driven deposition of material onto the existing seeds.

Relationship of Bottom-Up Fabrication to Biology

The basic building blocks of nanotechnology—nanoparticles, polymers, carbon nanotubes, and so forth, are built up from the addition of many atoms. In achieving greater functionality and sophistication, these building blocks are assembled into discrete structures with well-defined short-range order. These structures further integrate into systems of increasing complexity. The hierarchical ordering of atoms, particles, assembled subunits, and nanostructured systems mirrors almost exactly the world of proteins.

Proteins have primary, secondary, tertiary, and quaternary structure. Primary structure refers to the order of amino acids along the linear polypeptide chain. Amino acids are small molecules that form the basis of a polymer chain known as a polypeptide or protein. They are linked end to end in a precise sequence. The order is important because each amino acid has a set of specific physical properties that not only compel the chain to fold into a precise 3D shape, but also determine the function of the final structure. Amino acids can be positively charged, negatively charged, hydrophilic and neutral, hydrophobic, or include chemically active species.

The immediate impact of the amino acid sequence is to drive the assembly into a pair of relatively simple motifs referred to as the secondary structure. Secondary structure includes the alpha helix and beta sheet. As their names imply, the alpha helix is a coil-shaped configuration, and the beta sheet looks like a pleated two-dimensional (2D) sheet.

These two structural motifs then organize to create the overall 3D structure of the polypeptide. Typically taking the form of a compact globule, this 3D structure is referred to as the tertiary structure. Finally, quaternary structure describes the self-assembly of multiple globular proteins into a single supramolecular structure.

Proteins perfectly illustrate the concept of self-assembly because the blueprint for the final structure is encoded in the protein sequence. No intervention is required to create the final product once the individual building blocks have been synthesized. This feature has a number of desirable consequences for manufacturing technology.

Scalability: Self-assembly is a parallel process. 10^{23} objects self-assemble in the same time as a single object.

Cost: The costs for a self-assembly process lie entirely in the design and synthesis of the initial building blocks. Capital costs are comparable to those of a general wet chemistry laboratory. The costs of the required materials are relatively small from that point forward.

Low energy: Self-assembly frequently occurs best at standard temperature and pressure.

Hierarchical order: Although self-assembly does not provide the precision and control of top-down methodologies, living organisms demonstrate how self-assembly manages complexity and control across all size scales. Because forces do not scale linearly with size, hierarchically, organization can exploit the quirks of each size regime to obtain combinations of properties that generally do not go together.

It is insufficient to argue that bottom-up assembly is a promising manufacturing technique based merely on the success of biology in producing vast quantities of biomass. The point of bottom-up assembly is not the production of mass quantities. Rather, the promise lies in the ability to create devices and materials with unprecedented capabilities. Moreover, when one looks at the photonic properties of butterfly wings, the strength of spider silk, the toughness of mollusk shells, and the self-cleaning properties of the lotus leaf, one sees many examples where nature extracts extraordinary properties from ordinary materials.

The Physics of Self-Assembly

Thermal Energy

Self-assembly is generally possible at the nanoscale because the strength of most interactions drops close to the level of thermal energy below about 100 nm. The interaction forces must be relatively weak for the following reason: assembling objects must be allowed to try all possible combinations before settling into the desired configuration. In other words, the system cannot find the correct configuration before first attempting all possible mistakes. If the interactions are too strong, they become irreversible. The first random collision will cause assembly, and the probability that the first random choice is the desired choice is vanishingly small.

The system of assembling nanoscale objects must be at or near equilibrium for true self-assembly to occur. Equilibrium can only be achieved by balancing thermal energy with all possible attractive and repulsive forces. A few examples of the possible interactions are shown in Figure 9.2. They include electrostatic, electret, magnetic, van der Waals, steric repulsion, and hydrophobic forces.

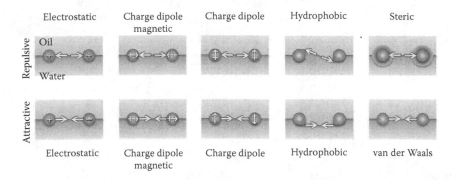

Figure 9.2 List of attractive and repulsive interactions common to building blocks in self-assembling systems.

Electrostatic Interaction

Perhaps the most familiar of these is the electrostatic interaction. It dictates that like charges repel each other and
opposite charges attract. It is normally a strong and long-
range force between charged objects. The situation changes
somewhat in a liquid medium like water. In water, the overall charge is neutral, so every charged object is surrounded
by a cloud of counterions. The total charge of the particle and
the surrounding cloud is zero, so no net attraction or repulsion occurs between pairs of charged particles at large distances. They can only "feel" the charge of the other particle
if they are close enough to penetrate the counterion cloud.
This distance decreases with increasing salt concentration.
In blood, it drops to 1 nm.[1]

A bizarre consequence of the counterion cloud is that it
controls the attraction and repulsion of the particle, and the
attraction is entropic in nature. A pair of oppositely charged
particles and their respective counterion clouds are depicted
schematically in Figure 9.3. The counterions balance their
attraction to the charged particle with the drive to maximize
their freedom of translation. When two oppositely charged
particles make contact, it is actually the entropy of liberating the counterions that causes the attraction, not the direct
electrostatic attraction between the particles. Essentially,
the system has traded particle/counterion bonds for par-
ticle/particle bonds. These enthalpic contributions cancel
out. However, the counterions that were originally between
the particles are now free to diffuse away in all possible
directions. The overall entropy of the system increases, and
assembly occurs.

Also arising from electrostatic interactions are dipolar
forces. An electric dipole, or electret, is basically a positively
charged object joined to a negatively charged object. It can
also be a molecule with an uneven electron distribution that
favors one side. All the above rules apply, except that orientation matters. A head-to-tail orientation is attractive, but a
head-to-head or tail-to-tail orientation is repulsive. Similarly,
antiparallel orientation (side-by-side but opposite direction) is
attractive, and parallel orientation is repulsive.

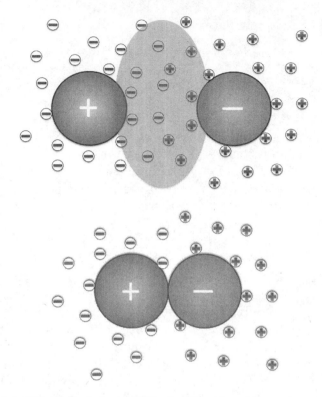

Figure 9.3 Electrostatic interactions in a liquid medium are controlled not by the charge of the interacting particles but by the intervening counterion cloud between them. The energy of attraction between a pair of charged particles comes from the entropy gained by liberating counterions to diffuse freely through solution.

Hydrogen Bonding

Even though the H_2O molecule is typically represented as a V-shaped molecule, it actually has a tetrahedral shape. The two hydrogen atoms are spaced by a 108° angle on one side of the oxygen atom, and two free electron pairs are tetrahedrally oriented on the opposite side. Because oxygen is more electronegative than hydrogen, the electrons are not shared equally with hydrogen. This leaves a partial positive charge on hydrogen and a partial negative charge on each electron pair. Naturally, the hydrogen atoms are attracted to the electron pairs of the neighboring water molecules. When these two come into close contact, it forms a hydrogen bond.

Hydrogen bonds are not distinct from a dipole/dipole bond, but because of their importance in biology and self-assembly, they are given their own classification. Next to a covalent bond, it is frequently the strongest bond in a nanoscale system. It is also responsible for many of the unique properties of water. One of these is described in the following subsection.

Hydrophobic Forces

The hydrophobic force is familiar to anyone who has ever shaken vinaigrette to disperse oil in vinegar. Following from the "like dissolves like" principle, it simply states that oily substances dissolve in other hydrophobic substances and phase separate from water. Polar or charged substances similarly dissolve in water and precipitate in oil. The underlying cause of this behavior is the polarity of the molecules. Polarity refers to how uneven the distribution of charge is within a molecule. Oils tend to be nonpolar and hydrophobic, whereas salts tend to be polar and hydrophilic.

The forces driving this segregation are much stronger than expected because of the internal structure of water. The water molecules arrange themselves tetrahedrally within the liquid to form as many hydrogen bonds as possible. This orientation is disrupted by the presence of a nonpolar particle. The water molecules are actually forced to form a cage around the hydrophobic object in order to preserve hydrogen bonds with their nearest neighbor. As in the previous examples, this increased order incurs an entropic penalty, which forces the water to minimize the contact area between itself and the hydrophobic particles. Entropy again drives the assembly of hydrophobic particles with each other in water.

van der Waals Forces

Also originating in the polarity of a material is the van der Waals attraction. Unlike electrostatic forces, van der Waals forces are always attractive. When molecules or electrons are free to redistribute themselves in the presence of neighboring materials, the dipoles will align to create the most favorable attraction (Figure 9.4). Even in nonpolar materials, temporary

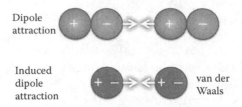

Figure 9.4 Dipolar interactions between electret pairs can be calculated from the sum of all electrostatic interactions. Induced dipolar interactions can also be calculated on this basis, but the charge pair that composes dipole is harder to define because it results from a temporary, time-varying fluctuation in the electron density of the particle or molecule.

dipoles form due to spontaneous fluctuations in their electron density. These temporary dipoles subsequently induce dipoles in neighboring materials, which then result in an attractive force.

The van der Waals interaction is normally tiny in comparison to electrostatic interactions. However, unlike the electrostatic interaction, it does not attenuate in water. In fact, it is frequently the largest force holding a nanoscale system together. It normally acts in concert with the hydrophobic interaction to assemble hydrophobic materials in water. The main difference is that the hydrophobic interaction is caused by interfacial tension, and van der Waals attraction arises from the bulk material.

Polymeric Steric Repulsion

Like electrostatic and hydrophobic forces, steric repulsion also has roots in the entropy of the system. It causes polymer chains to elastically resist compression. As shown in Figure 9.5, polymers are routinely attached to the surfaces of particles to prevent aggregation because of this property. They allow colliding particles to bounce off of each other rather than make intimate contact. The role of entropy is to force the polymer chains into a completely random coil. Because compression decreases the configurational disorder of the polymer chain, the entropic energy penalty creates a restoring force for the polymer to return to its original randomized state.

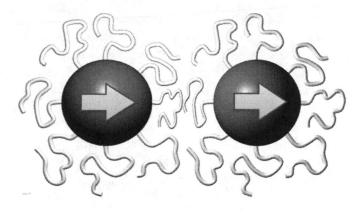

Figure 9.5 Two particles with magnetic dipoles that are attracted to each other but cannot come into close contact because of the steric repulsion of grafted polymer chains on their surfaces.

Magnetic Interactions

Magnetic fields are largely unaffected by either the liquid medium or the other materials present in the self-assembling system. However two important consequences arise from the reduced length scale. The first is that the material properties frequently diverge from their bulk values. In particular, the magnetization nanoparticles made from ferromagnetic materials behave in unexpected ways when exposed to an applied magnetic field. A bulk ferromagnetic material is composed of many magnetic domains. Because each domain has a random orientation, the net magnetic dipole cancels out. A nanoparticle, on the other hand, is too small to fit more than one magnetic domain. In fact, the enthalpic penalty of the domain walls outweighs the entropic gain from having randomly oriented domains in the same particle. The single domain nanoparticle, therefore, effectively possesses the saturation magnetization value even under zero field conditions.[12]

As the nanoparticle shrinks even further in size, another unexpected phenomenon occurs. The orientation of the single magnetic domain can rotate due to random thermal fluctuations. So although the particle does not rotate, the magnetic dipoles of the atoms within the particle make random, coordinated rotations at random intervals.[13] The alignment only remains fixed in a particular direction under the influence of an external field. Because only weak fields are required for alignment, the reinforcement of the external field by the nanoparticles can be substantial; so large, in fact, that this property merits its own classification. Such particles are said to be superparamagnetic. They have the high magnetization of a ferromagnetic material and the small hysteresis of a paramagnetic material, but with the advantage of reaching a large, constant magnetization for even very weak aligning fields.

The second important consequence of the small length scale is that magnetic nanoparticles do not respond strongly to magnetic field gradients. Unlike charges accelerated in a uniform electric field, magnetic dipoles are only accelerated by a magnetic field gradient. A uniform field merely rotates and aligns a magnetic dipole. Because nanoparticles are so small, it is nearly impossible to create a gradient from one side of the particle to the other. So while an external field may have a steep gradient, a particle will hardly experience that gradient across a few nanometers. Consequently, magnetic nanoparticles readily align with external fields but they do not readily accelerate toward the gradient. Alternatively stated, it is easy to apply a torque and difficult to apply a force.

Self-Assembly versus Directed Assembly

Hybrid Nanofabrication

Although bottom-up manufacturing is usually associated with self-assembly, it frequently occurs with the assistance of an external field or template. When external fields are present, nanoscale assembly is more accurately described as directed assembly. The most obvious example is the use of an applied electric field to assemble particles in electrophoresis. In this process, negatively charged particles migrate toward a positively charged electrode. Another common example is the use of evaporation to drive assembly. In this process, the liquid medium slowly evaporates, leaving behind a film of assembled particles. Rather than an electric field, it is helpful to think in terms of a directional surface tension field with a vector that points toward the center of the droplet.

Directed assembly sometimes takes a hybrid form that merges some aspects of top-down and bottom-up manufacturing. In these cases, one begins with a pattern on a solid substrate. The pattern could be alternating strips of different materials or simply rows of trenches etched into the substrate. In either case, particles follow the pattern by depositing preferentially in the trenches or on top of one particular material.

Symmetrical Self-Assembly

Useful self-assembly generally does not occur unless the building blocks possess broken symmetry, either in shape or in their interaction field. Perfectly symmetrical spheres can only aggregate into a handful of structures including disordered globules, diffusion-limited aggregates, and close-packed crystals.

One can switch between these three morphologies by modulating the strength of the interaction. In the limit of very strong binding relative to thermal energy, diffusion-limited cluster aggregation (DLCA) occurs.[14–16] Pairs of particles bind irreversibly upon first contact and will remain permanently fixed in their original relative orientation. The resulting structures are highly branched and loosely packed. This characteristic morphology emerges because the available binding sites near the center are shadowed by the outer branches and therefore grow more slowly.[17] At the other limit, the attractive potential energy is comparable with thermal energy. In this case, the adsorption and desorption of particles are at equilibrium. The condensed particle droplets will anneal over time, generally taking on a spherical shape to minimize the surface tension. The particles within the droplet will further densify into a closest packed crystalline structure so long as the friction at the particle junctions is low enough to allow rearrangement. For moderately strong, but not irreversible, binding energies, reaction-limited cluster aggregation (RLCA) occurs.[18] In this process, the probability of two particles sticking is lower than that for DLCA, so many contacts can be explored before the particles finally adhere.

Epitaxially Directed Growth

A simple way to break the symmetry of self-assembly and to create more complex structures is to assemble the particles on an asymmetric substrate. The deposition of particles upon a topographically patterned surface is known as graphoepitaxy.[19,20] The most common method for deposition is to slowly evaporate the solvent, leaving behind the solid particles on the surface. This process usually results in selective deposition of particles into depressions on the surface. Careful choice of size and geometry can lead to highly ordered assembly. For example, a trench with a width and depth equal to one particle diameter will promote end-to-end assembly into a straight line.

Another common way to template a surface is through chemical patterning. Chemical patterning can be achieved through standard photolithographic techniques or through a recently developed technique known as soft lithography, which essentially uses a rubber stamp with nanoscale embossing.[21] Once the pattern is transferred to the surface, particles will selectively adhere to the surface chemistry that is most compatible with its own surface. This compatibility is most frequently determined by either electrostatic or hydrophobic interactions.

Field-Aligned Growth

Externally applied fields are also helpful for introducing asymmetry to the bottom-up assembly process. A common example is the use of magnetic fields to drive the dipolar assembly of magnetic nanoparticles into one-dimensional (1D) chains. Figure 9.6 shows an example where an external magnetic field causes the alignment of 1D chains perpendicular to a surface. The synthesis and assembly of magnetic colloids have been extensively investigated as a bottom-up methodology to form self-organized mesostructures with 1D, 2D, or 3D ordering. Initially, magnetic assembly involved building blocks on the micrometer length scale.[22–24] The resulting structures had the added benefit of remaining

Figure 9.6 When no magnetic field is applied, ferromagnetic nanoparticles self-assemble into disordered chains and loops. When a magnetic field is applied, the chains align with the external field and are pulled toward the location with the steepest magnetic field gradient.

responsive to externally applied magnetic fields following assembly. Scaling of these systems below 1 µm was finally demonstrated by Singh et al. using polyelectrolyte-coated latex beads in the 500 to 800 nm range.[25,26] They demonstrated assembly of the beads into dispersed and surface-tethered magnetic chains spanning 30 to 50 µm in length.

The dipolar assembly of ferromagnetic nanoparticles with sizes below 50 nm has more recently been explored to prepare responsive 1D mesostructures. Once the synthesis became routine, these building blocks were used to form ordered 1D chains that resembled giant, mesoscopic polymer chains.[27] Keng et al. and Bowles et al. used polystyrene-coated ferromagnetic cobalt nanoparticles (diameter = 23.5 nm) to form micron-sized mesoscopic 1D assemblies when dispersed in cross-linkable organic solvents.[28,29]

Using an electric field to control the motion of charged particles is known as electrophoresis. Electrophoretic deposition has seen wide use as a means to coat ceramic surfaces with charged particles.[30,31] As expected, charged particles move in an electric field and eventually come to rest on an oppositely charged electrode. It allows a great degree of control as the voltage, particle concentration, and deposition time determine the degree of deposition. The deposition process is stochastic by nature, so the particles tend to adopt a random distribution on the surface. However, if the particles are coated with a surfactant or are otherwise prevented from coming into intimate contact with the surface, they may rearrange into a hexagonal lattice. Electrophoresis can be combined with graphoepitaxy to provide additional patterning if desired. The primary limitation of this technique is the requirement for a conductive substrate.[32]

Perhaps even more common than the previous techniques is the use of an evaporating solvent to drive particle assembly upon a surface.[33] Confined to the liquid medium, the particles retreat from the approaching liquid/air interface (Figure 9.7). As mentioned previously, the receding liquid/solid/air contact line defines the location and direction of this repulsion. It has

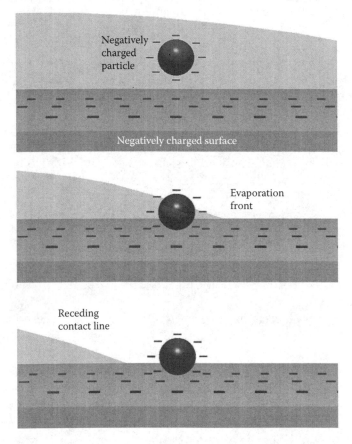

Figure 9.7 As the solvent evaporates, suspended particles retreat from the approaching liquid/air interface. The effective repulsion can overcome electrostatic repulsion to force the colloid in contact with the negatively charged surface.

the same effect as a field as it sweeps across the entire surface. The repulsion pushes the particles down to the surface. The repulsive nature of the capillary interaction drives the particles into the deepest areas of the surface topology. Grooves, trenches, and depressions are the preferred locations using this method.

Enhanced order arises from this hybrid top-down/bottom-up approach, because the top-down template restricts the possible degrees of freedom. This hybrid approach capitalizes on the strengths of the two methods: top-down fabrication works best at microscopic length scales, and bottom-up fabrication works best at nanometer length scales. Used individually, the degree of order will gradually break down at either small or large feature sizes. Used together, they achieve order continuously throughout the entire range of length scales.

Examples of Bottom-Up Assembly

Physical Chemistry Approaches

Bottom-up assembly frequently takes the form of nanoparticle aggregation through some form of phase separation. A technologically relevant example is the phase separation of fullerenes in polymer solar cells during spin casting. A common polymer solar cell design consists of fullerenes embedded in a matrix of conducting polymer. The polymer absorbs light to create electron-hole pairs known as excitons. The role of the fullerene is to separate the electrons from the holes by abstracting the electrons from the polymer as soon as they form. The electrons then hop from one fullerene to the other until they reach the cathode. The positively charged holes similarly conduct through the polymer to the transparent anode.

Charge separation only occurs very close to the polymer/fullerene interface because excitons only travel about 10 nm before they recombine.[34,35] If the solar cell contained only a single, planar interface, only a miniscule fraction of the material would participate in charge generation. The solution to this problem is to create a bulk heterojunction. Simply stated, a bulk heterojunction is an interpenetrating network where the polymer-rich phase and fullerene-rich phase are finely interwoven. The idea is to maximize the interfacial area per volume so that all locations within the material are no more than 10 nm away from an interface.

Because order does not affect the performance of the solar cell, the easiest way to create this morphology is through the natural phase separation of fullerenes from polymer/solvent solutions. Fullerenes and conducting polymers such as poly(3-hexylthiophene) (P3HT) are both soluble in organic solvents such as dichlorobenzene (DCB). However, fullerenes are not soluble in the pure polymer. Therefore, the fullerenes precipitate out of solution as the solvent evaporates. What one gets is a highly intercalated structure as shown in Figure 9.8.[36]

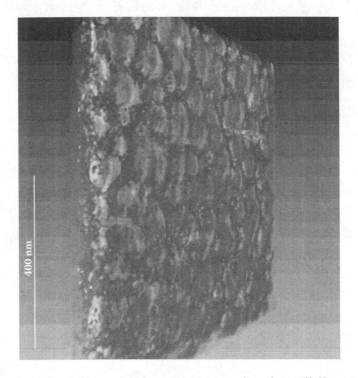

Figure 9.8 Three-dimensional reconstruction of a polymer/fullerene solar cell phase morphology obtained from a transmission electron microscope tomography tilt series. The more darkly colored phase represents the fullerenes, and the lighter phase is the conducting polymer. (From Yang, X., and Loos, J., *Macromolecules,* 40, 1353, 2007. With permission.)

This particular phase morphology occurs because the fullerenes only have limited time to diffuse away from the polymer before the solvent completely evaporates. During the final stages of evaporation, the polymer solidifies and diffusion essentially stops. Because the fullerenes have little time to move, the phase domains remain small. In addition to controlling the rate of evaporation, another method for controlling self-assembly is to add surfactant-like molecules that decrease the interfacial tension between the two phases. The lower the energy penalty for interfacial area, the more refined the structure becomes.[37]

Colloidal assembly in water and other liquids represents a large fraction of bottom-up assembly. As is typical with colloidal systems, the individual building blocks are typically polymer spheres with charged coatings that measure between 100 nm and 1 μm. One of the most active areas of research in colloidal, bottom-up fabrication involves photonic materials. A photonic material is defined as one that causes diffraction of light but is structured on length scales below the relevant wavelength. The name *metamaterial* denotes the fact that the composite behaves more like a homogeneous material as it interacts with light.

As the solvent slowly evaporates, the colloids in suspension gradually crowd closer together. Uniformly sized particles readily pack to form highly ordered crystals once the solvent has been completely removed. Because the periodicity of colloidal crystals can be easily controlled to be commensurate with the wavelength of light, they have been extensively investigated as photonic materials.[38] In fact, naturally occurring opals acquire their photonic properties in the same fashion. A man-made example in Figure 9.9 shows how spherical colloidal crystals can be formed by slowly removing water from the droplets of aqueous colloidal suspensions.[39] The ability to direct, trap, or guide light via photonic materials is a key enabler for optoelectronic devices such as broadband matrix switches and fiber optic relays.

The previous example perfectly illustrates how the spherical symmetry of colloids is reflected in their assemblies. To achieve variety beyond dense shapes, asymmetry is frequently introduced into the system by using a template. Channels and pits etched into a surface create an asymmetric repulsive potential that forbids spherical or globular packing. Though

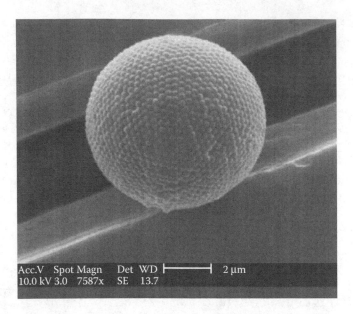

Figure 9.9 Hexagonally close-packed colloidal crystal with an overall spherical shape. The periodicity is comparable to visible light, creating a photonic bandgap. (From Yi, G.-R., Jeon, S.-J., Thorsen, T., Manoaran, V.N., Quake, S.R., Pine, D.J., and Yang, S.-M., *Synthetic Metals,* 139, 803, 2003. With permission.)

working mostly with microscale colloids, Younan Xia's group has demonstrated an exceptional variety of complex and controllable structures using template-assisted colloidal assembly.[40] Several examples of linear and even helical packing are given in Figures 9.10 and 9.11.

While templated colloidal assembly beautifully demonstrates the control possible from bottom-up assembly, the structures are not functional in their stand-alone form. The primary interest is to use these patterns as a mask for patterning microelectronic devices. Because nanoparticles have dimensions well below the wavelength of visible light, colloidal lithography can achieve smaller dimensional control relative to conventional photolithography.

The use of self-assembly for lithographic patterning has seen even greater interest in block copolymer systems.[41] Figure 9.12 illustrates an example of the processing steps involved when using a block copolymer as a lithographic mask. Essentially an amphiphile, a diblock copolymer is a linear

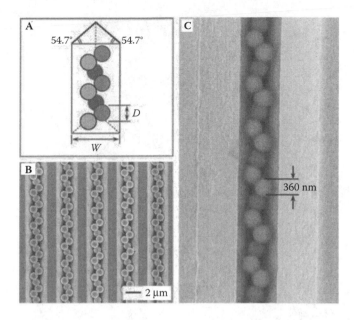

Figure 9.10 (A) Unit crystal showing helical packing of colloidal particles in a V-shaped trench. (B) Scanning electron micrograph of helical packing. (C) Helical packing of 360 nm particles in a V-shaped trench. (Xia, Y., Yin, Y., Lu, Y., and McLellan, J., *Adv. Funct. Mater.,* 13, 907, 2003. With permission.)

polymer chain of composition A covalently bound at one end to a chain of composition B. When the two blocks are mutually immiscible, they separate into A-rich and B-rich domains. The morphologies of these microdomains range from cylinders to interpenetrating tetragonal phases. The chemical differences between the two blocks are then exploited to preferentially etch one phase or the other. What is left behind is an impression of the original block copolymer phase morphology.[42] Depending on processing conditions, the degree of long-range order can be remarkable.[43] A notable example is given in Figure 9.13 where a block copolymer with a cylindrical morphology demonstrates nearly perfect hexagonal order over several micrometers.

Biological Approaches

In contrast to the majority of nanoscale building blocks, proteins, virus capsids, and other biological building blocks

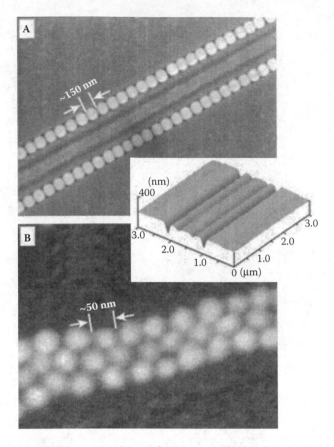

Figure 9.11 (A) Linear packing of 150 nm colloids along a 150 nm wide trench. (B) Ordered packing of 50 nm particles within a 150 nm trench. (Xia, Y., Yin, Y., Lu, Y., and McLellan, J., *Adv. Funct. Mater.,* 13, 907, 2003. With permission.)

encompass enormous heterogeneity in both shape and surface chemistry. An excellent example is the work of the Belcher group who used the M13 virus to template the growth of lithium ion battery electrodes. For this process, the virus templates were incubated in aqueous cobalt chloride solution to produce cobalt oxide (Figure 9.14). Coupled with the high surface area of the virus capsid aggregates, Co_3O_4 provided an extremely large reversible storage capacity. Transmission electron microscopy (TEM) images are shown in Figure 9.15.[44]

The use of biological building blocks has the additional advantage of precise chemical control. Proteins have identical

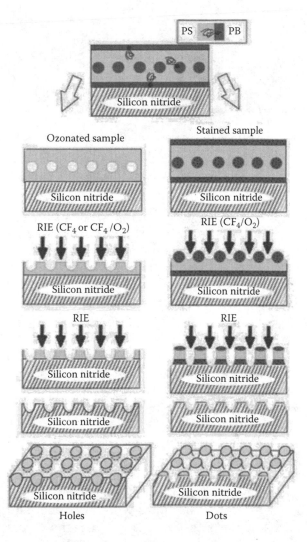

Figure 9.12 Processing steps required to form an impression of a block copolymer template onto the surface of a hard material such as silicon nitride (PS, polystyrene; PB, polybutadiene; RIE, reactive ion etching). (From Park, M., Harrison, C., Chaikin, P.M., Register, R.A., and Adamson, D.H., *Science,* 276, 1401, 1997. With permission)

size, shape, and chemical patterning according to the primary, secondary, and tertiary self-assembly of globular protein nanoparticles. This capability was beautifully exploited in the same study where the virus capsids were genetically

Figure 9.13 Atomic force microscope image of a polystyrene-poly(ethylene oxide) block copolymer that microphase segregated into hexagonally ordered, standing cylinders of poly(ethylene oxide). (Kim. S.H., Misner, M.J., and Russell, T.P., *Adv. Mater.*, 16, 2119, 2004. With permission.)

engineered to present gold binding thiol groups at regular intervals across the surface. Gold nanoparticles grew specifically from these sites, and the addition of gold nanoparticles to the Co_3O_4 nanoparticles increased the specific capacity by 30%.

On a smaller scale, oligopeptide chains, consisting of fewer than 100 amino acids in a specific sequence, can serve as building blocks for materials having unique viscoelastic and biological properties. A self-complementary peptide was designed with 20 alternating polar and nonpolar amino acids. This ~3 nm molecule first folded back upon itself and then self-assembled into β-sheet ribbons measuring 3 nm across with one hydrophobic face and one hydrophilic face. To protect the hydrophobic face from water, the ribbons then paired up to form bilayer strands with a buried hydrophobic interior. These strands continued to build up into a three-dimensional hydrogel network. An elastic solid at room temperature, this hydrogel rapidly shear thins and flows when it is pushed through a syringe. Perhaps more impressively, it regains its elastic properties almost immediately after it is dispensed.[45]

The fact that this material is composed of polypeptides makes it a prime candidate as a tissue scaffold or as a material for encapsulating stem cells for therapeutics. The ability to flow and rapidly set makes it especially attractive for implantation through a simple injection.

In a similar system, Stupp and coworkers have developed a peptide amphiphile that combines the β-sheet forming aspects of the molecule described above with surfactant behavior.[46] What results is a molecule that forms extraordinarily strong

Virus Biotemplating

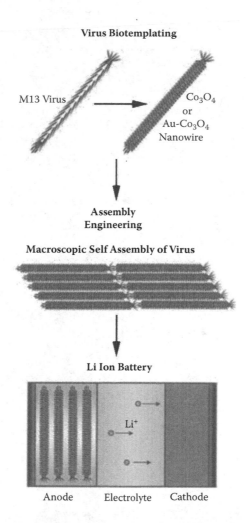

Figure 9.14 Virus templates used to fabricate the anode of a Li ion battery. (Nam, K.T., Kim, D.-W., Yoo, P.J., Chiang, C.-Y., Meethong, N., Hammond, P.T., Chiang, Y.-M., and Belcher, A M., *Science,* 312, 885, 2006. With permission.)

cylindrical micelles that measure 5 to 8 nm in diameter and more than 1 μm long (Figure 9.16A). The interactions that drive micelle formation include chiral dipole-dipole interactions, π-π stacking, hydrogen bonds, van der Waals interactions, hydrophobic interactions, electrostatic interactions, and repulsive steric forces. This group has demonstrated the nucleation of hydroxyapatite nanocrystals with *c*-axis orientation along

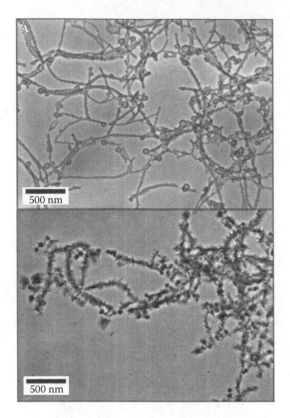

Figure 9.15 Transmission electron micrograph of viruses coated with cobalt oxide at two different thicknesses. (Nam, K.T., Kim, D.-W., Yoo, P.J., Chiang, C.-Y., Meethong, N., Hammond, P.T., Chiang, Y.-M., and Belcher, A M., *Science,* 312, 885, 2006. With permission.)

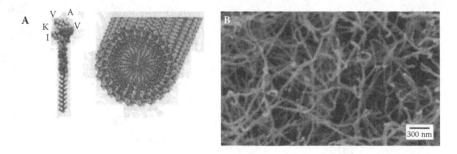

Figure 9.16 (A) Peptide amphiphile in monomer form and self-assembled into a cylindrical micelle. (B) Scanning electron micrograph of cylindrical micelles after removal of water. (Hartgerink, J.D., Beniash, E., and Stupp, S.I., *PNAS,* 99, 5133, 2002. With permission.)

the micelle axis, which mimics the geometrical relationship between apatite crystals and collagen fibrils in bone. Because the amino acids decorate the outer surface of the cylinder, they also provide excellent scaffolds for cell growth. By presenting the neurite-promoting laminin epitope with the amino acid sequence Ile-Lys-Val-Ala-Val (IKVAV), they created an artificial nanofiber scaffold (Figure 9.16B) that induced the rapid differentiation of neural progenitor cells into neurons while discouraging the development of astrocytes.[47] Examples like this point to the possibility of using nanostructured scaffolds to direct the growth of complex tissues and organs from various progenitor cells.

The final example of biologically driven bottom-up fabrication is given in Figure 9.17. The Montemagno group developed a cell-free artificial photosynthesis platform that embedded a nanoscale photophosphorylation system within foam formed from the Tungara frog surfactant protein Ranaspumin-2.[48] Essentially, they assembled vesicles from lipid molecules and incorporated various enzymes and proteins into the lipid bilayer. Requiring water to preserve their function, these vesicles were then protected from dehydration by encapsulating them within the walls of the foam. The foam resisted hydration much longer than conventional detergent foams due to the unique properties of the surfactant protein. This protein was first discovered because of the unusual resiliency of foams produced by the Tungara frog to protect its eggs. As a large protein, it has the additional advantage that it does not disrupt the lipid vesicles, unlike a detergent, which would immediately cause lysis. In addition to keeping the system hydrated, this long-lived foam provided a large surface area for light absorption and CO_2 diffusion.

Future Outlook of Bottom-Up Assembly

The continuing expansion of bottom-up assembly into mainstream technology depends largely on advances in chemical synthesis. The largest barrier to commercialization at this

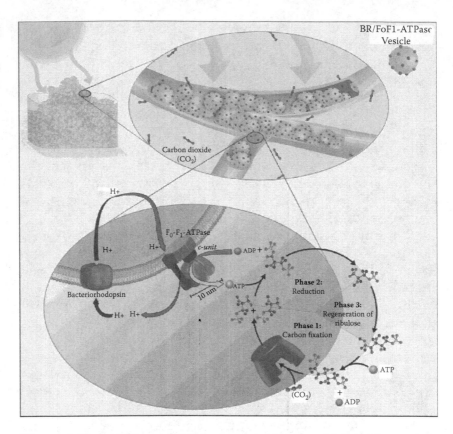

Figure 9.17 A cell-free artificial synthesis platform. The vesicles contain two proteins within their phospholipid bilayer membranes. Together with a water-soluble protein, they convert adenosine diphosphate into adenosine triphosphate, much like plants. The vesicles are protected within the watery confines of a long-lived foam formed from a protein that is produced by the Tungara frog. (From Wendell, D., Todd, J., and Montemagno, C., *Nano Letters* 2010, DOI: 10.1021/nl100550k. With permission.)

point is the limited ability to synthesize well-defined building blocks with structural complexity and chemical heterogeneity. The protein remains the gold standard. All copies of a protein are identical, they have specific chemical functionality at precise locations, and they routinely fit together in unique combinations via precise mating surfaces. Biology achieves this level of control through the complex chemical machinery of DNA and RNA, and it remains to be seen whether synthetic chemistry will ever match it.

In the roughly 20 years since nanotechnology became an active research area, the majority of nanomaterials are either spherical or cylindrical. Examples of tetrahedral, colloidal molecules, faceted polyhedra, rods, ellipses, and chemically patterned nanoparticles exist, but few have been included into functioning systems.[49]

Despite being some of the greatest achievements in this field, these sophisticated examples have yet to find an application because the research was not driven by need. Due to the great difficulty of achieving synthetic control in the first place, many chemists are understandably content to push their capabilities forward without pushing for a specific end use. The emphasis on technology push should not be concerning. It mostly follows from the relative early stage of technology development in this field. Tools such as the atomic force microscope have only recently become commonplace, and the foundation of basic science needed to mature these technologies is still being built. The end goal of self-assembly has always been to imbue engineered systems with the flexibility, adaptability, and robustness of natural systems. However, going from routine synthesis of nanocrystals to the shape-shifting robot from *Terminator 2* will take some time.

The slow trickle of mature nanotechnology emerging from self-assembly research should be taken in context. Although no examples of complete self-assembled systems exist outside biology, many self-assembled components have already crept into the marketplace. One has to look no further than L'Oreal, which currently ranks number six among nanotechnology patent holders in the United States. With 192 patents, it is ahead of General Electric, Motorola, and Eastman Kodak.[50] Procter & Gamble, Estée Lauder, Christian Dior, and Japan's Shiseido also incorporate nanoparticles into their products. They include cleansers, moisturizers, and other personal care products.

Initially, the safest and most practical uses of bottom-up assembly may appear relatively pedestrian from a technological standpoint. Sunscreen made from titanium dioxide nanoparticles was among the first nanotechnologies to see widespread consumer adoption. The real promise of self-assembled nanotechnology is the ability to combine the high performance of engineered systems with the adaptability and self-correction of natural systems. Although several exciting steps toward this goal have been discussed in this chapter, the realization of these goals and mainstream adoption still appear to be a decade or more away at this writing. That said, large leaps in properties and performance do not come without risk. The use of nanotechnology as the centerpiece of a critical system design carries the uncertainty that one normally finds with technologies at an early stage of development. Properties may not be preserved upon mass production, durability may not compare well with proven technologies, interfacing nanotechnologies with existing components may cause additional challenges, and a general lack of experience with nanotechnologies will naturally lead to unforeseen difficulties. Moreover, if the idea is to leverage bottom-up nanotechnology to exploit self-healing properties, adaptability, and scalability, then systems engineering may need to create entirely new concepts for handling uncertainty not only in the environment but in the state of the system as it evolves. These risks must be balanced against the real threat of falling behind the competition. Complex systems such as satellites and submarines involve enormous barriers to change. Conservatism, therefore, threatens these systems with obsolescence as the unpredictable world of bottom-up nanotechnology threatens their reliability.

Summary

The difference between top-down and bottom-up fabrication methods manifests in three ways. The first major difference is that bottom-up assembly generally involves a change in shape accompanied by an increase in volume, whereas top-down methods involve a decrease in volume. Occasionally, nanotechnology fabrication involves a hybrid of the two. One may then more broadly distinguish the two approaches by the fact that top-down fabrication techniques are usually associated with semiconductor manufacturing, and bottom-up fabrication relies primarily upon wet chemistry techniques. Finally, the third difference between the two manifests in the final appearance.

Top-down nanotechnology features Euclidian shapes, precision, and order. Bottom-up nanotechnology resembles biological structures and is characterized by fractal morphologies, characteristic length scales, and randomness.

Although top-down nanotechnology is more mature than bottom-up nanotechnology, the latter holds the promise of providing the best features of biological systems—adaptability, scalability, and self-correction. A comparison of top-down and bottom-up nanotechnology is in many ways analogous to the comparison of man-made and natural systems. A fighter jet flies much faster than the fastest bird, but birds are more energy efficient; harvest energy from their local environment; heal their own wounds; and adapt extremely well to a variety of wind, temperature, and barometric conditions.

Nanotechnology assembled from the bottom-up holds these possibilities because it operates according to the same physical principles as cells, proteins, and DNA. At these small size scales, electrostatic, magnetic, van der Waals, capillary, and hydrophobic interactions are roughly comparable to thermal energy. Nanoscale systems can therefore spontaneously switch on and off under operating conditions. The spontaneous action triggered by thermal fluctuation underlies the adaptability of these seemingly fragile systems. However, the system is not fragile because the size of the ensemble creates a state of thermodynamic equilibrium. The dynamically changing properties of a nanosystem are essentially the result of thermodynamic phase changes. And because the entropy term of the free energy often dominates in self-assembled systems, these changes can be quite complex and unlike bulk materials in their behavior.

Examples of this unexpected behavior can be found in the exquisite structure of bulk heterojunction solar cells, the surprising mechanical strength of peptide amphiphile cylindrical micelles, and the large reversible storage capacity of virus-templated batteries. The beauty of each example is that the number of manufacturing steps is greatly reduced relative to traditional top-down manufacturing. Once the chemicals have been synthesized, assembly can involve little more than adding water. So long as regularity and precision do not affect the final properties, equal or better performance can be achieved through self-assembly.

References

1. Israelachvili, J. *Intermolecular and Surface Forces*, Academic Press, San Diego, CA, 1992, p. 315.
2. Kramers, H.A. "Brownian motion in a field of force and the diffusion model of chemical reactions," *Physica* 1940, *7*, 284.
3. Hänggi, P.; Talkner, P.; Borkovec, M. "Reaction-rate theory: fifty years after Kramers," *Rev. Mod. Phys.* 1990, *62*, 251.
4. Nohynek, G.J. "Grey goo on the skin? Nanotechnology, cosmetic and sunscreen safety," *Crit. Rev. Toxicol.* 2007, *37*, 251.
5. Daniel, M.C. "Gold nanoparticles: assembly, supramolecular chemistry, quantum-size-related properties, and applications toward biology, catalysis, and nanotechnology," *Chem. Rev.* 2004, *104*, 293.
6. Ulman, A. "Formation and structure of self-assembled monolayers," *Chem. Rev.* 1996, *96*, 1533.
7. Chen, S.H.; Kimura, K. "Synthesis and characterization of carboxylate-modified gold nanoparticle powders dispersible in water," *Langmuir* 1999, *15*, 1075.
8. Mulvaney, P. "Not all that's gold does glitter," *MRS Bull.* 2001, *26*, 1009.
9. Lamer, V.K.; Dinegar, R.H. "Theory, Production and Mechanism of Formation of Monodispersed Hydrosols," *J. Am. Chem. Soc.* 1950, *72*, 4847.
10. Privman, V.; Goia, D.V.; Park, J.; Matijevic, E. "Mechanism of Formation of Monodispersed Colloids by Aggregation of Nanosize Precursors," *J. Colloid Interface Sci.* 1999, *213*, 36.
11. Overbeek, J.T.G. "Monodisperse colloidal systems, fascinating and useful," *Adv. Colloid Interfac. Sci.* 1982, *15*, 251.
12. Frankel, R.B.; Moskowitz, B.M. *Biogenic Magnets*, in *Magnetism: Molecules to Materials, Vol. 4* (Eds.: J.S. Miller, M. Drillon), Wiley-VCH Verlag KGaA, Weinheim, 2003, p. 485.
13. Jakubovics, J.P. *Magnetism and Magnetic Materials, 2nd ed.*, The Institute of Materials, Cambridge, 1994.
14. Stauffer, D.; Stanley, H.E. *From Newton to Mandelbrot: A Primer in Theoretical Physics with Fractals for the Macintosh*, Springer, New York, 1996, ch. 3.
15. Ossadnik, P. "Multiscaling analysis and width of the active zone of large off-lattice DLA," *Physica A* 1993, *195*, 319.
16. Vandewalle, N.; Ausloos, M. "Magnetic diffusion-limited aggregation," *Phys. Rev. E* 1995, *51*, 597.

17. Witten, T.A.; Sander, L.M. "Diffusion-limited aggregation, a kinetic critical phenomenon," *Phys. Rev. Lett.* 1981, *47*, 1400.
18. Forrest, S.R.; Witten, T.A. "Long-range correlations in smoke-particle aggregates," *J. Phys. A* 1979, *12*, L109.
19. Segalman, R.A.; Yokoyama, H.; Kramer, E.J. "Graphoepitaxy of spherical domain block copolymer films," *Adv. Mater.* 2001, *13*, 1152.
20. Segalman, R.A.; Hexemer, A.; Kramer, E.J. "Effects of lateral confinement on order in spherical domain block copolymer thin films," Macromolecules 2003, 36, 6831.
21. Woodson, M.; Liu, J. "Functional nanostructures from surface chemistry patterning," *Phys. Chem. Chem. Phys.* 2007, *9*, 207.
22. Furst, E.M.; Suzuki, C.; Fermigier, M.; Gast, A.P. "Permanently linked monodisperse paramagnetic chains," *Langmuir* 1998, *14*, 7334.
23. Furst, E.M.; Gast, A.P. "Micromechanics of dipolar chains using optical tweezers," *Phys. Rev. Lett.* 1999, *82*, 4130.
24. Dreyfus, R.; Baudry, J.; Roper, M.L.; Fermigier, M.; Stone, H.A.; Bibette, J. "Microscopic Artificial Swimmers," *Nature* 2005, *437*, 862.
25. Singh, H.; Laibinis, P.E.; Hatton, T.A. "Synthesis of flexible magnetic nanowires of permanently linked core-shell magnetic beads tethered to a glass surface patterned by microcontact printing," *Nano Lett.* 2005, *5*, 2149–2154.
26. Singh, H.; Laibinis, P.E.; Hatton, T.A. "Rigid, superparamagnetic chains of permanently linked beads coated with magnetic Nanoparticles. Synthesis and rotational dynamics under applied magnetic fields," *Langmuir* 2005, *21*, 11500–11509.
27. Zhou, Z.H.; Liu, G.J.; Han, D.H. "Coating and Structural Locking of Dipolar Chains of Cobalt Nanoparticles," *ACS Nano* 2009, *3*, 165.
28. Keng, P.Y.; Shim, I.; Korth, B.D.; Douglas, J.F.; Pyun, J. "Synthesis and self-assembly of polymer-coated ferromagnetic nanoparticles," *ACS Nano* 2007, *1*, 279.
29. Bowles, S.E.; Wu, W.; Kowalewski, T.; Schalnat, M.C.; Davis, R.J.; Pemberton, J.E.; Shim, I.; Korth, B.D.; Pyun, J. "Magnetic assembly and pyrolysis of functional ferromagnetic colloids into one-dimensional carbon nanostructures," *J. Am. Chem. Soc.* 2007, *129*, 8694.
30. Hamaker, H.C. "The influence of particle size on the physical behaviour of colloidal systems," *Trans. Faraday Soc.* 1940, *36*, 279.

31. Sarkar, P.; Nicholson, P.S. "Electrophoretic Deposition (EPD): Mechanisms, Kinetics, and Applications to Ceramics," *J. Am. Ceram. Soc*. 1996, *79*, 1987.
32. Zhang, Q.; Xu, T.; Butterfield, D.; Misner, M.J.; Ryu, D.Y.; Emrick, T.; Russell, T.P. "Controlled Placement of CdSe Nanoparticles in Diblock Copolymer Templates by Electrophoretic Deposition," *Nano Lett*. 2005, *5*, 357.
33. Xia, Y.; Yin, Y.; Lu, Y.; McLellan, J. "Template-Assisted Self-Assembly of Spherical Colloids into Complex and Controllable Structures," *Adv. Func. Mater*. 2003, *13*, 907.
34. Yoshino, K.; Hong, Y.X.; Muro, K.; Kiyomatsu, S.; Morita, S.; Zakhidov, A.A.; Noguchi, T.; Ohnishi, T. "Marked Enhancement of Photoconductivity and Quenching of Luminescence in Poly(2,5-dialkoxy-p-phenylene vinylene) upon C60 Doping," *Jpn. J. Appl. Phys., Part 2* 1993, *32*, L357.
35. Halls, J.J.M.; Pichler, K.; Friend, R.H.; Moratti, S.C.; Holmes, A.B. "Exciton diffusion and dissociation in a poly (p-phenylenevinylene)/C60 heterojunction photovoltaic cell," *Appl. Phys. Lett*. 1996, *68*, 3120.
36. Yang, X.; Loos, J. "Toward high-performance polymer solar cells: the importance of morphology control," *Macromolecules* 2007, *40*, 1353.
37. Peet, J.; Kim, J.Y.; Coates, N.E.; Ma, W.L.; Moses, D.; Heeger, A.J.; Bazan, G.C. "Efficiency enhancement in low-bandgap polymer solar cells by processing with alkane dithiols," *Nat. Mater*. 2007, *6*, 497.
38. Stein, A.; Schroden, R.C. "Colloidal crystal templating of three-dimensionally ordered macroporous solids: materials for photonics and beyond," *Curr. Opinion in Solid State Mater. Sci*. 2001, *5*, 553.
39. Yi, G.-R.; Jeon, S.-J.; Thorsen, T.; Manoaran, V.N.; Quake, S.R.; Pine, D.J.; Yang, S.-M. "Generation of uniform photonic balls by template-assisted colloidal crystallization," *Synthetic Metals* 2003, *139*, 803.
40. Xia, Y.; Yin, Y.; Lu, Y.; McLellan, J. "Template-Assisted Self-Assembly of Spherical Colloids into Complex and Controllable Structures," *Adv. Funct. Mater*. 2003, *13*, 907.
41. Hawker, C.J.; Russell, T.P. "Block copolymer lithography: Merging "bottom-up" with "top-down" processes," *MRS Bull*. 2005, *30*, 952.
42. Park, M.; Harrison, C.; Chaikin, P.M.; Register, R.A.; Adamson, D.H. "Block copolymer lithography: periodic arrays of ~ 10^{11} holes in 1 square centimeter," *Science* 1997, *276*, 1401.

43. Kim, S.H.; Misner, M.J.; Russell, T.P. "Solvent-Induced Ordering in Thin Film Diblock Copolymer/Homopolymer Mixtures," *Adv. Mater.* 2004, *16*, 2119.

44. Nam, K.T.; Kim, D.-W.; Yoo, P. J.; Chiang, C.-Y., Meethong, N.; Hammond, P.T.; Chiang, Y.-M.; Belcher, A. M. "Virus-enabled synthesis and assembly of nanowires for lithium ion battery electrodes," *Science* 2006, *312*, 885.

45. Haines-Butterick, L.; Rajagopal, K.; Branco, M.; Salick, D.; Rughani, R.; Pilarz, M.; Lamm, M.S.; Pochan, D.J.; Schneider, J.P. "Controlling hydrogelation kinetics by peptide design for three-dimensional encapsulation and injectable delivery of cells," *PNAS* 2007, *104*, 7791.

46. Hartgerink, J.D.; Beniash, E.; Stupp, S.I. "Peptide-amphiphile nanofibers: a versatile scaffold for the preparation of self-assembling materials," *PNAS* 2002, *99*, 5133.

47. Silva, G.A.; Czeisler, C.; Niece, K.L.; Beniash, E.; Harrington, D.A.; Kessler, J.A.; Stupp, S.I. "Selective differentiation of neural progenitor cells by high-epitope density nanofibers," *Science* 2004, *303*, 1352.

48. Wendell, D.; Todd, J.; Montemagno, C. "Artificial photosynthesis in ranaspumin-2 based foam," *Nano Letters* 2010, DOI: 10.1021/nl100550k.

49. Glotzer, S.C.; Solomon, M.J. "Anisotropy of building blocks and their assembly into complex structures," *Nature Mater.* 2007, *6*, 557.

50. Matlack, C.; Carey, J. "Nano, nano, on the wall," *Bloomberg Businessweek*, December 12, 2005.

PART 3

Systems Engineering Process Elements

Component Advanced Development ...
The fundamental objectives of this stage of development are to accomplish risk-reduction activities as required to establish confidence that the building blocks of the system are sufficiently well defined, tested, and demonstrated to provide confidence that when integrated into higher-level assemblies and subsystems, they will perform reliably.

Part 3 discusses key activities in all phases of micro- and nanoscale technology development that support and run parallel to systems engineering verification and validation and risk management activities. Tools and techniques that are unique to micro- and nanoscale technology development are described. Issues integral to the conduct of a systems engineering effort are discussed, from planning to consideration of broader management issues (Chart III.1). These issues include the differences in process flow, need for strict configuration management, risk management of technology readiness, management of interfaces across scales, selection of verification methods, need for prototyping (Chart III.2), and quality management. The importance of defining decision gates and conducting technical reviews is discussed (Chart III.3). Throughout the part, the role of multidisciplinary subject matter experts and "product" systems engineers in providing data for key decision points and risk mitigation techniques is described.

Chapter 10: Modeling and Simulation in the Small World, *Morgan Trexler and John Thomas*

Chapter 11: Interfaces at the Micro- and Nanoscale, *Jennifer Breidenich*
Chapter 12: Systems Reliability, *O. Manuel Uy*
Chapter 13: Test and Evaluation Techniques from Very-Large-Scale Integration (VLSI) to New Developments in Micro- and Nanoscale Technology (MNT), *William Paulsen*
Chapter 14: Developing and Implementing Robust Micro- and Nanoscale Technology Programs, *Janet L. Barth*

CHART III.1
Systems Analysis and Control—Sampling of Tools

Work Breakdown Structure

The Work Breakdown Structure (WBS) is a means of organizing system development activities based on system and product decompositions. These product architectures, together with associated services (e.g., program management, systems engineering, etc.) are organized and depicted in a hierarchical tree-like structure that is the WBS.

Configuration Management

Configuration management permits the orderly development of a system, subsystem, or configuration item. A good configuration management program ensures that designs are traceable to requirements, that change is controlled and documented, that interfaces are defined and understood, and that there is consistency between the product and its supporting documentation. Configuration management provides documentation that describes what is supposed to be produced, what is being produced, what has been produced, and what modifications have been made to what was produced.

Data Management

Data management documents and maintains the database reflecting system life cycle decisions, methods, feedback, metrics, and configuration control. It directly supports the configuration status accounting process. Data Management governs and controls the selection, generation, preparation, acquisition, and use of data imposed on contractors.

CHART III-1 *(Continued)*
Systems Analysis and Control—Sampling of Tools

Interface Management

Interface Management consists of identifying the interfaces, establishing working groups to manage the interfaces, and the group's development of interface control documentation. Interface Management identifies, develops, and maintains the external and internal interfaces necessary for system operation. It supports the configuration management effort by ensuring that configuration decisions are made with full understanding of their impact outside of the area of the change.

Interface Identification

An interface is a functional, physical, electrical, electronic, mechanical, hydraulic, pneumatic, optical, software, or similar characteristic required to exist at a common boundary between two or more systems, products, or components.

Trade Studies

Trade studies identify desirable and practical alternatives among requirements, technical objectives, design, program schedule, functional and performance requirements, and life-cycle costs are identified and conducted. Choices are then made using a defined set of criteria. Trade studies are defined, conducted, and documented at the various levels of the functional or physical architecture in enough detail to support decision making and lead to a balanced system solution.

Modeling and Simulation

A model is a physical, mathematical, or logical representation of a system entity, phenomenon, or process. A simulation is the implementation of a model over time. A simulation brings a model to life and shows how a particular object or phenomenon will behave. It is useful for testing, analysis or training where real-world systems or concepts can be represented by a model.

Metrics in Management

Metrics are measurements collected for the purpose of determining project progress and overall condition by observing the change of the measured quantity over time. Management of technical activities requires use of three basic types of metrics:

- Product metrics that track the development of the product,
- Earned Value that tracks conformance to the planned schedule and cost, and
- Management process metrics that track management activities.

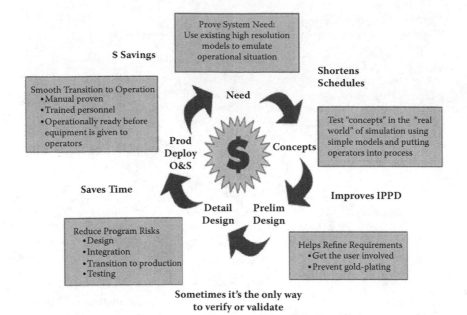

$ Savings

Prove System Need:
Use existing high resolution
models to emulate
operational situation

**Shortens
Schedules**

Smooth Transition to Operation
• Manual proven
• Trained personnel
• Operationally ready before
  equipment is given to
  operators

Need

Test "concepts" in the "real
world" of simulation using
simple models and putting
operators into process

**Prod
Deploy
O&S**

Concepts

$

Saves Time

Improves IPPD

**Detail
Design**

**Prelim
Design**

Reduce Program Risks
• Design
• Integration
• Transition to production
• Testing

Helps Refine Requirements
• Get the user involved
• Prevent gold-plating

**Sometimes it's the only way
to verify or validate**

Chart III.2 Advantages of Modeling and Simulation

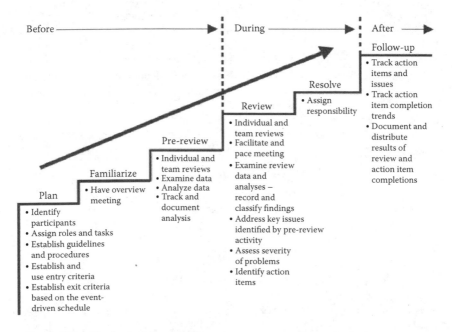

Before ————————————————→ During ——————————→ After ——→

Follow-up
• Track action
  items and
  issues
• Track action
  item completion
  trends
• Document and
  distribute
  results of
  review and
  action item
  completions

Resolve
• Assign
  responsibility

Review
• Individual and
  team reviews
• Facilitate and
  pace meeting
• Examine review
  data and
  analyses –
  record and
  classify findings
• Address key issues
  identified by pre-review
  activity
• Assess severity
  of problems
• Identify action
  items

Pre-review
• Individual and
  team reviews
• Examine data
• Analyze data
• Track and
  document
  analysis

Familiarize
• Have overview
  meeting

Plan
• Identify
  participants
• Assign roles and tasks
• Establish guidelines
  and procedures
• Establish and
  use entry criteria
• Establish exit criteria
  based on the event-
  driven schedule

Chart III.3 Technical Review Process

10

Modeling and Simulation in the Small World

Morgan Trexler
John Thomas

In search of the perfect model.

Contents

Introduction

Modeling and simulation tools are an integral part of the design process utilized by the systems engineer and are also useful to the nanotechnologist for gaining understanding of nanoscale phenomena. However, modeling and simulation in these two contexts can differ greatly in form and function. This chapter will serve to explain how modeling and simulation are used by the systems engineer and conversely by the nanotechnologist. Figure 10.1 presents a framework for the entire chapter. Overviews of theories used for modeling at the nanoscale will

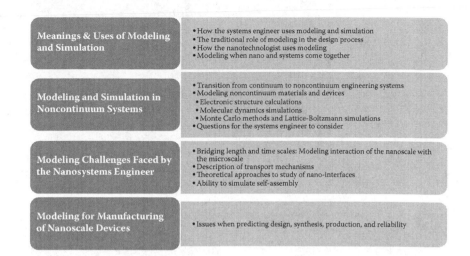

Meanings & Uses of Modeling and Simulation	• How the systems engineer uses modeling and simulation • The traditional role of modeling in the design process • How the nanotechnologist uses modeling • Modeling when nano and systems come together

Modeling and Simulation in Noncontinuum Systems	• Transition from continuum to noncontinuum engineering systems
	• Modeling noncontinuum materials and devices
	  • Electronic structure calculations
	  • Molecular dynamics simulations
	  • Monte Carlo methods and Lattice-Boltzmann simulations
	• Questions for the systems engineer to consider

Modeling Challenges Faced by the Nanosystems Engineer	• Bridging length and time scales: Modeling interaction of the nanoscale with the microscale • Description of transport mechanisms • Theoretical approaches to study of nano-interfaces • Ability to simulate self-assembly
Modeling for Manufacturing of Nanoscale Devices	• Issues when predicting design, synthesis, production, and reliability

Figure 10.1 Chapter framework.

be given, and challenges faced in trying to capture the important nanoscale phenomena will be reviewed. As nanoscale components merge into the systems engineering framework, up to 10 orders of magnitude are spanned. To predict system performance, material properties and component behavior must be translated from computations at the atomistic scale to the nanoscale and then up to the micro- and macroscales without losing important effects or propagating error by oversimplifying behaviors or making excessive assumptions. This chapter will explore how nanoscale theory, modeling, and simulations fit into the traditional systems engineering framework and what obstacles need to be overcome to enable the systems engineer's traditional approach to design.

Meanings and Uses of Modeling and Simulation

Modeling and simulation can be used to perform many different functions in the context of systems engineering versus nanotechnology. Modeling denotes a mathematical approach for solving a problem in the context of nanotechnology, whereas systems engineers use experimental and qualitative models in addition to mathematical models. Different types of models employ frameworks and theory with varying degrees of accuracy, physical fidelity, and assumptions. For each type of model, the objectives and expectations for results range from a qualitative understanding of a process to a detailed quantitative understanding of phenomena. Each application of modeling and simulation has a specific purpose in systems engineering or nanotechnology. These will be reviewed here along with thoughts on how the two could merge together.

How the Systems Engineer Uses Modeling and Simulation

Modeling and simulation tools are used by the systems engineer throughout the development process [1] and are characterized on scales of low to high physics fidelity and low to high situational relevance [2]. A chart summarizing types of models and where they lie on these scales is presented in Figure 10.2. Models that fall into each of the four quadrants are important in the development process, but it is critical to recognize the utility of each. Some models incorporate high physics fidelity, whereas others are useful for replication of an environmental scenario, for example. Systems engineers use models that range from schematic diagrams to mathematical equations and physical scale models or prototypes. Schematic models are used to represent a system element or process. Although schematic diagrams present an abstract and limited view, these are an indispensable means of communication because they can be easily drawn and changed. Mathematical models represent relationships or functions via equations. Physical models directly reflect some or most of the physical characteristics of a system but are often simplified in scale or scope. In systems engineering terms, a simulation is a type of modeling often utilizing numerical computations to study the dynamic behavior of a system or its components. Simulations are used at each step of system development and are critical to understanding the effects of the environment and projected

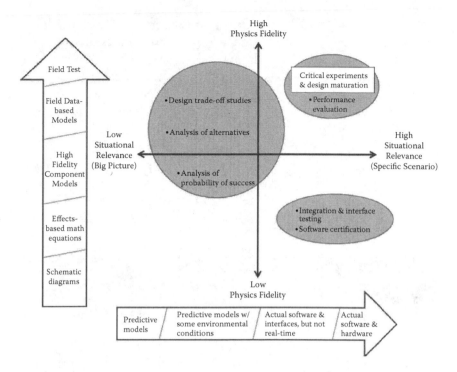

Figure 10.2 The modeling and simulation activities needed to mature a complex system of systems design plotted as a two-dimensional grid of physics fidelity versus situational relevance. (Adapted from Bath, W. and Miller, G., 2011, "Systems of Systems Network Engineering," to be published in Johns Hopkins Applied Physics Laboratory Technical Digest, Vol. 29).

performance of the final system. The various types of models and simulations are heavily relied upon during the design and development of complex systems. In order for the final system to achieve the intent of the designers and systems engineers, high fidelity models are required.

Kossiakoff and Sweet [1] state that mathematical models are "most useful where systems elements can be isolated for purposes of analysis and where their primary behavior can be represented by well-understood mathematical constructs" (p. 418). They further state that "an important advantage of mathematical models is that they are widely understood" (p. 419). Although models play a critical role in systems development, the statements made by Kossiakoff and Sweet indicate that mathematical models must be highly accurate and the theory well understood, which is a challenge when working with new and complex technologies such as nanoscale materials. Given that models in all four quadrants of Figure 10.2 are important to systems engineers, it is likely that challenges will be faced when developing a system for which there are no models that lie comfortably in the upper-right quadrant (high physics fidelity, high situational relevance). Alternatives to modeling or development of more accurate nanoscale models may be necessary for the traditional systems engineering approach for nanosystems to be utilized. This chapter will address strictly computational and mathematical modeling at the nanoscale and will not address other types of modeling used by systems engineers.

The Traditional Role of Modeling in the Design Process

Consider the materials science tetrahedron given in Figure 10.3. This tetrahedron illustrates links between processing, properties, structure, and performance of a given material. This particular view of the tetrahedron is looking down the axis of performance. This emphasizes the distinct relationship between the other three points—processing, structure, and properties—all of which contribute to performance. It follows that if you can understand and model each of these three regions of the tetrahedron and how they relate to one another, modeling material performance will follow.

The traditional approach to materials synthesis consists of iterative physical sample processing to optimize a specific property or microstructure. In recent years, a paradigm shift has occurred due to development of predictive modeling capabilities [3]. Materials modeling that links processing to structure to properties to performance has allowed for faster development of materials for specific applications. Modeling can be used to identify materials that will theoretically meet specified performance criteria, and then its structure and a corresponding processing route can be elucidated to realize the "perfect" material for the given application. Development

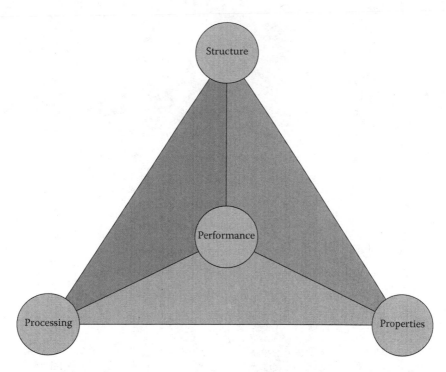

Figure 10.3 Materials science tetrahedron linking processing, properties, structure, and performance. This is a view down the performance axis, showing that there is a clear relationship between the other three points, all of which contribute to performance.

of physics-based modeling tools and predictive capabilities, as opposed to empirical relationships fit to experimental data, is a growing effort and will allow computational materials design to become more widely used.

A flowchart describing the materials design cycle [3] that begins with the systems engineer's description of performance requirements is shown in Figure 10.4. This can again be thought of as a view down the performance axis of the tetrahedron with the addition of an outer ring connecting processing, structure, and properties. This outer ring describes modeling that can be performed to relate each of these. Performance requirements can serve as inputs for multiscale models ranging from the continuum down to atomistic scale, as appropriate. The outputs from these models are the material properties required for performance in a specific application. These

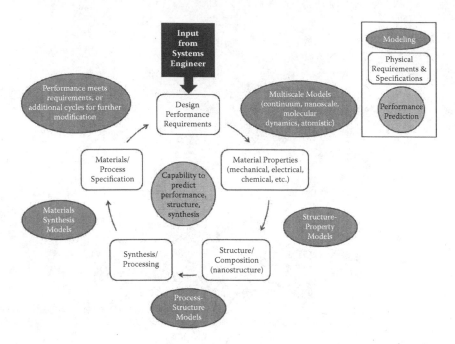

Figure 10.4 Flowchart illustrating predictive capability allowing information flow from performance to structure to synthesis. (Adapted from Gray, G.I., Journal of Materials, 62(3), 9–10, 2010.)

properties can then be translated into structure-property models, which will yield the ideal composition and (nano) structure. Next, the synthesis procedure of the desired material must be determined. Process-structure models and materials synthesis models are used to determine process specifications. This cycle is repeated until the designed material meets the requirements set by the systems engineer.

Beyond the individual material and component level, modeling must be performed to integrate components and predict device behavior. Figure 10.5 shows a flowchart outlining the multiple stages and scales of modeling necessary for modeling nanoscale systems. In comparison to a similar flowchart for microscale systems [4], the physical level of modeling requires multiscale modeling and the conveyance of information between scales such that nanoscale component properties and behaviors can be captured at the device scale. Traditionally, the physical-level behavior of components is modeled in a three-dimensional continuum using techniques including finite element/finite

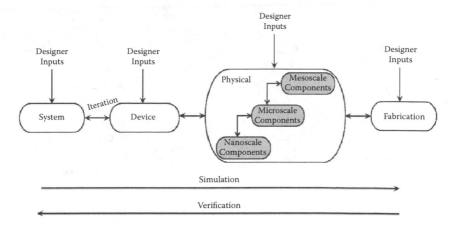

Figure 10.5 The multiple levels and scales of modeling necessary for design, fabrication, and implementation of nanoscale systems. Double-headed arrows indicate iterative information exchange between levels and scales of modeling. Multiscale modeling is required for physical modeling when working with nanoscale systems. (Adapted from Senturia, S.D., *Microsystem Design*, Kluwer Academic, Boston, 2001.)

difference/boundary element models. However, the existence of many length scales is yet another challenge faced when modeling behavior of nanosystems. To capture accurate physical behavior of nanoscale structures, other approaches including atomistics and molecular mechanics are necessary. However, these atomistic and molecular modeling approaches can handle only limited amounts of atoms while keeping computation time reasonable. Multiscale approaches must be used to bridge the physical properties of the nanostructures to the microscale and mesoscale finite element models. Meshed continuum models are often too cumbersome to capture whole device behavior. To transition from the microscale and mesoscale models to the device level, macromodels or reduced-order models are needed to capture the essential physical behavior of components and simultaneously be compatible with the system-level description. Determining which nanoscale phenomena and properties are relevant to component behavior, and then translating that information to the model of the next higher scale, is a critical step in accurately modeling systems containing nanoscale components.

Modeling of materials processing is especially important because details of the synthesis and processing sequence can dictate material properties, as illustrated by Figure 10.4. An example of the importance of process-level modeling for synthesis of nanoscale structures is growth of carbon nanotubes. Many commonly used synthesis techniques cannot controllably grow nanotubes with a specific type of conductivity (metallic versus semiconducting); however, varying the noble gas ambient during thermal annealing of the catalyst while also in the presence of oxidative and reductive species can yield 91% metallic conducting nanotubes [5]. The ability to model and predict the resulting differences in material properties when synthesis parameters (ambient environment, temperature, humidity, etc.) are altered would be a significant tool for design of nanoscale systems; however, these effects are not as well characterized, understood, or studied as in the case of larger-scale materials.

How the Nanotechnologist Uses Modeling

A flowchart illustrating the general sequence required for modeling materials can be seen in Figure 10.6 [6]. Once a physical problem of interest has been identified, this problem must be

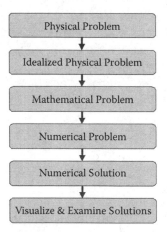

Figure 10.6 Sequence of steps required to model a nanomaterials problem. (See Ramesh, K., *Nanomaterials: Mechanics and Mechanisms,* Springer, Dordrecht, 2009.)

idealized and simplified. A set of mathematical equations can then be derived to represent the idealized physical problem. These mathematical equations can be solved using a numerical approach. Numerical solvers range in accuracy, thus introducing numerical approximations. Similar to the verification and validation performed by the systems engineer for traditional systems, modelers must ask if the correct equations are being solved and if the equations are being solved correctly. This can be difficult when studying behavior of nanoscale components, because quantitative understanding of phenomena occurring at the nanoscale is lagging behind experimental advances. Ultimately, the outputs and predictive capability of a model are only as good as the inputs and theory used. This should be noted especially at the nanoscale, because the theory is not as well developed as for other technologies. Therefore, modeling results should be utilized cautiously, and the degree of assumptions required to progress through this modeling sequence should be taken into consideration. However, if implemented effectively and accurately, simulations of nanoscale systems can potentially do the following [7]: (1) compare and evaluate various molecular-based theoretical models, (2) evaluate and direct experiments that may not otherwise be possible, and (3) replace an experiment provided that accurate intermolecular potentials or other appropriate descriptors are available.

An example demonstrating the effects of multiple assumptions and the need for improved nanoscale modeling can be found in efforts to model the strength of carbon nanotubes. Carbon nanotubes are nominally 10 times the strength of steel, making them of interest for many applications including structural composites, sporting equipment, and body armor. Experimental fracture stress data range from ~10 to 100 GPa. The wide range of these data illustrates the importance in understanding the role of defects. Modeling has been used to investigate the theoretical strength of carbon nanotubes and effects of defects on strength. Figure 10.7 shows data obtained from three different types of theory—density functional theory (DFT), a semiempirical quantum mechanical method known as PM3, and a reactive bond-order molecular mechanics potential function known as MTB-G2—compared with experimental data [8]. The calculated theoretical fracture

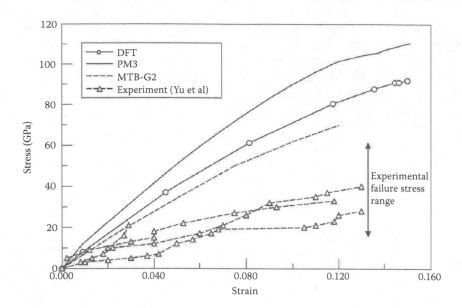

Figure 10.7 Comparison between experiments and modeling of the strength of carbon nanotubes. (See Schatz, G.C., in *Proceedings of the National Academy of Science*, 104(17), 6885–6892, 2007.)

stresses differ by 15% indicating the theoretical uncertainties, and experimental scatter is also considerable. However, these discrepancies are small compared to the difference between the theoretical and experimental data. If the performance of a nanoscale device is dependent on the mechanical strength of a carbon nanotube, it will be nearly impossible to predict with any accuracy at what stress the device would fail. This example illustrates not only the need for better developed models for nanoscale materials, but also the need to first better understand nanoscale phenomena.

Modeling When Nano and Systems Come Together

The current lack of computational models to accurately describe matter at the nanoscale and predict behavior of devices and systems composed of nanoscale components is a barrier to progress in this area. Although models exist to describe matter interactions at the atomic scale, tools to predict behavior of systems containing nanoscale components have yet to be developed. It is unlikely that the electronics, medical, or other industries will risk deploying devices based on nanoscale technology, even if they can be produced, unless the technology is thoroughly understood [9]. Computational tools for understanding and predicting behavior of nanotechnology are necessary before these devices can be employed.

When designing nanoscale materials for a system application, it is critical that the systems engineer and materials engineer recognize the degree of assumptions necessary when modeling at the nanoscale, and that these models cannot be relied upon for design and prediction of behavior as are models for more established technology regimes. Predictive modeling capability would be a great asset to the systems engineer working with nanoscale components; however, there are many challenges that must first be overcome before simulations of nanoscale phenomena can be utilized in these ways.

Modeling and Simulation in Noncontinuum Systems

In-depth theoretical details of various theoretical methods used for nanoscale modeling can be found elsewhere [6,7,10], but the general approaches, advantages, and drawbacks will be covered here. Also presented is a list of questions the systems engineer should ask him- or herself when working with nanoscale systems.

Transition from Continuum to Noncontinuum Engineering Systems

Consider the gold ingot illustrated in Figure 10.8, which has a well-characterized set of electrical, thermal, mechanical, and optical properties. If we cleave the ingot into two equal-sized pieces, the intensive thermophysical properties (e.g., thermal

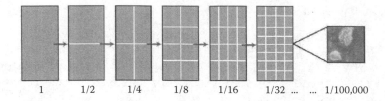

Figure 10.8 Repeated bisection of a gold ingot. Thermophysical properties (e.g., thermal conductivity, electrical conductivity, etc.) that are intensive at the macroscopic level become extensive with decreasing fragment size.

conductivity, electrical conductivity, elastic modulus, etc.) of the two smaller gold pieces will be equal to those of the full-sized ingot. If we repeat this cleaving process such that the original ingot is split into four quarter-sections, these smaller fragments will again possess the same intrinsic thermophysical properties as the larger fragments. As we continue this cleaving process, nothing noteworthy will happen until about the 18th cleaving when the fragment size is about approximately 1:100,000 the size of the original ingot. Then the measured material properties will begin to deviate from those of the larger samples. Although these smaller gold particles will remain electrically conducting, their conductivity will become an extensive property that depends on the size and geometry of the fragment. Similarly, the thermal conductivity and the thermal diffusivity will become increasingly dependent on the fragment size and local environment interacting with the exposed surfaces. The transition from macroscopic engineering regimes, wherein materials properties are intensive and well defined, to microscopic engineering regimes, wherein material properties are extensive and geometry dependent, must be recognized when designing small-scale systems and engineering devices.

Insight into the physical mechanisms governing this transition from bulk-like behavior to small-scale behavior comes from thinking about materials at the atomic and molecular level. In a traditional finite-element based engineering analysis, as illustrated in Figure 10.9, we virtually partition the sample into a large number of volumetric elements that are small compared to the system size but large compared to the atomic structure. The size of the elements (dx) is taken to be

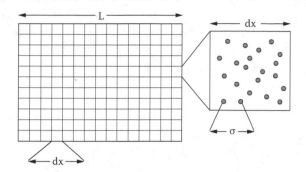

Figure 10.9 Analysis of macroscopic engineering systems typically begins by discretizing the system (L) into a series of infinitesimal volumetric elements (dx) that each contain a statistically large number of molecules. This analysis is valid only when the mean-free path of the molecules (σ) is small compared to dx. If the mean-free path is comparable to dx or larger, this finite element-based approach is inappropriate, and systems must be analyzed at the atomic or molecular level.

small compared to the system size (L) such that gradients across the element can be linearized. Concurrently, the sizes of the elements are taken to be large compared to the atomic length scale (σ) such that the element contains a statistically sufficient number of atoms and exhibits well-defined material properties [11,12]. As the system length approaches the atomic length scale, as shown in Figure 10.9, these two criteria become impossible to satisfy simultaneously, and the notion of a finite element with well-defined material properties is no longer correct. Instead, within this regime, the behavior and performance of the element must be explicitly correlated to the properties of the constituent atoms and molecules [13,14].

A key metric used to characterize the transition from continuum-level transport to noncontinuum transport with decreasing system size is the nondimensional Knudsen number [15]. The Knudsen number, as illustrated in Figure 10.9, is defined as the ratio of a representative atomic length scale to a representative system length scale. The representative atomic length scale typically corresponds to an intrinsic molecular property, such as a crystal lattice constant, the diameter of a molecule, or the mean-free path of an electron or gas molecule, and so forth. The representative system length scale is a system dimension that best characterizes the transport domain. When the Knudsen number is less than 0.1, such that the system length scale is large compared to the atomic length scale, materials and devices may be modeled as a continuum and differential calculus can be used to model transport through a system. When the Knudsen number is larger than 0.1, the continuum assumption becomes less valid and the role of boundary scattering on systems/materials behavior becomes increasingly important. Predictions from continuum-based transport prediction models become less reliable, and system behavior must instead be analyzed at the atomic level using statistical mechanics-based models.

Different materials and systems experience a transition from continuum to noncontinuum behavior at different length scales. For example, the mean-free path of water molecules in a condensed-phase liquid state is around 0.2 nm (which is comparable to the effective molecular diameter). This value suggests that a continuum description of liquid water will be valid in systems with a wetted diameter larger than 2 nm. In smaller systems, where the Knudsen number is greater than 0.1, the effective water viscosity becomes correlated to the confined liquid structure, and a noncontinuum description of transport becomes necessary. In crystalline silicon, where the longest phonon mean-free path at room temperature is around 1 micrometer, a continuum-based description is valid only in samples thicker than 10 micrometers. With decreasing system length, the applicability of bulk-like (macroscopic) material properties becomes increasingly questionable as performance becomes increasingly size and geometry dependent [16].

Modeling Noncontinuum Materials and Devices

For reasons discussed above, tools for predicting the performance of noncontinuum systems must describe the behavior of materials and devices from at the atomic or molecular level. Such bottom-up modeling approaches do not, in general, use a prespecified set of material properties when predicting the performance of engineering systems. Rather, these properties naturally come from the atomic/molecular interactions and dynamics. In this sense, molecular-level modeling provides a

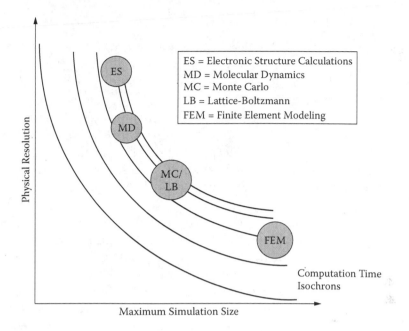

Figure 10.10 Compromises between physical resolution versus maximum simulation size. Increasing the simulation size (physical resolution) of a high physical resolution (simulation size) model typically requires an increase in computational resources.

more accurate description of behavior of materials and systems. Because the complexity of the bottom-up approaches increases quickly with increasing system size, careful consideration is required when selecting an appropriate modeling framework and building a modeling system.

In this section, we overview four popular molecular modeling and simulation tools: electronic structure methods, molecular dynamics methods, Monte Carlo methods, and Lattice–Boltzmann methods. We discuss the strengths and limitations of each modeling method and, as illustrated in Figure 10.10, discuss these models in the context of physical resolution, maximum simulation size, and computational demand. The objective of this discussion is to apprise systems engineers of the available modeling tools and the associated strengths and limitations. Technical specifics, implication details, and extended discussion are available in cited references.

Electronic Structure Calculations

Electronic structure calculations are based upon the laws of quantum mechanics and use a variety of mathematical transformations to solve the multielectron Schrödinger equation [17]. These solutions are typically generated without reference to experimental data, such that the calculations are considered to be ab initio (a Latin phrase meaning "from the beginning"). In general, electronic structure methods offer the highest system resolution and require the fewest number of initial assumptions. However, the large computational costs associated with the calculations make them computationally prohibitive for systems containing more than a few hundred atoms. Thus, although the predictions obtained from electronic structure calculations may be physically correct, the situational relevance may be limited.

As illustrated in Figure 10.11, three parameters must be specified prior to running an electronic structure calculation: (i) an electron basis set, (ii) an electron-electron correlation model, and (iii) an initial atomic configuration. These inputs are used to approximate solutions to the Schrödinger equation, which takes the form of an eigenvalue equation and describes the discrete energy levels accessible to the system. The output of the calculation is a three-dimensional wave function, which is related to the spatial variation in the electron density. The

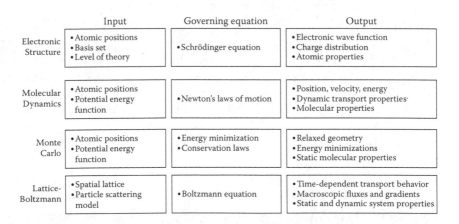

	Input	Governing equation	Output
Electronic Structure	• Atomic positions • Basis set • Level of theory	• Schrödinger equation	• Electronic wave function • Charge distribution • Atomic properties
Molecular Dynamics	• Atomic positions • Potential energy function	• Newton's laws of motion	• Position, velocity, energy • Dynamic transport properties • Molecular properties
Monte Carlo	• Atomic positions • Potential energy function	• Energy minimization • Conservation laws	• Relaxed geometry • Energy minimizations • Static molecular properties
Lattice-Boltzmann	• Spatial lattice • Particle scattering model	• Boltzmann equation	• Time-dependent transport behavior • Macroscopic fluxes and gradients • Static and dynamic system properties

Figure 10.11 Overview and comparison of molecular simulation modeling methodologies.

electronic wave function can be manipulated to predict a number of molecular and crystalline properties, such as the optimized molecular geometry, the atomic vibration spectra, chemical reaction paths, optical absorption properties, and electronic band structures [18].

The electron basis set describes the number and type of functions used to build the multielectron wave function. Increasing the number of basis functions increases the dimensionality and accuracy of the predicted multielectron wave function. Just as grid resolution tests are required when running a finite element-based model, convergence tests are required to ensure that the basis set used to describe the electrons is sufficiently large. Along with the absolute number of basis functions used to build the multielectron wave function, the form and shape of selected basis functions (e.g., diffuse, polarizable, etc.) must be balanced and physically motivated [19].

The electron-electron correlation model describes how interactions between electrons are modeled within the Schrödinger equation. In general, electron-electron interactions are complex and cannot be solved analytically. For example, in a two-electron atom, predicting the equilibrium position distribution of electron A requires knowing the equilibrium position distribution of electron B. However, to predict the equilibrium position distribution of electron B, we must already know the equilibrium position distribution of electron A. A variety of perturbation-based and functional-based methodologies for handling these electron-electron interactions are available in the literature. The accuracy of the multielectron wave function improves with increasing level of sophistication.

The initial atomic configuration defines the relative location of each electron orbital and the potential energy landscape experienced by the electrons within the system. Because the mass of the atomic nuclei are multiple orders of magnitude greater than the mass of the electrons, their motion is generally neglected when predicting the electronic wave function. However, because nuclei are charged, they still interact with electrons through Coulomb's law [17].

For some systems, the Schrödinger equation used to calculate the electronic wave function can be simplified using parameters derived from experimental data or predictions from previous calculations. These "semiempirical" methods typically incorporate the charge of the core electrons into the nuclear charge and treat only the valence electrons explicitly, thereby reducing the computational load. Semiempirical methods are much faster than traditional electronic structure methods and are currently the only computationally practical method for predicting the quantum mechanical characteristics of large (~1,000 atoms) systems. Semiempirical methods can also be used as a quick first step toward predicting the equilibrium atomic positions of systems with unknown geometries. The same approximations that reduce the simulation runtimes also reduce the accuracy and precision of the electronic wave function. Thus, efforts should be made to validate predictions from semiempirical methods using more sophisticated simulation techniques [20].

Molecular Dynamics Simulations

Molecular dynamics (MD) simulations are based upon Newton's laws of motion and use a prespecified set of interatomic potential functions to predict the time evolution of atomic positions and velocities [21]. From the ensuing atomic interactions and trajectories, one can extract the thermophysical properties of atoms (e.g., temperature, pressure, density, thermal conductivity, etc.) and transport processes (e.g., heat conduction, fluid flow, fracture propagation) through nano- and microscale engineered systems. Because the nature and form of the interatomic potential function are specified a priori, MD simulations are much (several orders of magnitude) faster than electronic structure methods and can handle systems containing hundreds of thousands of atoms. The ability to handle larger systems means MD simulations can have better situational relevance than electronic structure calculations. However, the assumptions required to realize this speed increase result in the physical resolution of MD simulation being lower than electronic structure calculations.

Potential functions are algebraic expressions that describe how the energy between two or more atoms is related to the interatomic separation distance. A simple potential function would be a harmonic spring connecting the two atoms of a molecular dipole. More complex potential functions are often multibodied (meaning they depend on the positions of three or more atoms) and, in addition to a harmonic-like stretch term, contain higher-order torsion, dihedral, and bending terms. The predictive accuracy of an MD simulation is governed primarily by the correctness of the prespecified interatomic potential functions in reproducing the actual quantum-mechanical interatomic interactions. Although the form of the potential function may be motivated by experimental data, predictions from electronic structure calculations, and analytical predictions grounded in physical chemistry, some interactions are typically neglected in order to maximize computational efficiency. Thus, as with any other simulation methodologies, efforts must be made to choose potential functions that appropriately balance accuracy and computational efficiency [22].

Dynamical data obtained from an MD simulation are transformed into more meaningful thermophysical properties using tools grounded in statistical mechanics [23]. For example, consistent with ideas grounded in kinetic theory, the temperature of a collection of atoms is proportional to its total kinetic energy following the equipartition theorem. Likewise, the pressure of a collection of atoms is related to the virial expansion of the ideal gas law. In an equilibrium simulation, transport coefficients can be evaluated using ensemble-averaged autocorrelation functions (e.g., diffusion coefficient from the autocorrelation function of the velocity and thermal conductivity from the autocorrelation of the heat current). In nonequilibrium, a gradient can be applied to a system and transport coefficients evaluated directly from the resultant fluxes (e.g., predicting the thermal conductivity by applying a temperature gradient and measuring the resulting heat flux).

Although MD simulation sizes (100,000's of atoms) and simulation runtimes (0.1 to 0.5 of microseconds) may be smaller than actual operating conditions, they are typically sufficient to obtain converged thermodynamic properties or the steady-state response to an applied gradient. As suggested by the central-limit theorem, increasing the number of particles within the system reduces the simulation uncertainty. Likewise, increasing the simulation runtime provides a better sampling of the system phase space and more accurate reproduction of actual engineering materials and devices. Further improvements in the situational relevance are realized by choosing to simulate within a device-appropriate thermodynamic ensemble. For example, by modifying the atomic equations of motion, MD simulations can accurately reproduce the behavior of materials and systems in constant energy, constant temperature, constant pressure, constant enthalpy, and constant volume ensembles.

It is important to recognize that within an MD simulation, electrons are not modeled explicitly. Although the effects of the electron–electron and electron–ion interactions are incorporated into the potential function (these interactions define the potential function), MD simulation cannot be used to model excited electronic states, chemical reactions, or electron transport phenomena. Moreover, because the dynamics of the atoms and molecules is governed by Newton's law of motion, their movement and interactions are classical, meaning all quantum degrees-of-freedom are assumed to be fully populated. For liquid systems, wherein quantum effects become negligible at temperatures greater than a few Kelvin, this classical treatment is appropriate. For solids, where quantum effects remain relevant in systems with operating temperatures in the hundreds of Kelvin, the implications of this assumption must be considered more carefully [13].

For materials and systems that do not have well-defined potential functions, MD simulation can be hybridized with electronic structure calculations to perform ab initio MD (AIMD) simulations. In AIMD, the intermolecular potential energy is calculated on-the-fly from the instantaneous atomic configuration using electronic structure calculations. Although computationally expensive, this additional step eliminates the need to specify a preparameterized potential function and can improve the correctness of the predicted thermophysical properties. Although the interaction energy is modeled quantum mechanically, the positions and velocities still evolve classically according to Newton's laws of motion.

Monte Carlo Methods and Lattice–Boltzmann Simulations

Monte Carlo methods are a general class of simulation tools used to predict the thermophysical behavior of engineering systems at the atomic, molecular, and transport carrier (e.g., photon, phonon, and electron) level. Named after the famous casino in Monaco, Monte Carlo methods use random sampling to investigate the multidimensional potential energy landscape of a material or device. Like MD simulations, the theory behind Monte Carlo methods is grounded in statistical mechanics, and interactions between particles are typically modeled using potential functions. However, unlike MD simulation, Monte Carlo methods do not predict the dynamical properties of particles. Thus, although Monte Carlo methods are generally much faster than MD, they are not suitable for investigating time-dependent processes or properties that depend on velocities (e.g., temperature, pressure) [19].

As an example of one Monte Carlo technique, consider predicting the minimum energy configuration of a long (multiple degree-of-freedom) atomic chain. To begin, we calculate the initial potential energy of our molecule (which is saved as a reference value for the next calculation) and generate a large set of random numbers. Next, for each atom in the chain, we randomly perturb the components of each position vector using a pre-generated random number sequence. We then recalculate the potential energy of the molecule and compare this new energy to the reference value. If the energy of the molecule is lower than the previously calculated value (meaning we have moved closer to the minimum energy configuration), we accept the positional changes to the molecule, set reference potential energy value to this new potential energy, and repeat the process. If the energy of the molecule is greater than the reference value (meaning we have moved away from the minimum energy configuration), we return the atoms to their previous positions (without updating the reference energy) and try a new set of perturbations. This process is repeated until perturbations of the atomic positions yield no further decrease in the molecular potential energy. The magnitude of the perturbations (e.g., the simulation step size) can be tuned to balance the required solution accuracy with a realistic computational runtime.

Beyond this simple structure relaxation example, numerous Monte Carlo-based techniques have been developed for describing photon, fluid, and phonon transport processes through nanoscale and bulk-like systems [21,24,25]. Other Monte Carlo-based techniques for finding the roots of a complex, multidimensional design space are also available. In addition to these engineering-related applications, Monte Carlo-based statistical sampling methods are used in traffic flow simulations, computational finance, and computational mathematics. The attractiveness of these methods stems from their ability to easily handle very complex systems using relatively simple algorithms. The challenges with these methods arise from ensuring that the sampling process has sufficiently sampled the available design space. Thus, as with any computational method, convergence and resolution tests must be employed to assess the validity and robustness of the simulation predictions.

Lattice–Boltzmann (LB) simulation is a hybridized continuum/noncontinuum numerical modeling technique for predicting the macroscopic behavior of systems and materials from the microscopic behavior of atoms, molecules, and transport carriers [26]. Like MD simulation and Monte Carlo methods, materials and systems within an LB simulation are described in terms of molecule-like particles that evolve through space and time. These particles evolve according to the Boltzmann transport equation and scatter according to a prespecified set of kinetic collision operators. Like finite element-based schemes, particles within the LB simulation can only step from lattice site to lattice site along a prespecified grid. However, by localizing particles' dynamics and using a prespecified interaction operator, LB simulations are typically multiple orders of magnitude faster than MD simulations of the same size. Not surprisingly, this increase in speed is associated with a reduction in physical fidelity through the use of a prespecified set of collision operators as compared to a potential function.

The theoretical underpinnings of LB simulations are grounded in the Boltzmann equation, which describes the time evolution of a particle through position and velocity phase space [27]. In simplified terms, the Boltzmann equation requires that the rate of change in the number of particles within a specific region of phase space is equal to the number of particles scattered into that region minus the number of particles scattered out of the region. Thus, within an

LB simulation, the particles that collide and move about are not merely physical particles moving through real space. In a more subtle sense, the particles also represent the time evolution of each single-particle distribution function through the phase space.

Owing to the local nature of the interactions and particle movement, LB simulations are particularly successful at predicting transport through systems with complex or cluttered geometries. Moreover, due to the inherent particle-based description of systems, LB simulations can be used to investigate transport through both continuum and noncontinuum systems and are easily parallelizable. Although the first LB modes were developed to model gas and fluid transport, robust techniques now exist for predicting phonon and electron transport through crystals. Thus, although LB simulations have lower physical resolution than electronic structure calculations or MD simulations, they can exhibit better situational relevance.

Questions for the Engineer to Consider

Although systems engineers may not be directly responsible for building or running simulations, they must be able to assess critically both the appropriateness of the modeling framework and the validity of the model predictions. In this section, we provide a list of questions to aid the systems engineer in this assessment process. These questions are somewhat open-ended and meant to encourage discussion. Technical experts and the scientists running the models should participate in this dialogue. Of course, during implementation and data collection periods, additional task-specific technical questions will arise. The systems engineer must remain abreast of these issues and understand their possible implications.

"What is the Knudsen number of my system; do I need to be concerned about noncontinuum material properties?" Remember that although the distance between atoms is quite small (~1 angstrom), the mean-free path of electrons, phonons, and fluid molecules in many systems can be quite large (>1 micrometer). If the

Knudsen number is greater than unity, noncontinuum effects must be considered explicitly, and a molecular transport model should be used. With a decreasing Knudsen number, noncontinuum effects become less important, and finite-element approaches become increasingly valid.

"What are the relevant system/material properties; will my model correctly capture these properties?"

Different types of modeling frameworks capture different types of transport processes. For example, MD simulations do not describe electron transport phenomena. Electronic structure calculations, which model the electronic wave function explicitly, must instead be used to predict electronic transport phenomena. Likewise, Monte Carlo energy minimization methods provide little insight into dynamic molecular relaxation processes. Molecular dynamics simulations, which include a time dimension, must instead be used to predict such time-dependent properties.

"Does my model have an appropriate balance between simulation size and physical resolution?"

In general, as illustrated in Figure 10.10, increasing the physical fidelity of a simulation requires compromises in the situational relevance. Increases in computational resources (e.g., more runtime or more processors) may be required to achieve sufficient accuracy in both of these requirements.

"Is my model fully converged; have I sufficiently sampled the system phase space?"

When running any type of numerical simulation, one must carefully perform basis-set, grid size, and sample length convergence tests. For example, when using MD simulation, multiple simulations (from differential initial conditions) should be used to assess the robustness of the predicted behavior and the variance in the simulation results. Likewise, tests must be performed to ensure that the simulation runtime is sufficient to observe fully converged (e.g., steady-state or equilibrium) behavior.

"How does my model compare to an analytical solution; is experimental reference data available?"

Every model must be validated; the mere ability to run a simulation does not guarantee that the simulation predictions are correct. For simple geometries or simple materials, the properties predicted from the simulation should be compared directly to analytical solutions. If possible, the simulation predictions should be compared directly to relevant experimental data.

Modeling Challenges Faced by the Nanosystems Engineer

Despite the significant progress of theory, modeling, and simulation at the nanoscale, many challenges remain. Additional obstacles may be faced by the systems engineer looking to integrate nanoscale components, because the use of theory and predictive modeling capability in this regime is not as straightforward as with technologies with larger characteristic dimensions. Several of the challenges called out by "Theory and Modeling in Nanoscience," a report of the workshop conducted by the Basic Energy Sciences and Advanced Scientific Computing Advisory Committees to the Office of Science in the Department of Energy in 2002 [9], as well as other challenges, are addressed in this section.

Bridging Length and Time Scales: Modeling Interaction of the Nanoscale with the Microscale

Nanoscience bridges multiple length and time scales and combines classes of materials and molecules traditionally investigated in different subdisciplines. This translates to a need for combination and cooperation of theory, modeling,

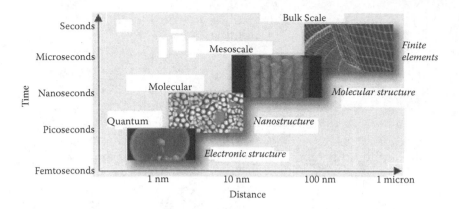

Figure 10.12 The range of time and length scales and corresponding types of models relevant for systems containing nanoscale components. (See Mize, S., *Toward Nanomaterials by Design: A Rational Approach for Reaping Benefits in the Short and Long Term,* Foresight Institute, Palo Alto, CA, 2004.)

and simulation tools that were previously developed and used separately or for the invention of new computational tools to bridge length and time scales [9,28]. Because no single type of theory is appropriate for all nanoscale scenarios, different approaches must be used to understand phenomena occurring at different length and time scales [8]. The range of time (femtoseconds to seconds) and length (1 nanometer to 1 micrometer) scales and corresponding types of theory are shown in Figure 10.12.

Approaches using purely atomistic or purely continuum theory have limitations for modeling nanoscale properties [8]. Hybrid and mesoscale properties often play an important role in bridging the gap between the two approaches. These approaches have begun to be utilized but are still in development. Elhard [30] outlined the need for "development and coherent integration of a suite of modeling tools dedicated to the design and fabrication of atomically precise functional nanosystems" (p. 1). It is further stated that the need "involves developing computational software tools that link nano-meso-macro-scales (time, length, and energy)" (p. 1) and requires methods that "simultaneously incorporate atomistic (1 to 10 μm) and mesoscale (1 to 10 μm) and have uniform bridging scales across 9 [orders of magnitude] in length, 12 [orders of

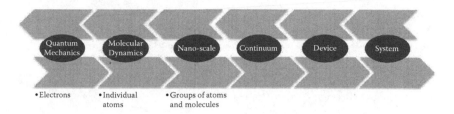

Figure 10.13 Flowchart depicting levels in multiscale modeling. It can be necessary to upscale or downscale during the modeling process to attain sufficient understanding of the physics at each level.

magnitude] in time (and energy flow) and integrate with self-consistent scaling laws."

Multiscale modeling is the idea of computing material behavior beginning at the smallest scale and then successively passing the resulting information to a model at a larger scale [28], leading ultimately to device and then system behavior, or vice versa, as depicted in the flowchart in Figure 10.13. The ultimate goal of a multiscale approach is to predict macroscopic behavior of an engineering process from first principles. In the modeling flow, quantum mechanics calculations include information about electrons, which is then conveyed to molecular dynamics models, which focus on individual atoms. Next, nanoscale models describe groups of atoms, and information obtained here is passed to continuum, device, and system-level models. Table 10.1 summarizes atomistic and continuum approaches for modeling various properties of materials. Also included are hybrid, or mesoscale, approaches that can be used to bridge the gap between length and time scales. Development of these hybrid techniques has progressed, but there is still a long way to go to accurately predict behavior of nanostructures and, ultimately, to integrate the behavior into models of entire devices and systems.

An example of a hybrid computational approach to modeling mechanical properties of carbon nanotubes can be seen in Figure 10.14 [8]. The nanotube shown has a two-atom vacancy but is otherwise perfect. To describe fracture, quantum mechanics (QM) is used to model the breaking of bonds in the area around the defect. QM is limited by being computationally expensive and by the fact that due to strong size-dependence

TABLE 10.1
Atomistic, Continuum, and Hybrid Theory Approaches
for Modeling of Various Nanoscale Properties

Property	Atomistic Level Theory	Hybrid/Mesoscale Theory	Continuum Theory
Structural, thermal	Quantum mechanics, empirical potentials	Coarse-grained molecular dynamics, grand canonical Monte Carlo on lattices	Static structure models, thermodynamics
Mechanical	Quantum mechanics, empirical potentials	Coarse-grained models, quantum mechanics/ molecular mechanics/ continuum mechanics	Elasticity theory
Optical	Quantum mechanics	Multipole coupling of particles, quantum mechanics/ electrodynamics	Continuum electrodynamics

Source: From Schatz, G.C., *Proceedings of the National Academy of Science*, 104(17), 6885–6892, 2007. With permission.

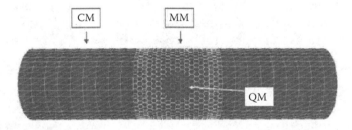

Figure 10.14 Example of multiscale modeling—mechanical properties of a carbon nanotube. This hybrid approach combines quantum mechanics (QM), molecular mechanics (MM), and continuum mechanics (CM). (Adapted from Schatz, G.C., *Proceedings of the National Academy of Science*, 104(17), 6885–6892, 2007.)

of mechanical properties, it is not an accurate way to predict the properties of nanotubes that are large enough to be tested experimentally. To circumvent these limitations, a molecular mechanics (MM) force field is applied around the patch described by QM. Further, MM, which is still atomistic, is interfaced with continuum mechanics (CM) to extend the stress field to distances far away from the original defect. For any hybrid or multiscale approach, interfaces (boundaries between types of models) require precise definition to accurately convey information from one level to the next. This is still the subject of research and remains a challenge to be addressed. The development of hybrid and mesoscale techniques to bridge length and time scales has progressed significantly in recent years, but a need remains for self-validating and benchmarking methodologies in which modeling at larger scales is always validated against more detailed calculations at smaller length scales, because experimental validation is often not possible.

Description of Transport Mechanisms

The science of transport mechanisms in nanostructures is essential to many applications of nanotechnology, as summarized in Table 10.2 [9]. Technological advances in fabrication, characterization, and control at the nanoscale level have enabled the manufacturing of a variety of new organic-inorganic nanostructured devices. Therefore, new simulation approaches are required because the inherent quantum-mechanical physics involved must be treated properly, and the exact nature of the transport mechanisms in many of such systems has yet to be explained [31]. As mentioned earlier, it is unlikely that devices of this nature will be commercialized unless their behavior is fully understood. This challenge can perhaps be tackled by the systems engineer and nanotechnologist together.

Electron transport is a topic of great interest for nanotechnologists studying nanotubes and nanowires, molecular electronics, and transport in semiconductors, metals, and thermoelectric devices. Di Carlo et al. [31] are working to develop theory to describe transport in carbon nanotubes as field-effect transistors (FETs) and field emitters for TeraHertz (THz) sources. Despite their differing transport issues, quantum transport in nanotransistors, nanowires, and molecular electronic devices has been described by Darve et al. [33] using the Nonequilibrium Green's Function. Lundstrom and Ren [34] employed two-dimensional (2D) simulations and a simple conceptual view of the nanoscale

TABLE 10.2
Summary of Types of Transport Occurring in Nanostructures and Nanoscale Devices and the Technology That Rely on This Transport

Type of Transport	Technology to Which This Transport Is Fundamental
Electron[a–d]	Molecular electronics, nanotubes, nanowires, semiconductors, metals, etc.
Spin[e–h]	Spintronics-based devices such as spin valves and spin qubits
Molecule[i,j]	Chemical and biological sensors, molecular separations/membranes, nanofluidics, photons, phonons

[a] See also Di Carlo, A. et al., Modelling of Carbon Nano Tube-Based Devices: From nanoFETs to THz Emitters, SPIE, San Diego, CA, 2006.

[b] See also Sangiorgi, E. et al., Solid-State Electronics, 52(9), 1414–1423, 2008.

[c] See also Darve, E., Li, S., and Teslyar, Y., Calculating Transport Properties of Nanodevices, SPIE, Philadelphia, PA, 2004.

[d] See also Lundstrom, M., and Ren, Z., Electron Devices, IEEE Transactions on, 49(1), 133–141, 2002.

[e] See also Sanyal, B., and Eriksson, O., Physica Status Solidi (A) Applications and Materials, 204(1), 33–43, 2007.

[f] See also Wang, C. et al., IEEE Transactions on Nanotechnology, 6(3), 309–315, 2007.

[g] See also Saikin, S., Pershin, Y.V., and Privman, V., IEE Proceedings: Circuits, Devices and Systems, 152(4), 366–376, 2005.

[h] See also Nikonov, D.E., and Bourianoff, G.I., Journal of Superconductivity and Novel Magnetism, 21(8), 479–493, 2008.

[i] See also Zhou, S.-A., and Uesaka, M., International Journal of Applied Electromagnetics and Mechanics, 24(1), 51–67, 2006.

[j] See also Bohn, P.W., Annual Review of Analytical Chemistry, 2(1), 279–296, 2009.

Note: Some references on modeling of each type of transport are given.

Source: Adapted from *Theory and Modeling in Nanoscience. Report of the Workshop Conducted by the Basic Energy Sciences and Advanced Scientific Computing Advisory Committees to the Office of Science, Department of Energy,* 2002.

transistor to describe transport in metal-oxide semiconductor field-effect transistors (MOSFETs). Monte Carlo simulations have been used to provide insight and understanding of nano-MOSFET device physics [32]; however, many transport issues remain unaddressed.

Use of the spin property of electrons in semiconductor electronic devices (semiconductor spintronics) may revolutionize electronics [38]. Devices for digital logic based on spintronics [41] are still a topic of research [37] and require insight from modeling and simulation for further development. Sanyal and Eriksson [35] have stated that "it is quite clear that a fundamental understanding of the materials involved in spintronics research is absolutely essential. This will not only explain the existing experimentally observed results (mutually contradictory in many cases) but also help to predict new materials with desired properties" (p. 35). This example illustrates the need for better theoretical models of spin transport to enable prediction of device behavior by the systems engineer.

The Monte Carlo simulation approach is a method commonly used to study characteristics of transport beyond quasi-equilibrium approximations, such as drift diffusion or linear response approximations, and has been widely used for charge carrier transport in semiconductor structures and modern devices [37]. This approach can accommodate scattering mechanisms, specific device design, material properties, and boundary conditions in the simulation [37]. However, there is once again a trade-off between accuracy and computation time and resources, so simplified simulation schemes are often utilized, but clearly cannot be relied upon for prediction of performance of a critical device or system. In contrast, Wang et al. [36] used the nonequilibrium Green's function to calculate conductance in spintronic gates and have shown useful operations for "multiterminal" logic gates. Sanyal and Eriksson [35] employed ab initio density functional electronic structure calculations of several spintronic materials to investigate the nature of exchange interactions, electron correlation effects, influence of defects, effects of disorder, and volume dependence of exchange interactions. The variety of approaches to modeling spin transport suggests that a commonly accepted approach has yet to be agreed upon. Although no single modeling approach is currently superior to others across the board, perhaps approaches could be integrated or combined to yield a more accurate and consistent approach.

The ability to understand and control molecular transport by structures of nanometer scale is critical to applications including chemical and biological sensors, molecular separations/membranes, and nanofluidics [40]. These applications often incorporate mass-limited samples in nanometer-scale structures, so precision and accuracy are critical. Employing an atomistic simulation or MD analysis of ion permeation on large time and length scales is not possible due to computational limitations. Continuum models have been used extensively to investigate ion permeation and related transport processes. However, some questions still exist, and disputes in the applications of these continuum models, which are mostly related to the dielectric (polarization) properties of the medium and ion channels, are not taken into account properly in the classical continuum theory [39]. Zhou and Uesaka [39] have formulated a continuum theory for the study of transport phenomena of ions and polarizable molecules based on the basic laws in nonequilibrium (irreversible) thermodynamics and continuum electrodynamics. Their approach considers effects of electrodiffusion, heat conduction, thermomechanical motion, polarization, and polarization relaxation. Approaches along these lines are necessary to overcome computational limitations while still incorporating important phenomena; however, simplification of the theory must be done with caution to ensure that the model will still be a valuable predictive tool for the systems engineer. Most of the approaches utilized to model transport thus far are quite simplified, and although useful for understanding device behavior and optimizing design, they cannot yet be relied upon for performance predictions by the systems engineer.

One challenging aspect of modeling transport properties is the abundance of available scattering mechanisms that cannot be characterized within a given sample. For example, even in silicon MOSFETs, there is interface roughness scattering, impurity scattering from many different types of impurities, oxide scattering from trapped oxide charges, scattering from various types of lattice defects, and scattering from phonons. Scattering from phonons is the only one in this list that can be modeled from first principles. For the rest, the theorist needs to know the density of impurities or trapped oxide charges, and so forth, and this is often impossible. To circumvent this limitation, theorists commonly use empirical fits to experimental data to figure out what the relevant scattering intensities from different scattering sites are. This approach may or may not have value in designing new systems and predicting their behavior. An awareness of what is possible is also important when attempting to model nanoscale.

Theoretical Approaches to Study of Nano-Interfaces

Modeling and simulation have been used to study behavior of interfaces including grain boundaries and metal/ceramic interfaces [42]. Specifically, the response of atomic bonds at or near interfaces to applied stresses or strains can be elucidated via ab initio calculations [42]. The capability to simulate accurately complex nanostructures involving many molecular and atomic species, and the combination of soft biological or organic materials with hard inorganic materials, would be a beneficial development [9]. However, before that can be accomplished, accurate modeling of nano-interfaces must be achieved. Nano-interfaces are highly complex, heterogeneous in shape and substance, and often composed of dissimilar classes of materials [9]. Because of the high surface area to volume ratio, interfacial properties can dominate at the nanoscale and thus must be highly understood. Theory and simulation approaches to understand and predict nano-interface behavior are crucial to the design of structural materials and nanoscale devices. Further, simulation of interfaces between hard and soft materials will play a critical role in biomolecular nanomaterials and associated applications. Challenges associated with interfaces at the nanoscale will be discussed further in Chapter 11.

Ability to Simulate Self-Assembly

As discussed in Chapter 9, self-assembly typically spans several length and time scales and is the key to large-scale production of novel structures based on nanotechnology [9,43]. Although biological systems have mastered self-assembly, scientists are only beginning to learn and understand the process for application to nonbiological systems. Predicting self-assembly behavior is extremely difficult [8]. The potential impact of nanoscale devices and systems has generated interest in better understanding the self-assembly process. Initial theoretical studies have been aimed at understanding self-assembly mechanisms and designing and controlling properties of the resulting materials via selection of initial conditions and structure and chemical properties of precursor molecules [43].

Tsonchev et al. [43] describe one approach for modeling molecular self-assembly in which they first reduced the complex macromolecules to simpler structures, retaining their overall shape, symmetry, and charges. Next, they used these simplified molecules to predict the shape and structure of the final assembly. MD simulations were performed on the self-assemblies to elucidate their detailed structure and stability and the role of dipole-dipole interactions and hydrogen bonding. The authors state that although this method is less rigorous than traditional Monte Carlo or MD methods (which are not computationally capable of solving this problem), the lack of rigor is compensated for by the sampling of a very large number of initial conditions at each stage of the assembly to ensure that the most stable final structure is obtained. They expect that it should be possible to extend this technique to predict and study the self-assembly of many new materials; however, one can imagine there are limitations to this simplified approach. This example illustrates the challenges associated with modeling molecular self-assembly. As more and more devices incorporate nanoscale components, self-assembly will become increasingly more utilized as a manufacturing methodology. For these techniques to be utilized effectively, the fundamental mechanisms must be better understood.

Modeling for Manufacturing of Nanoscale Devices: Issues When Predicting Design, Synthesis, Production, and Reliability

As is true for all types of modeling, technology computer-aided design (TCAD) tools for simulating processes and devices are accurate for predicting behavior only if the technology is sufficiently mature or the model has been calibrated appropriately, which can be a lengthy process [44]. This prompts the question of whether modeling devices composed of new technology and materials is relevant and worthwhile [45]. For example, properties of nanoscale electronic devices are extremely sensitive to the presence of defects or impurities. They are also very geometry dependent, so predicting nanodevice behavior is extremely complicated because it seems impossible to capture all of these variables accurately. However, even with less than ideal model accuracy, TCAD can still be used to provide guidelines for optimization, aid in explanation of characterization results, and provide insight into transport mechanisms. Additionally, modeling and design can guide fabrication by allowing one to perform computational experiments on structures, devices, and systems [46]. These uses for modeling should be considered by the systems engineer when developing a nanoscale system. As long as the modeling outputs are relied upon only as much as the theory they are built on, nanoscale modeling can be used as a guide for the systems engineer, although it is not a definitive predictor of behavior.

A contrasting viewpoint is that theory and computation can lead the way in nanostructure design and synthesis because experimental tools often provide an incomplete picture of the structure and function of nanomaterials [8]. In this case, theory can be thought of as a way to fill in missing information and features critical to understanding experimental characterization results. However, as seen in the case of the strength of carbon nanotubes in Figure 10.7, it is not yet possible to fully understand experimental results at the nanoscale or reach conversion of experimental data and theory. Perhaps modeling and simulation could aid in elucidation of the origin of the experimental scatter. Again, it is important to recognize that the role of the modeling is to act as a guide for the designer.

As size scales decrease and experimental methods of evaluating different production methods become increasingly difficult, modeling becomes an important tool for optimizing production and fabrication. As a part of the design process depicted in Figure 10.4, process modeling is a useful and necessary tool and will become increasingly critical as system components decrease in size to the nanoscale. For example, this type of process modeling has been used to predict initial fabrication parameters to enable design of a specific microfluidic microlens profile [47]. In another case, thermomechanical modeling was used to optimize the nano-imprint forming process for different materials used in components such as mini-fluidics and biochemical systems, opto-electronics, photonics, and health usage monitoring systems [48]. Using modeling and simulation to optimize these processes prior to beginning fabrication can save time and lead to better-quality production. Linked with modeling of performance and reliability, the entire lifetime of a system could potentially be predicted computationally.

Use of nanoscale components introduces new challenges for reliability given increasing complexity of materials, interfaces, and failure mechanisms. Miniaturization leads to materials that have size-dependent properties including the occurrence of failure in smaller geometries [49]. Due to this added complexity, reliability must be approached from multiple scales and approaches, as depicted in Figure 10.15, especially because each component's individual properties, requirements, and failure mechanisms must be considered. Additionally, nanospecific simulation methods must be utilized to capture structure-property relations at the relevant scale where failure may occur. In their work on lifetime modeling, Wunderle and Michel [49] described the emerging field of "nano-reliability" as one that will "encompass research on the properties and failure behavior of materials and material interfaces under explicit consideration of their micro- and nano-structure and the effects hereby induced" (p. 799). These authors believe that a coupled experimental and simulative approach is necessary to develop and use nanoscale components in systems and to ensure reliability of these small-scale parts. Further, nanoreliability will use modeling to explain failures from first principles. It is evident that with the error inherent in both experiments and theory with nanoscale components, modeling reliability is a particular challenge that has yet to be addressed.

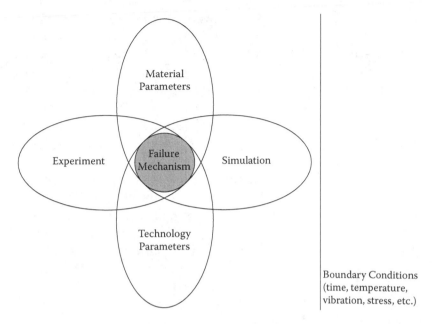

Figure 10.15 A physics-of-failure lifetime reliability modeling approach. (See Wunderle, B., and Michel, B., *Microsystem Technologies*, 15(6), 799–812, 2009.)

Prospects for the Future

Projected advances in modeling at the nanoscale and projected applications of various modeling methods, as they mature, are given in tables 4.1 through 4.5 in the publication entitled, *"Productive Systems: A Technology Roadmap"* [46]. A wide range of projected applications is presented, but many challenges are still ahead for the nanoscale modeling community. As modeling of nanoscale phenomena continues to develop, "the key is to understand (a) precisely what is the scientific objective of the calculation, (b) precisely what is being computed, and (c) when can the computation provide reasonable results" (p. 302) [6]. Existing modeling tools can be quite useful if restricted to their intended purpose. However, none are broadly applicable for nanoscale modeling; thus, a combination of many modeling tools or development of new tools is required. A new set of numerical techniques is necessary for addressing different scales in the same simulation in order to handle many realistic multiscale problems [28]. For some applications, it may also become necessary to design a truly multiscale code (i.e., one that integrates multiple codes developed to address different length scales and time scales) [50].

As discussed in Chapter 7, the four generations of nanotechnology have been defined by Roco [51] as passive nanostructures, active nanostructures, three-dimensional nanosystems and systems of nanosystems, and heterogeneous molecular nanosystems. Many technological challenges exist as we attempt to progress through these generations, and modeling and simulation are among these challenges. Currently available modeling and simulation tools lack the accuracy necessary to be relied upon for prediction and device behavior; nonetheless, they are incredibly useful for learning about phenomena at the nanoscale. As computational capabilities continue to advance, modeling and simulation also progress, and calculations that were once impossible are becoming feasible. As time goes on, we will continue to learn more about nanoscale phenomena and how to implement multiscale models relevant to design and behavior prediction for nanosystems. Experimental and modeling capabilities will drive each other to advance as we learn from both. Modeling and simulation will likely provide much of the fundamental understanding necessary to progress through the generations of nanotechnology.

References

1. Kossiakoff, A., and W. Sweet, *Systems Engineering Principles and Practice*, ed. A. Sage. 2003, Hoboken: Wiley Interscience.
2. Ajji, A. et al., Amorphous orientation of poly(ethylene terephthalate) by X-ray diffraction in combination with Fourier transform infra-red spectroscopy. *Polymer*, 1995. 36(21): 4023–4030.
3. Gray, G.I., Predictive capability for inverting information flow from performance to structure to processing: An evolving paradigm shift in MSE. *Journal of Materials*, 2010. 62(3): 9–10.

4. Senturia, S.D., *Microsystem Design*. 2001, Boston: Kluwer Academic.
5. Harutyunyan, A.R. et al., Preferential growth of single-walled carbon nanotubes with metallic conductivity. *Science*, 2009. 326(5949): 116–120.
6. Ramesh, K., *Nanomaterials: Mechanics and Mechanisms*. 2009, Dordrecht: Springer.
7. Esfarjani, K., and G. Mansoori, Statistical mechanical modeling and its application to nanosystems, in *Handbook of Theoretical and Computational Nanotechnology*, eds. M. Rieth and W. Schommers. 2006, Valencia, CA: American Scientific.
8. Schatz, G.C., Using theory and computation to model nanoscale properties. *Proceedings of the National Academy of Science*, 2007. 104(17): 6885–6892.
9. Theory and Modeling in Nanoscience. Report of the Workshop Conducted by the Basic Energy Sciences and Advanced Scientific Computing Advisory Committees to the Office of Science, Department of Energy. 2002.
10. Delerue, C., and M. Lannoo, *Nanostructures Theory and Modeling*. 2004, Berlin: Springer.
11. Welty, J., C. Wicks, and R. Wilson, *Fundamentals of Momentum, Heat and Mass Transfer*. 3rd ed. 1984, New York: Wiley & Sons.
12. White, F., *Viscous Fluid Flow*. 3rd ed. 2006, Boston: McGraw-Hill.
13. Pathria, R., *Statistical Mechanics*. 2nd ed. 1996, Oxford: Elsevier.
14. White, F., *Fluid Mechanics*. 5th ed. 2003, Boston: McGraw-Hill.
15. Shen, C., *Rarefied Gas Dynamics*. 2nd ed. 2005, Berlin: Elsevier.
16. Thomas, J.A., Water flow and thermal transport through carbon nanotubes, PhD Thesis, Pittsburgh, PA: Carnegie Mellon University.
17. McQuarrie, D., and J. Simon, *Physical Chemistry: A Molecular Approach*. 1997, Herndon, VA: University Science Books.
18. Martin, R., *Electronic Structure: Basic Theory and Practical Methods*. 2004, Cambridge: Cambridge University Press.
19. Jensen, F., *Introduction to Computational Chemistry*. 2006, New York: Wiley and Sons.
20. Foresman, J., and A. Frish, *Exploring Chemistry with Electronic Structure Methods*. 1996, Pittsburgh, PA: Gaussian.
21. Allen, M., and D. Tildesley, *Computer Simulation of Liquids*. 1987, Oxford: Clarendon Press.
22. Rapaport, D., *The Art of Molecular Dynamics Simulation*. 2004, Cambridge: Cambridge University Press.

23. Frenkel, D., and B. Smit, *Understanding Molecular Simulation: From Algorithms to Applications*. 2nd ed. 2002, San Diego: Elsevier.
24. Binder, K., and D. Heermann, *Monte Carlo Simulation in Statistical Physics*. 2010, Berlin, Heidelberg: Springer.
25. Modest, M., *Radiative Heat Transfer*. 2nd ed. 2003, San Diego: Academic Press.
26. Sacci, S., and S. Succi, *The Lattice Boltzmann Equation for Fluid Dynamics and Beyond*. 2001, New York: Oxford University Press.
27. Sukop, M., and D. Thorne, *Lattice Boltzmann Modeling: An Introduction for Geoscientists and Engineers*. 2006, New York: Springer.
28. Liu, S. et al. *Several Modeling Issues in LED, 3D-SiP, and Nano Interconnects*. 2009, Delft, Netherlands: IEEE Computer Society.
29. Mize, S., *Toward Nanomaterials by Design: A Rational Approach for Reaping Benefits in the Short and Long Term*. 2004, Palo Alto, CA: Foresight Institute.
30. Elhard, J., *Productive Nanosystems: Multi-Scale Modeling and Simulation, in Productive Nanosystems: A Technology Roadmap*, K. Eric Drexler et al., eds. Columbus, OH: Battel Memorial Institute. *Tech Report* 26, p. 1, 2007.
31. Di Carlo, A. et al. *Modelling of Carbon Nanotube-Based Devices: From nanoFETs to THz Emitters*. 2006, San Diego, CA: SPIE.
32. Sangiorgi, E. et al., The Monte Carlo approach to transport modeling in deca-nanometer MOSFETs. *Solid-State Electronics*, 2008. 52(9): 1414–1423.
33. Darve, E., S. Li, and Y. Teslyar, *Calculating Transport Properties of Nanodevices*. 2004. Philadelphia, PA: SPIE.
34. Lundstrom, M., and Z. Ren, Essential physics of carrier transport in nanoscale MOSFETs. *IEEE Transactions on Electron Devices*, 2002. 49(1): 133–141.
35. Sanyal, B., and O. Eriksson, *Ab-initio* modeling of spintronic materials. *Physica Status Solidi (A) Applications and Materials*, 2007. 204(1): 33–43.
36. Wang, C. et al., Modeling multiterminal spintronic devices. *IEEE Transactions on Nanotechnology*, 2007. 6(3): 309–315.
37. Saikin, S., Y.V. Pershin, and V. Privman, Modelling for semiconductor spintronics. *IEE Proceedings: Circuits, Devices and Systems*, 2005. 152(4): 366–376.
38. Nikonov, D.E., and G.I. Bourianoff, Operation and modeling of semiconductor spintronics computing devices. *Journal of Superconductivity and Novel Magnetism*, 2008. 21(8): 479–493.

39. Zhou, S.-A., and M. Uesaka, Modeling of transport phenomena of nano-particles in biological media in electromagnetic fields. *International Journal of Applied Electromagnetics and Mechanics*, 2006. 24(1): 51–67.

40. Bohn, P.W., Nanoscale control and manipulation of molecular transport in chemical analysis. *Annual Review of Analytical Chemistry*, 2009. 2(1): 279–296.

41. Datta, S., and B. Das, Electronic analog of the electro-optic modulator. *Applied Physics Letters*, 1990. 56(7): 665–667.

42. Ogata, S., Y. Umeno, and M. Kohyama, First-principles approaches to intrinsic strength and deformation of materials: Perfect crystals, nano-structures, surfaces and interfaces. *Modelling and Simulation in Materials Science and Engineering*, 2009. 17(1).

43. Tsonchev, S. et al., All-atom numerical studies of self-assembly of zwitterionic peptide amphiphiles. *Journal of Physical Chemistry B*, 2004. 108(39): 15278–15284.

44. Law, M.E., Jones, K.S., Radic, L., Crosby, R., Clark, M., Gable, K., and Ross, C., 2004, "Process Modeling for Advanced Devices," MRS Online Proceedings Library, 810.

45. Iannaccone, G., Perspectives and challenges in nanoscale device modeling. *Microelectronics Journal*, 2005. 36(7): 614–618.

46. Drexler, K.E. et al. (eds.), *Productive Nanosystems: A Technology Roadmap*, Columbus, OH: Batelle Memorial Institute and Foresight Nanotech Institute, 2007.

47. O'Neill, F.T., C.R. Walsh, and J.T. Sheridan. *Microfluidic Microlenses for Nano-Technology Applications: Production, Modeling and Fabrication Techniques*. 2004. Denver, CO: SPIE.

48. Stoyanov, S. et al. *Modelling the Nano-Imprint Forming Process for the Production of Miniaturised 3D Structures*. 2008. Freiburg im Breisgau, Germany: IEEE Computer Society.

49. Wunderle, B., and B. Michel, Lifetime modelling for microsystems integration: From nano to systems. *Microsystem Technologies*, 2009. 15(6): 799–812.

50. Maiti, A., Multiscale modeling with carbon nanotubes. *Microelectronics Journal*, 2008. 39(2): 208–221.

51. Roco, M., Nanoscale science and engineering: Unifying and transforming tools. *AIChE Journal*, 2004. 50(5): 890–897.

11

Interfaces at the Micro- and Nanoscale

Jennifer Breidenich

Hazards of blind interfacing.

Contents

Introduction

Special attention to interfaces is imperative when engineering micro- and nanotechnology (MNT) systems. At the micro- and nanoscale, minute changes in processing procedures have large effects on nanomaterial structure and properties. As a result, working at the micro- and nanoscale sizes dictates that the interactions between components are difficult to predict reliably. In order to solve this problem from a systems engineering point of view, MNT interfaces must be accounted for during all phases of the design, development, and implementation process.

This chapter will discuss how to address the characterization and modeling challenges that arise when working in the micro- and nanoscale. A review of fundamental systems engineering (FSE) concepts related to interfaces indicates that the problems posed by the unique properties of systems at the nano- and microscale can be approached using FSE methodologies during the design stage. Stand-alone MNTs, such as lab on a chip and the accelerometer, will be used throughout the chapter to further illustrate the importance of implementing systems engineering concepts during the early planning stages of technology development. It will be shown that properly accounting for interfaces during the design stage can rapidly advance the sophistication of these systems by allowing for the early integration of self-regulation and feedback components that will monitor the interface throughout the life cycle of the technology. Further, as one moves from stand-alone MNTs to a multiscale system, complexity increases.

This chapter will summarize fundamental systems engineering concepts related to interface controls and adapt those concepts to MNTs while addressing the challenges related to MNT interfaces.

Background on Interfaces in Fundamental Systems Engineering (FSE)

Properties of Interfaces

An interface has both functional and physical characteristics and exists at a common boundary between two or more systems, products, or components. The functional characteristics of an interface relate to actions that the interface carries out, such as connecting one component to another, isolating a component, and converting information. The physical characteristics describe the actual electrical, mechanical, and hydraulic machinery needed to complete the functional allocation of the interface (Table 11.1).

In FSE, interfaces are discrete, well defined, and predictable. When systems are designed, their connections are considered simultaneously with the design of the components.

TABLE 11.1
Examples of Interface Elements

Type	Electrical	Mechanical	Hydraulic	Man/Machine
Interaction Medium	Current	Force	Fluid	Information
Connectors	Connector Cable Switch	Joint Coupling	Pipe Valve	Display Control Panel
Isolator	RF shield Insulator	Bearing Shock Mount	Seal	Cover Window
Converter	Antennae A/D Converter	Gear Train Piston	Reducing Valve Pump	Keyboard

Source: Kossiakoff, A., and W.N. Sweet, *Systems Engineering: Principles and Practices.* 2003, Hoboken, NJ: John Wiley & Sons. Reprinted with permission of John Wiley & Sons, Inc.

A "golden rule" of systems engineering is that when connectivity is huge, it should be hidden away, encapsulated by that system element [2]. The interface is treated as a part of each component and rarely as its own entity.

Interfaces in FSE Design

In FSE, interfaces are identified and managed during the design stage. At the beginning of the design stage, the system boundaries are defined, and external interfaces are identified. Once the big picture inputs and outputs of the system have been outlined, the components of the system needed to complete the overall functions of the system are identified, and the internal interfaces of those components are defined. Next, each interface is analyzed in depth so that the functional and physical allocations can be accounted for at each interface. Once the functional and physical allocations of each interface are managed and documented, the interface portion of the design is essentially locked in or frozen.

Although the interface is actively defined in both functional and physical terms during the design stage, the definition becomes fixed and is rarely altered following the design phase. Once defined and designed, the interface is seen as event driven. The interface is the train track that moves the passenger from one city to another, serving as merely a means to move between actions or events, while the medium passing through (an electrical current, fluid, force, or information) is the dynamic portion that drives system events. In FSE, the interface does not change, and actions that take place across it are accurate and repeatable.

This view of frozen interfaces is not acceptable when working at the micro- and nanoscale. Due to their scale, MNT systems and system components have properties that are unpredictable, leading to a need for a dynamic interface that responds to the changing needs of the system. For a more seamless integration of components, the MNT system interface should be treated as a stand-alone component or a subsystem. The design parameters for the interface should be set up and treated as one would treat the rest of the system. The interface should be open to changing requirements throughout the entire life cycle of the system to account for the dynamic properties of the MNT system.

Characterizing Micro- and Nanotechnology (MNT) System Interfaces

Components and internal interfaces of MNT systems are in the micro- and nanoscale. Lack of control over material synthesis at these size scales makes MNT interfaces difficult to fabricate repeatedly and test individually. Although many analysis techniques of nanostructured materials exist in the literature and in practice, industry standardized characterization techniques and modeling theory related to nanomaterials are lacking. Additionally, MNT systems face the challenge of interfacing across several scales as they externally interface with the macroworld to receive and transmit information.

Analytical Techniques for Nanoscale Characterization

The nanostructure of a material affects the mechanical, thermal, electrical, and chemical properties of the material. Tiny

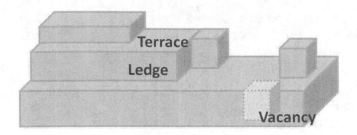

Figure 11.1 Defects that may occur on a solid surface and could affect material properties and behavior.

manipulations at the nanoscale have a profound effect on material properties. On solid surfaces, several types of defects with different dimensionalities can form as a result of processing temperature, humidity, time, and so forth. Vacancies of atoms form point defects, and line defects form ledges in the material at the atomic level (Figure 11.1). These ledges form steps that have interesting electronic properties that change with the material type. A step orientation that is different when compared with the adjacent planar (terrace) region will cause different chemical bonds to be exposed and affect the material properties. Specifically, in semiconductors with strong covalent bonds, the steps modify the electronic energy levels of atoms in close proximity to the defect, and in metals, the steps can cause a formation of dipole moments [3]. Although defects do not as commonly affect the properties of soft materials, their surface properties change with processing techniques and still have an effect on the functionality of the nanomaterial.

The American Society for Testing and Materials (ASTM) is an international standards organization that is often looked to for an accepted industry-wide method of characterizing a material. Yet, standard ASTM methods have not been developed and cannot be readily applied to test and characterize the system interfaces at these size scales. The first ASTM standard related to nanotechnology, ASTM E2456-06, was published in 2006 on the Terminology of Nanotechnology. Since then, standards from the ASTM Committee on Nanotechnology have been developed to standardize the analysis of nanoparticle size (ASTM E2490-09) and toxicity (ASTM E2526), but less progress has been made to standardize interface property characterization.

The literature illustrates many advances in nanomaterial characterization, but many steps need to be taken to achieve standardization of nanomaterial interfaces. Although industry-wide nanostructure characterization standards are not currently in place, there are several analytical methods used to determine the physical properties and image atomic defects of a nanostructured interface. The most common methods used to characterize properties at nanoscale interfaces are scanning probe microscopy (SPM), scanning electron microscopy (SEM), transmission electron microscopy (TEM), and x-ray diffraction (XRD).

Scanning probe microscopy is a class of techniques that investigates the interactions between a tip (probe) and the material of interest. During the experiment, the probe is scanned over the material of interest, and an image representative of the surface of the material is generated. Electron tunneling interactions are measured using scanning tunneling microscopy. Scanning the probe over the surface provides information on the roughness and general morphology of the material at the atomic level [4].

Another SPM technique, atomic force microscopy (AFM), uses the contact forces between the probe and material to characterize the material. AFM has been shown in the literature to provide information about nanomaterial properties, including electric properties, magnetic properties, surface properties, and adhesion. For example, Nguyen et al. at the National Institute of Standards and Technology (NIST) characterized the nanoscale surface coating degradation of car paints using AFM [5]. Mendez-Vilas quantified the adhesive strength of bacteria by attaching a *Staphylococcus epidermidis* bacterial cell to their AFM tip and analyzing the force between the bacteria-coated AFM tip and an adhesive substrate [6]. Olrich et al. used AFM to electrically characterize their silicon dioxide coatings on a nanometer-length scale by measuring tunneling currents after applying a voltage between the AFM tip and the silicon substrate [7].

Scanning electron microscopy and transmission electron microscopy both give information on the surface of materials. SEM use an electron beam to scan the surface of a material and collect an image, and TEM uses transmission of the

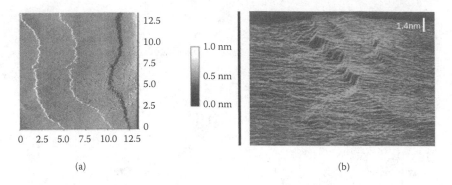

Figure 11.2 Examples of nanostructured surface characterization techniques of zinc and zinc oxide that are studied for the effect of nanostructure on bulk electronic properties. (a) Atomic force microscopy (AFM) image of terraced planes on zinc. (See Sziraki, L. et al., *Electrochimica Acta*, 46, 3743–3754, 2001.) (b) Scanning tunneling microscopy (STM) of polycrystalline zinc oxide surfaces. (See Rohrer, G.S., and Bonnell, D.A., *J. Am. Ceram. Soc.*, 73(10), 3026–3032, 1990.)

electron beam through the sample to generate an image. TEM is a much higher resolution characterization method and can sometimes be used to distinguish individual atomic layers. Figure 11.2 shows examples of SEM and TEM imaging.

X-ray diffraction is an analytical technique that relies on the interaction between x-rays and crystalline structure through diffraction. Although x-ray diffraction does not result in images of the material surface, it does provide information on nanomaterial phase composition, lattice strain, crystallite size, and crystallographic orientation.

Modeling Nanoscale Interfaces

Although characterization techniques help to elucidate the structure of nanomaterials, these methods are expensive, time consuming, and often do not give a direct measurement of the nano-interface. This lack of standard experimental methods for nanoscale systems drives the need for new theory and simulation in these systems. More extensive theory is necessary to provide the information needed to get better quantitative data from experimental techniques.

Braatz summarized the challenges of modeling molecular and multiscale systems as being [8]

- Uncertainties in physicochemical mechanisms
- Dynamically coupled model structures and high computational costs for model simulation
- A lack of online measurements at the molecular scale
- A lack of controllable variables during processing

Nano-interfaces are highly complex, heterogeneous in shape and substance, and often composed of dissimilar classes of materials [9]. Nanodimensional features alter the interface properties making the physiochemical mechanisms at the interface very difficult to define. Interfacial properties can dominate at the nanoscale because of the high surface area to volume ratio; hence, it must be highly understood. Theory and simulation approaches to understand and predict nano-interface behavior are crucial to device design. Further, simulation of interfaces between hard and soft materials play a critical role in biomolecular nanomaterials and associated applications.

The interface properties also have a profound effect on failure mode, which makes nanosystem reliability difficult to predict. The failure mode, most often solder fatigue, interface delamination, wire bond fatigue, cracking, fracture, and so forth, is dependent on the structure and processing conditions of the nanomaterial. The lack of controlled variables during processing and the lack of methods to characterize the structure illustrates the need for theory and modeling advances that will allow for the prediction of the lifetime of the nanoscale components [10]. See Chapter 10 for further discussion on the challenges of modeling and simulation of micro- and nanotechnologies.

The lack of online measurements and feedback mechanisms within nanoscale systems also contributes to the inability to gauge nanoreliability and lifetime. Methodologies that relate to the control of MNT systems have yet to be completely established. Stand-alone MNT systems illustrate the need for systems engineering at the conceptual stage. Their size lends itself to little or no flexibility in making physical design changes following fabrication. The interfaces of nanomaterials can play a key role in providing feedback and diagnostic data and should be incorporated into the design from the onset to increase the robustness of the system.

Examples of Stand-Alone MNT Systems

Stand-alone MNT systems stay within the micro- and nanoscale. Therefore, they have macroscale inputs and outputs; however, all components and interfaces reside in the micro- and nanoscale. The accelerometer and lab on a chip (LOAC) are two prominent examples of stand-alone MNT systems. The interfaces are difficult to fabricate repeatedly and test individually because their components remain at the nano- and microscale. The accelerometer and LOAC will be used as examples throughout the next section of the chapter to illustrate some of the challenges of designing MNT systems.

The Micro Electro Mechanical Systems (MEMS) accelerometer was first demonstrated in 1979 at Stanford University [11], and lab on a chip was first developed in 1975. Although these microscale systems have been around for decades, they are in very different development stages. The accelerometer and LOAC will be used throughout this chapter to illustrate the application of FSE principles to MNTs.

Accelerometer

The accelerometer measures acceleration. It is commonly used in many technologies ranging from medical applications to building technologies to measure vibrations and changes in acceleration. One of the most common uses is in automobiles to sense vibrations resulting from impact.

An accelerometer behaves as a damped mass on a spring. As a MEMS device, the accelerometer consists of a cantilever beam that acts as a spring and a proof mass (seismic mass). Once the device is accelerated, the proof mass is displaced from its neutral position. The displacement is measured by analog or digitally. Often, the capacitance difference between the fixed beams and the beams attached to the proof mass is measured to quantify acceleration. An alternative method is to integrate piezoresistors into the beam to detect displacement. Another, far less common, type of MEMS-based accelerometer contains a small heater at the bottom of a very small dome, which heats the air inside the dome to cause it to rise. A thermocouple on the dome determines where the heated air reaches the dome, and the deflection off the center is a measure of the acceleration applied to the sensor.

Lab on a Chip

Lab-on-a-chip (LOAC) devices employ nano- and microfluidics for biomedical applications. LOAC systems can be used for clinical analysis, DNA analysis, immunoassays, toxicity monitoring, and forensic analysis applications. There is great potential for using LOAC technology for point-of-care diagnostics, especially in developing countries. LOAC is small, relatively cheap, and requires little technical skill or machinery to operate, making it a very sustainable option in developing countries.

The design of lab-on-a-chip devices is very diverse. Yet, generally they require input of a biological fluid. The fluid is transported through a microfluidic channel that holds several wells of reacting chemicals intended to trigger some sort of response from the biological fluid. The response of the biological fluid is detected by an instrument or human and can provide information on the health of the patient.

Designing MNT Interfaces: Adapting Macro-Interface Design to Fit MNTs

Micro- and nanoscale interface design and control principles cannot be directly transcribed from fundamental systems engineering due to the dynamic nature of the interface at small size scales. Yet, fundamental systems engineering principles that dictate interface treatment can be built upon to devise a plan for the treatment of MNT system interfaces. There are four basic practices from macro-interface control that can be adapted and applied to MNTs:

1. Define system boundaries
2. Identify the internal and external interfaces

3. Map out the functional and physical allocations at each interface
4. Manage the interfaces and incorporate necessary feedback mechanisms

In order to efficiently manage the system interfaces, the central activity of the system must first be identified and always remain the top priority as each component of the system is broken down and planned. The central activity of the system dictates the design requirements of the system. Design decisions regarding interfaces should be consistent with what directly or indirectly contributes to the central activity of the system.

Define System Boundaries

The initial step to developing interfaces, whether one is working with a macrosystem, a stand-alone MNT system, or a MNT system that must interface with other MNT systems, a larger system, the environment, and humans is to define the system boundaries. Defining the system boundaries involves a big picture analysis of what the system will do, what interactions need to take place to further the central activity of the system, and what uncontrolled interactions automatically take place that need to be considered.

The central activity or purpose of the system must be established to define the system boundaries. Specifically, defining system boundaries involves identifying both elements that are under the control of the central activity of the system as well as elements that are outside of the control of the central activity of the system but interact with the system at a higher level. As shown in Figure 11.3, outputs, inputs, enablers, controls, and activities must all be mapped to accurately define system boundaries [12].

For example, when identifying system boundaries of an MNT system, such as an accelerometer that resides in an automobile, the primary input is going to be the acceleration frequency of the automobile. The accelerometer needs to be designed to be sensitive to large changes in the frequency of the automobile and respond allowing the device to perform its

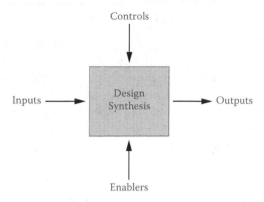

Figure 11.3 Defining the system boundaries not only includes inputs and outputs but also system controls and system enablers. (See College, D.S.M., ed., *Systems Engineering Fundamentals,* Defense Acquisition University Press, Fort Belvoir, VA, 2001.)

central activity. Yet, the packaging must be sturdy enough to disregard the normal operating frequency of the car and count it as background noise. These overall frequency requirements should be outlined in the initial design step.

Lab-on-a-chip design faces more of the problems associated with MNT systems. The primary LOAC input is a nanoliter volume of biological fluid. In addition to the difficulty involved in measuring out that volume of fluid accurately, the samples require pretreatment including the sampling procedure, extraction, filtration, preconcentration, and dilution just to be initially interfaced with the LOAC [13]. Some groups have centered their entire LOAC design on the goal of minimizing the sample preparation steps [14], highlighting the complexity of the input interface for LOAC. Equally complex is the output for lab on a chip. This nano-/macroscale device needs to be capable of outputting information that can be detected reliably by humans. Signal detection by humans has been completed by methods such as fluorescence, absorbance, and conductance [13]. These methods have worked but often require that the LOAC be paired with another, macroscale system creating a multiscale system. Factors such as the environment and other system enablers and controllers also need to be taken into account when defining the boundaries of the LOAC before identifying the interfaces.

Identify Internal/External Interfaces

Identifying internal and external interfaces takes defining system boundaries one step further, looking specifically at each interface and its functional allocation. Internal interfaces are those that address components and subsystems inside the boundaries established for the system addressed. These interfaces are generally identified and controlled by the designer responsible for developing the system. External interfaces, on the other hand, are those that involve relationships outside the established boundaries and are often uncontrollable [12].

In FSE the central function of the system can be broken down into subfunctions assigned to each component or set of components. Once subfunctions and components are identified, the functional allocation of each interface is determined. During this step, tools such as a schematic block diagram (SBD) and a work breakdown structure (WBS) can be used to map out each component and its function. The SBD and WBS can help to organize the components and sets of components and can aid in minimizing the amount of interfaces necessary in a system by structuring the functional allocations in the most efficient manner (functional partitioning). When identifying the internal and external interfaces, the performance design requirements of the system must constantly be revisited and then translated to each interface within the system.

While the SBD and the WBS serve as static representations of the system, a functional flow block diagram (FFBD) can be used to identify the dynamic actions that the system needs to take to perform its primary function. A FFBD illustrates the responses that the system will initiate when it is faced with specific environmental, internal, and external stimuli. In order to develop a system properly, it is necessary to sketch a FFBD to model the system functions, how they are derived, and how they are related to one another [1].

Special attention needs to be paid to the physical environment that the device will reside in as well as the packaging material of the device, because the accelerometer external interface is the entire device. These two factors will play a key role in determining the interactions at the interface. Internal interfaces will include the interface between the physical architecture of the seismic mass, the resonating beam, and the piezoresistors and other circuitry needed to measure the frequency change.

The LOAC external interfaces would include the interface between capillary or syringe needle needed to inject the biological fluid and the LOAC as well as the LOAC and amplification device interface needed to allow the signal to be detected by humans. Internal interfaces as simple as the microfluidic well to biological fluid interface also need to be defined. Because properties change so significantly at the nanoscale, all interfaces where any interaction takes place need to be mapped out and well characterized.

Map Out Functional and Physical Allocations at Each Interface

Once interfaces have been identified, their functional and physical allocations can be defined. The functional allocation of an interface is what the interface actually does, such as transferring information, providing power, and giving feedback, whereas the physical allocation is the actual infrastructure of the interface, such as a wire, a microfluidic channel, an adhesive, and so forth. This step requires a quantitative evaluation of the inputs and outputs (see Figure 11.4) at each interface.

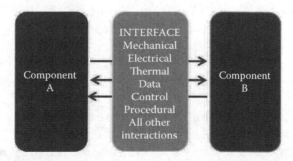

Figure 11.4 Interfaces at the nanoscale should be treated with equal consideration as the components of the system throughout the entire design and development process.

In FSE, interfaces are given the same consideration as components during the design process but are often frozen once the design is complete. For MNT systems, interfaces should remain a component of the design throughout the entire process. Their trade space should remain large to account for the dynamic nature of micro- and nanoscale systems.

Within large systems, failures often occur at the interfaces. Minimizing the number of interfaces decreases the probability of failure and can minimize the disruption of the central activity of the system. The practice of configuration management is often used to systematically develop large-scale systems with multiple levels of organization and control the interfaces. A "configuration" consists of the functional, physical, and interface characteristics of existing or planned hardware, firmware, software, or a combination thereof as set forth in technical documentation and ultimately achieved in a product [12]. Configuration management programs often include guidance on topics including the following:

- Confirming that designs are traceable to requirements
- Controlling and documenting change
- Defining and understanding interfaces
- Ensuring consistency between products and supporting documentation

Configuration management plans focus on providing documentation describing what is supposed to be produced, what is being produced, what has been produced, and what modifications have been made to what was produced [12]. Although a configuration management program is not required when dealing with MNTs, we can apply configuration management to MNTs when integrating them into larger systems and when designing an MNT with multiple functions.

One aspect of configuration management that directly relates to all systems, including MNTs, is interface management. Interface management encompasses identifying, managing, controlling, and documenting interfaces. Interface management "identifies, develops, and maintains the external and internal interfaces necessary for system operation. It supports the configuration management effort by ensuring that

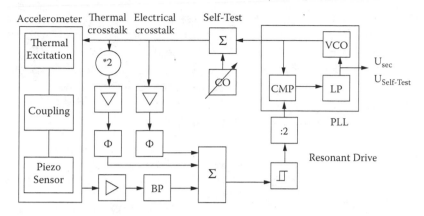

Figure 11.5 An example of mapping the functional allocations of a micro- and nanoscale technology (MNT) system. A depiction of the electronic circuitry of the resonant sensor including its self-test in an accelerometer. (See Aikele, M. et al., *Sensors and Actuators A*, 92(1–3), 161–167, 2001.)

configuration decisions are made with full understanding of their impact outside of the area of the change" [12 (p. 97)].

The practice of mapping out functional and physical allocations of an interface can be applied to even to simple stand-alone MNTs such as the accelerometer (see Figure 11.5). The physical allocations of interfaces in most accelerometers are the silicon connections between the resonant beam and the piezoresistors. The functional allocations would include the maximum and minimum electrical conductance and resistance that can travel across those connectors. Material properties and frequency response properties of the external interface also need to be characterized.

The LOAC functional and physical interactions are more intricate because several types of materials play a role to allow the LOAC to work. Interactions such as mixing between the biological fluid and the chemical reaction media within the LOAC need to be considered. Physical interaction between LOAC channel material and the biological fluid needs to be analyzed and characterized according to the channel material properties. For example, in even more complex, multilayer biochips, thermoplastic fusion bonding is key to constructing interfaces needed for design. Ahn looked at bond strength as a function of surface contact angle at these interfaces and showed that bond strength increases with hydrophilicity and that adhesion changes with both microscopic roughening and surface chemistry [15]. Characterization of the interfaces needs to be extremely in depth to be able to accurately predict interactions. The LOAC physical and functional interfaces illustrate that need to treat the interfaces and connectors as their own components throughout the entire process.

Manage the Interfaces and Incorporate Necessary Feedback Mechanisms to Determine System Reliability

Once interfaces have been identified and defined, they must be constantly managed and optimized so that they will continue to contribute to the central activity of the system. Their primary control mechanism needs to be self-regulation that can often be employed at the interfaces, because these MNT systems exist at the micro- and nanoscale and cannot be perturbed at the macroscale.

Stand-alone MNT systems require only one macroscale input and provide only one macroscale output, leaving little physical room in the system components to allow for real-time process monitoring sensors. These systems must have feedback mechanisms already designed into their infrastructure to inhibit the system's function once a system failure has been detected. For this reason, systems engineering concepts must be employed even during the conceptual stage of the design of a MNT system.

The accelerometer technology is at a stage where self-test mechanisms are employed absolving the need for a world to MNT interface to determine system reliability. Aikele et al. developed an accelerometer (Figure 11.6) that conducted an ongoing self-test based on the simultaneous excitation of the seismic mass and the resonating beam above the background frequency but below the triggering frequency [16]. The device would be able to self-diagnose any problems.

For LOAC technology, it remains difficult to determine reliability. Often long-term performance tests of microcomponents are conducted to determine reliability [15]. Once the device is out in the field and being used, the only way to assess reliability is to conduct a control test.

Figure 11.6 Scanning electron microscope (SEM) of the vibrating beam, the hinge, and the U-structure for excitation and detection. (See Aikele, M. et al., *Sensors and Actuators A*, 92(1–3), 161–167, 2001.)

Interfaces of Multiscale, Complex Systems

Many MNTs are not stand-alone systems. They are subsystems or components of larger, multiscale systems. A multiscale system spans size scales of several orders of magnitude. Interfaces within a multiscale system may take a variety of forms bridging the nanoscale components to the macroscale components or connecting components stepwise, through the scales, from nano to micro to meso to macro. Interfaces woven throughout multiscale systems are more difficult to define, characterize, and design than interfaces of stand-alone MNTs because the physical and functional interactions that occur between scales are more complex (Figure 11.7).

A multiscale system is often characterized as a complex system because of its hierarchical nature. In general terms, a complex system is nonlinear and dynamic. Complex systems are often characterized as nonequilibrium systems because

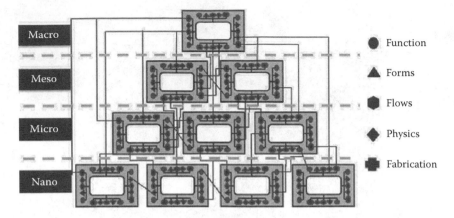

Figure 11.7 Scales and complexity of a multiscale system. (Image and concept courtesy of Prof. Martin Culpepper. Taken from his class notes for Multi-scale System Design, Massachusetts Institute of Technology. See Culpepper, M. *Lecture 3: Macro/Meso-Scales Components and Characteristics* 2004 [cited 2010; Available from: http://ocw.mit.edu/courses/mechanical-engineering/2-76-multi-scale-system-design-fall-2004/).

they change over time (Figure 11.8). The issues that arise when working with these multiscale, complex systems include the following [17]:

- Phenomena at different scales are difficult to correlate.
- The dominant mechanism is difficult to identify.
- Spatial and temporal structural changes cannot be coupled easily.
- Critical phenomena will occur in complex systems.

An averaging or reductionism approach cannot be taken when referring to multiscale, complex systems due to the many phenomena that occur at different size scales (Figure 11.9). A strategy used often by mathematicians and physicists to manage multiscale systems is "from the particular to the general." Each component is evaluated and characterized in depth individually, then the components are fitted into the system, and a more macroscale analysis is completed. This approach works well in conjunction with micro- and nanoscale systems that are often dynamic in nature and change over time and according to environmental factors [17].

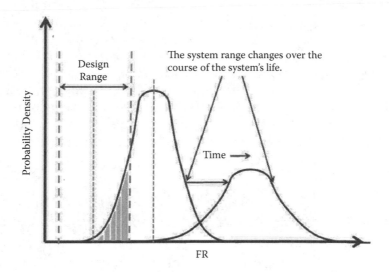

Figure 11.8 The effect of the dynamic nature of interfaces of micro- and nanotechnology—nonequilibrium systems change over the long term. (Adapted from Suh, N.P., *Axiomatic Design: Advances and Applications*, Oxford University Press, New York, 2001.)

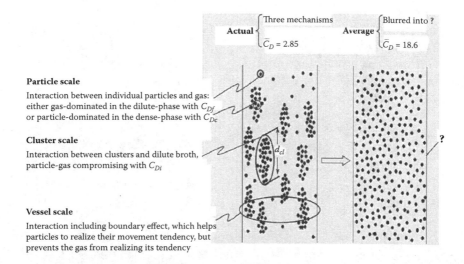

Figure 11.9 Dominant mechanisms change as scale changes, and the average approach does not account for this. (See Li, J.Z., *Chemical Engineering Science*, 1687–1700, 2004.)

The averaging approach also does not work for multiscale systems because it does not take into account the changes in interaction mechanisms as the scale changes. As one moves from scale to scale, the dominant mechanism needs to be identified in order to predict the behavior of the systems. According to Li, the formulation for identifying the dominant mechanism in each multiscale system is unique to that system and cannot be generalized across systems, emphasizing the need for the "particular to the general" strategy.

Future Challenges

When looking forward, the interface control definitions for the micro- and nanoscale world need to be reconstructed to address problems integrating MNTs into other systems. Nano- and macroscale interfaces are not well defined, physically or theoretically, making these systems unreliable and difficult to predict. Incorporating better control elements into the processing steps as well as feedback mechanisms into the design of MNT systems during the conceptual stage will serve to increase reliability and efficiency of the systems.

Improved process control and characterization of MNTs are needed to solve the problem of variability within the same process [18]. According to the IEEE Workshop on Control of Micro and Nano Systems, improved characterization is necessary to

- Clarify physical phenomena that are dominant at the micro- and nanoscale.
- Provide basic input to physically based models—a lot of basic fluid/solid properties are still largely unknown.
- Validate both physically based and reduced-order models.

Reliability of MNT systems could also be increased by the incorporation of sensors or feedback mechanisms into the physical architecture of the MNT systems. Sensors need to be designed into the MNT system to monitor the function and health of the MNT system, because these interfaces are dynamic and require a feedback mechanism that will adapt to changing conditions quickly. The IEEE Workshop on Control of Micro and Nano Systems identified several challenges of integrating these types of mechanisms into MNT systems, including the following:

- Incorporation of sensors within a very limited space
- Sensor accessibility (getting signals in and out)
- Limiting the addition of any sensor fabrication steps that may reduce system yield [18]

One core function of interface controls within systems engineering is to establish infrastructure that will allow for the monitoring and evaluation of the health of a given system. Systems analysis and control activities for macrosystems include establishing a work breakdown structure, determining the configuration of the system, auditing progress, employing trade studies, and measuring performance through metrics. All of these processes may not directly apply to micro- and nanotechnologies (MNTs), but interface control mechanisms still need to be integrated into MNT systems to allow the user to obtain feedback from the system.

The need for better process control and feedback mechanisms points to the utilization of a systems engineering approach early in the conceptual stage of the technology development process. Existing traditional systems engineering planning tools and definitions for interfaces need to be expanded to approach interfaces within MNTs successfully.

References

1. Kossiakoff, A., and W.N. Sweet, *Systems Engineering: Principles and Practices*. 2003, Hoboken, NJ: John Wiley & Sons.
2. Boardman, J., and B. Sauser, Systems of Systems—the meaning of. *Proceedings of the IEEE / SMC International Conference on Systems of Systems Engineering*, 2006: 118–123.
3. Luth, H., *Solid Surfaces, Interfaces, and Thin Films*. Graduate Texts in Physics, ed. W.T. Rhodes, H.E. Stanley, and R. Needs. 2010, Heidelberg, Germany: Springer.

4. Ramesh, K.T., *Nanomaterials: Mechanics and Mechanisms.* 2009, New York: Springer.
5. Nguyen, T., M. VanLandingham, R. Ryntz, D. Nguyen, and J. Martin, *Nanoscale Characterization of Coatings Surface Degradation with Tapping Mode AFM*, in *Adhesion Fundamentals: From Molecules to Mechanisms and Modeling. Proceedings, 26th Annual Meeting of the Adhesion Society.* February 23–26, 2003: Myrtle Beach, SC, 508–510.
6. Mendez-Vilas, A., A.M. Gallardo-Moreno, and M.L. González-Martín, Nanomechanical exploration of the surface and subsurface of hydrated cells of *Staphylococcus epidermidis. Antonie van Leeuwenhoek*, 2006: 89(3–4), 373–386.
7. Olrich, A.E., B. Ebersberger, and C. Boit, Conducting atomic force microscopy for nanoscale electrical characterization of thin SiO_2. *Applied Physics Letters,* 1998: 73(21).
8. Braatz, R.D., A multiscale systems approach to microelectronics. *Computers and Chemical Engineering*, 2006: 30(10–12), 1643–1656.
9. McCurdy, C.W., *Theory and Modeling in Nanoscience,* Department of Energy (DOE), Editor. 2002. Berkeley, CA: Lawrence Berkeley National Laboratory.
10. Wunderle, B., and B. Michel, Lifetime modeling for microsystems integration: From nano to systems. *Microsystems Technology*, 2009: 15(6), 799–812.
11. Maluf, N., *An Introduction to Microelectromechanical Systems Engineering.* 2000, Norwood, MA: Artech House.
12. College, D.S.M., ed. *Systems Engineering Fundamentals.* 2001, Defense Acquisition University Press: Fort Belvoir, VA.
13. Chin, C.D., V. Linder, and S.K. Sia, Lab-on-a-chip devices for global health: Past studies and future opportunities. *Lab Chip*, 2007: 7, 41–57.
14. Liu, J., C. Hansen, and S.R. Quake, Solving the "world-to-chip" interface problem with a microfluidic matrix. *Analytical Chemistry*, 2003: 75(18), 4718–4723.
15. Ahn, C.H. and Choi, J.-W. et al., Disposable smart lab on a chip for point of care clinical diagnostics. *Proceedings of the IEEE*, 2004: 154–173.
16. Aikele, M., et al., Resonant accelerometer with self-test. *Sensors and Actuators A*, 2001: 92(1–3), 161–167.
17. Li, J.Z., Multi-scale methodology for complex systems. *Chemical Engineering Science*, 2004: 59(8–9), 1687–1700.
18. Shapiro, B., *Control and System Integration of Micro- and Nano-Scale Systems.* 2004, National Science Foundation.

19. Sziraki, L. et al., Study of the initial stage of white rust formation on zinc single crystal by EIS, STM/AFM and SEM/EDS techniques. *Electrochimica Acta*, 2001: 46, 3743–3754.
20. Rohrer, G.S., and D.A. Bonnell, Electrical properties of individual zinc oxide grain boundaries determined by spatially resolved tunneling microscopy. *Journal of the American Ceramic Society*, 1990: 73(10), 3026–3032.
21. Culpepper, M., *Lecture 3: Macro / Meso-Scales Components and Characteristics* 2004 [cited 2010; Available from: http://ocw.mit.edu/courses/mechanical-engineering/2-76-multi-scale-system-design-fall-2004/.
22. Suh, N.P., *Axiomatic Design: Advances and Applications*, New York: Oxford University Press, 2001.

12
Systems Reliability

O. Manuel Uy

"When Erroneous meets Arrhenius
the results go nonlinear!"

Contents

Systems Reliability

Reliability is the probability of a unit surviving within its expected lifetime during normal use. When a unit has reached a period in which it has a constant failure rate, its reliability can be expressed by a simple exponential function, $e^{-\lambda t}$, where λ is the constant failure rate, and t is the operating time. This reliability function also applies to situations where the units are mixed in age and are replaced as they fail regardless of the cause of failures.

However, we really wish to evaluate the reliability of systems, which are often composed of complex arrangements of many units or components. To do this, we must have knowledge of the reliabilities of all of the components in the system. These component reliabilities are obtained from actual measurements that yield information about their respective failure rates. Thus, system reliability is based on the reliability of the individual components and the calculation of the reliability of the combination of these components. These calculations are carried out by basic yet exact probability rules.

Reliability and Failure Rate

Suppose that a large quantity of repairable devices, such as iPods, is observed over an extended period of time, and that records on the time of occurrence of each failure are maintained. From these data, a *bar chart* could be constructed showing the number of failures per hour per device during this operating period. Dividing the number of failures by the total number of devices under observation and the length of interval would yield the number of failures per hour per device during this interval. Typically, such a plot could be expected to appear as a *bathtub* curve, as shown in Figure 12.1.

If a smooth curve is drawn through the tops of the bars, the resulting curve is called the *failure rate* characteristic of the devices (more commonly known as the *hazard rate*). Because of the shape, it is also commonly called the bathtub curve. This curve logically divides into three age phases, which are identified in Figure 12.1. During the "infant mortality" phase, a decreasing failure rate is observed as the latent defects are

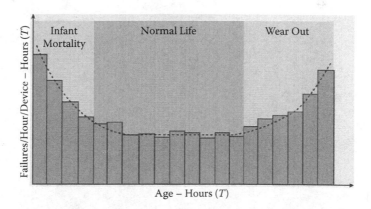

Figure 12.1 Failure per hour per device as a function of age in hours.

replaced with normal parts. During the "normal" phase, the failure rate remains relatively constant, and the failures are said to occur randomly (i.e., failures are equally likely to occur regardless of age). Finally, the failure rate increases with age, and this is referred to as the "wear-out" phase.

It would be expected that the failure rate, $h(T)$, for this device would be related to its reliability function. It can be shown that the reliability of the device is given by

$$R(t_1, t_2) = e^{-\int h(T)dT}$$

where T is the age of the device. The period starts at $T = t_1$ and ends at $T = t_2$, so that the operating time is $t = t_2 - t_1$. It is observed that the device reliability is a function of the area under the hazard curve during the interval t_1 to t_2 because $\int h(T)dT$ is the area under the hazard curve between t_1 and t_2. The above equation simplifies to

$$R(t_1, t_2) = e^{-A}$$

where A is the area under the curve.

This is shown in Figure 12.2. For a given operating (or mission) time, the reliability of the devices in general changes with the age of the device (i.e., the area, A) changes as the interval t_1–t_2 is moved along the T-axis. However, during the normal life phase when the failure rate is essentially constant

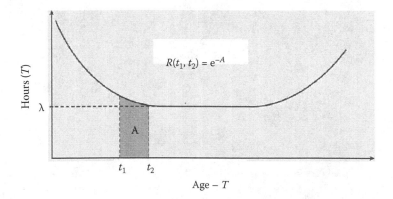

Figure 12.2 Area under the curve A between times t_1 and t_2.

over time, $h(T) = \lambda t$, the area A remains fixed for a given time interval. Thus, during this normal life phase, or constant failure rate phase, the reliability reduces to

$$R(t) = e^{-\lambda t}$$

This is the most commonly used model for reliability and is frequently called the *exponential reliability law*.

Basic Rules of Probability

Before going too much further, it is essential that we get into the basic rules of probabilities so that we can translate these rules into estimating reliabilities:

Rule 1: If A and B are two independent events with the probabilities $P(A)$ and $P(B)$, the probability that both events will occur is the product

$$P(AB) = P(A) * P(B)$$

Rule 2: If the two events can occur simultaneously, the probability that either A or B or both A and B will occur is

$$P(A \cup B) = P(A) + P(B) - P(A) * P(B)$$

Rule 3: If the two events are mutually exclusive such that when A occurs, B cannot occur, Rule 2 simplifies to

$$P(A \cup B) = P(A) + P(B)$$

Rule 4: If the two events A and B are complementary in addition to being mutually exclusive (i.e., if A does not occur, B must occur, and vice versa), we obtain

$$P(A) + P(B) = 1$$

The last rule is easily translated into the language of reliability because we normally designate reliability as R and its complement, unreliability, as Q such that

$$R + Q = 1$$

We can now expand the above probability rules to calculate the reliability of a system with two or more components. In these examples, we will use the case with components having a constant failure rate, which is $e^{-\lambda t}$ (more correctly known as the *Poisson distribution*).

Ancillary Rules of Reliability

Ancillary Rule 1: In a series connection, if a component has a reliability, R_1, and another component has a reliability, R_2, the probability that both components will survive an operating time t is

$$R_s(t) = R_1(t) * R_2(t) = \exp(-\lambda_1 dt) * \exp(-\lambda_2 dt) = \exp(-\Sigma \lambda dt)$$

where λ_1 and λ_2 are the failure rates of the components and can be a time variable or a constant.

Ancillary Rule 2: In a series connection, the probability that either one or both components will fail is

$$
\begin{aligned}
Q_s(t) &= Q_1(t) + Q_2(t) - Q_1(t) * Q_2(t) \\
&= [1 - R_1(t)] + [1 - R_2(t)] - [1 - R_1(t)] * [1 - R_2(t)] \\
&= 1 - R_1(t) * R_2(t) \\
&= 1 - R_s(t)
\end{aligned}
$$

Ancillary Rule 3: In a parallel connection, the probability that either one or both of the components will survive is

$$R_p(t) = R_1(t) + R_2(t) - R_1(t) * R_2(t)$$

and for the exponential failure rate,

$$R_p(t) = \exp(-\lambda_1 dt) + \exp(-\lambda_2 dt) - \exp(-\Sigma\lambda dt)$$

Ancillary Rule 4: The probability that both parallel components will fail is

$$\begin{aligned}
Q_p(t) = Q_1(t) * Q_2(t) &= [1 - R_1(t)] * [1 - R_2(t)] \\
&= 1 - R_1(t) - R_2(t) + R_1(t) * R_2(t) \\
&= 1 - R_p(t)
\end{aligned}$$

Ancillary cases 1 and 2 are complementary events, so $R + Q = 1$ because the complementary event to both components surviving is the event of both *not* surviving with the latter including three possibilities: one or the other fails or both fail. We call the R and Q for these two cases the reliability and unreliability, respectively, of a series connection.

Ancillary cases 3 and 4 are also complementary events such that $R + Q = 1$, because the complementary event to both components failing is that either one or both survive. We call the R and Q of these cases the reliability and unreliability, respectively, of parallel connections or a parallel redundant system. If one component fails, the other component will continue to perform the required function in parallel and, therefore, has not failed.

When nonexponential components are considered, failure rates are not constant with time but are a function of the age T of these components. Therefore, for an operating time, t, for which the reliability is to be computed, component failure rates corresponding to the component age at the given operating period must be used. With the aid of computers and modern statistical software, this is not as difficult as it used to be. Fortunately, in most cases, we can do very well with the exponential distribution for systems when system time is taken as the basis for observing component failures or when components operate only within their useful lifetimes.

In some cases, less reliable components in a system are backed up by parallel components to increase system reliability by using parallel redundancy. Such parallel arrangements of two or more components can be considered as a single unit in a series within the system so that, if the unit fails as a whole, the system fails. In this way, we again arrive at a series arrangement of reliabilities or a series system. Therefore, the reliability of every complex system can be expressed as the product of the reliabilities of all of those components and units on whose survival the system depends. For n components in a series, the system reliability is given by

$$R_s = R_1 * R_2 * R_3 * \ldots R_n = \prod R_n$$

where $n = 1$ to n, and where R can be exponential or not.

When all Rs are exponential, the system's reliability becomes simple—that is,

$$R_s = \exp(-\lambda_1 dt) * \exp(-\lambda_2 dt) * \ldots * -\exp(-\lambda_n dt)$$
$$= \exp(-\Sigma \lambda dt)$$

Thus, all we need to do is add up all of the constant failure rates of all of the series components in the system, multiply this sum with the operating time t, and obtain the value of R_s.

Example Reliability Calculation

Example of a Series System

As a simple example, let us consider an electronic circuit consisting of 10 silicon diodes, 10 transistors, 10 resistors, and 10 ceramic capacitors arranged in a series with failure rates per hour of 0.000002, 0.00001, 0.000001, and 0.000002, respectively. To calculate the reliability of this circuit, one needs to add up all the failure rates:

$$\Sigma \lambda = 10 * (0.000002) + 10 * (0.00001)$$
$$+ 10 * (0.000001) + 10 * (0.000002)$$
$$= 0.00015$$

To estimate the reliability of this circuit at 100 hours of operation, we obtain

$$R_s = \exp(-\Sigma \lambda dt) = \exp(-0.00015 * 100) = 0.985$$

Thus, the circuit is expected to operate on the average without failure in 100 hours 98.5% of the time or there is a chance of a failure at 100 hours of 1.5%. Another way to define reliability is to calculate its mean time between failures (MTBF), which is defined here as the reciprocal of the reliability, or

$$MTBF = 1/(\Sigma\lambda) = 1/(0.00015) = 6700 \text{ hours}$$

For the exponential distribution, this MTBF means that the system has a chance to survive 6700 hours on the average 37% of the time or a chance to fail 63% of the time. MTBF is important for all distributions because it can be measured quantitatively and defines the reliability of the system during its useful life.

Example of a Series and Parallel System

Suppose that the system consists of five components arranged in a series and parallel system as shown in Chart 12.1.

The first thing to calculate is the combined reliability of Parts 4A and 5A. From ancillary rule 1,

$$R_{4A/5A} = 0.6 * 0.5 = 0.03$$

The same is performed for Parts 4B and 5B:

$$R_{4B/5B} = 0.6 * 0.5 = 0.03$$

The arrangement shown in Chart 12.1 can now be simplified as shown in Chart 12.2.

Next we can combine the two parallel components using ancillary rule 3:

$$R_p = 0.3 + 0.3 - 03 * 0.3 = 0.51$$

As shown in Chart 12.3, we now further simplify the system.

Finally, we can treat this as a purely series system:

$$R_s = 0.9 * 0.8 * 0.7 * 0.51 = 0.250$$

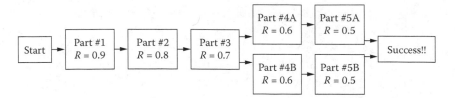

Chart 12.1 A system of five components arranged in a series and parallel system.

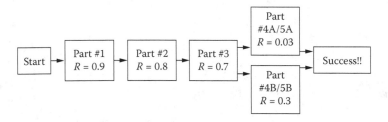

Chart 12.2 Simplified series and parallel system.

Chart 12.3 Series and parallel system simplified to a purely series system.

Derating for Higher System Reliability

The estimated MTBF of the example above may be too short of the system requirements, which might be as high as 100,000 hours. What can the designer do to upgrade the reliability of the circuit in order to meet requirements? The answer lies in derating the components.

Component failure rates apply to definite operating stress conditions such as operating voltage, current, temperature, and mechanical stresses such as shock and vibration. Failure rates usually change drastically with the operating stress levels. For instance, the failure rate of a component may drop to one-twentieth of the failure rate when the temperature decreases from 50°C to 30°C.

Thus, to upgrade the reliability of the circuit, one needs to reduce the stresses acting on the component or use components that are rated for higher stresses. Most component manufacturers supply failure rate derating curves from which failure rates can be directly read for the various stress levels. Large gains in reliability can also be achieved by reducing the number of parts used. Thus, the designer should always keep three things in mind:

1. Operate the components well below their rated stresses (i.e., voltage, temperature, etc.).
2. Design with as few components as possible.
3. With some low-reliability components, design with parallel redundancy.

Mean Time Between Failure

If a device has a constant failure rate, λ, the reciprocal of this failure rate is the MTBF for the device:

$$MTBF = 1/\lambda$$

Thus, if a device has a constant failure rate of 0.0001 failures per hour, its MTBF is 10,000 hours. For this example, a common misconception is that 50% of the failures will occur below 10,000 hours and the other 50% will occur over 10,000 hours. Rather, this is an exponential function, so by substituting t as MTBF in the exponential equation for reliability,

$$R(t) = e^{(-\lambda\,t)} = e^{(-\lambda\,(MTBF))} = e^{-1} = 0.37$$

Thus, only 37%, not 50%, of the devices will survive for a given operating time equal to their MTBF. It can also be inferred that a high-reliability device must have an MTBF that is much greater than its mission or operating time. In fact, to add another "9" to the reliability of a device, its MTBF must be increased tenfold as shown in Table 12.1.

TABLE 12.1
Relationship between Mean
Time Between Failures
(MTBF) and Reliability (R)

MTBF	$R(t)$
10 t	0.9
100 t	0.99
1000 t	0.999

In the usual cases where $t <$ MTBF, the reliability function is closely approximated by

$$R(t) \sim 1 - t/\text{MTBF}$$

This approximation is usually satisfactory for $t < 0.1$ MTBF. Thus, if the device with an MTBF = 10,000 hours is operated at 100 hours, its reliability is

$$R(100) \sim 1 - 100/10{,}000 = 1 - 0.01 = 0.99$$

Alternatively, if the device must have a reliability of 0.999 (for $t < 0.1$ MTBF), then it should only be operated for

$$t = (1 - R) * (\text{MTBF}) = (1 - 0.999) * (10{,}000 \text{ hours}) = 10 \text{ hours}$$

Although MTBF is measured in time scales (hours, months, years), it does not in general indicate the true life of a device. Rather, it is a reliability index of a device for a given operating or mission time. Device life, for all practical purposes, is the time when the increasing failure rate phase begins (i.e., the "wear-out" phase).

It is clear from Figure 12.3 that knowledge of MTBF gives no information about *life*, and vice versa, because *life and MTBF are not related*. This leads us to some strange but true relationships; for example, a new increased-life device may have a lower MTBF than an older device or a new design gives a higher MTBF but lower life. Because this chapter is devoted mainly to reliability estimation, we shall not delve further into maintenance concepts or cost trade-offs.

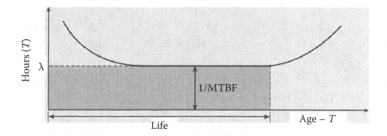

Figure 12.3 Why mean time between failure (MTBF) and life are not related.

Reliability Growth

As data from various test programs become available during the equipment design, manufacturing, and initial product introduction, it can be expected that certain changes or corrective actions will be made that should result in improved reliability. This is called *reliability growth*. This reliability growth has now become almost commonplace for commercial software, which is often labeled with a version number. Notice the continuous improved versions of the word-processing software from Microsoft. It is therefore advantageous to have techniques available for monitoring reliability growth trends. An early technique is one described by Duane.[1] Duane's method is to plot the cumulative observed failure rate against the cumulative test time on log–log paper or plot their exponents on a linear scale. The result is usually a reasonable straight line with the slope being the rate of reliability growth.

Acceptance and Accelerated Testing

Measuring MTBF under actual-use conditions usually takes a long time. Past experience has indicated that a meaningful MTBF requires the accumulation of durations that are sufficiently long enough to generate at least six failures. Thus, a device with an MTBF of 10,000 hours would take 60,000

TABLE 12.2

Test Length Guide Table from National Institute of Standards and Technology (NIST) Section 8.3.1.1

Number of Failures Allowed	Factor for Given Confidence Levels					
r	50%	60%	75%	80%	90%	95%
0	0.693	0.916	1.39	1.61	2.30	3.00
1	1.68	2.02	2.69	2.99	3.89	4.74
2	2.67	3.11	3.92	4.28	5.32	6.30
3	3.67	4.18	5.11	5.52	6.68	7.75
4	4.67	5.24	6.27	6.72	7.99	9.15
5	5.67	6.29	7.42	7.90	9.28	10.51
6	6.67	7.35	8.56	9.07	10.53	11.84
7	7.67	8.38	9.68	10.23	11.77	13.15
8	8.67	9.43	10.80	11.38	13.00	14.43
9	9.67	10.48	11.91	12.52	14.21	15.70
10	10.67	11.52	13.02	13.65	15.40	16.96
15	15.67	16.69	18.48	19.23	21.29	23.10
20	20.68	21.84	23.88	24.73	27.05	29.06

hours of test time to obtain a good estimate of MTBF. More recently, section 8.3.1.1. in the *National Institute of Standards and Technology (NIST) Engineering Statistics Handbook* presented a more statistical basis for determining the test time needed to prove a required (or warranted) MTBF assuming that the system has a constant failure rate.[2] This reliability test is often called the *product reliability acceptance test* (PRAT), qualification test, or simply acceptance test. This reference is based on the chi-square distribution given the number of failures allowed and the confidence level $(1 - \alpha)$ that the consumer is willing to accept. The reference is also provided in Table 12.2, where r equals the number of failures allowed for acceptance, and the columns from 50% to 95% are the confidence levels for this test.

For example, to confirm a 200-hour MTBF objective at 90% confidence allowing up to four failures on the test, the test length must be $200 \times 7.99 = 1598$ hours. If this is unacceptably long, try allowing only three fails for a test length of 200 × 6.68 = 1336 hours. The shortest test would allow no fails and last 200 × 2.3 × 460 hours. All of these tests guarantee a 200-hour MTBF at 90% confidence when the equipment passes. However, the shorter tests are much less "fair" to the supplier in that they have a large chance of failing a marginally acceptable piece of equipment. The recommended procedure is to iterate on r = the number of allowable fails until a larger r would require an unacceptable test length. For any choice of r, the corresponding test length is quickly calculated by multiplying M (the objective) by the factor in Table 12.3 below corresponding to the rth row and the desired confidence level column.

Often, the shortest test time required in Table 12.2 from NIST Reference 8.3.1.1 is still too long for a particular program. To further reduce the test time required, accelerated test techniques can be considered. That is, some stress known to increase failures such as voltage, current, or operating temperature, is chosen at higher values than normal for testing. At a higher stress, 1 hour may be equivalent to 10 hours at normal stress. The ratio is called the *accelerating factor*. Two major factors to consider before conducting an accelerated test are

1. There should not be a change in failure modes or mechanism between the normal and accelerated stress.
2. The magnitude of the acceleration factor between accelerated stress level and the normal level should not be too high. In general, this should be kept at <50 per an ISO standard.[3]

The first caveat comes into the heading of physics of failures (i.e., one can only accelerate within one mode of failure mechanism). The second caveat minimizes the chance of violating the first caveat and minimizes the extrapolation errors that generally increase exponentially with the regression or predictive models used.

The most common acceleration model used for decreasing test time is the Arrhenius model, although other useful models are also available, such as the Eyring model, the (Inverse) Power Rule for Voltage, the exponential voltage model, two temperature/voltage models, the electromigration model, three stress models (temperature, voltage, and humidity), and the Coffin–Manson mechanical crack growth model.

The Arrhenius equation takes the form of

$$AF = \exp\,[(\Delta H/k)(1/T_1 - 1/T_2)]$$

with T denoting temperature measured in degrees Kelvin at the point at which the failure process takes place. k is Boltzmann's constant (8.617×10^{-5} in eV/K), ΔH is the activation energy in electron volts, and AF is the acceleration factor.

The value of ΔH depends on the failure mechanism and the material involved and typically ranges from 0.3 to 1.5 eV. Using the value of k given above, this can be written in terms of T in degrees Celsius as

$$AF = \exp\left\{\Delta H \times 11605 \times \left[\frac{1}{(T_1 + 273.16)} - \frac{1}{(T_2 + 273.16)}\right]\right\}$$

Note that the only unknown parameter in this formula is ΔH.

Example

An electronic microprocessor had been shown to fail by opened contacts because of electromigration of the gold pad materials to the aluminum wires used to connect the silicon chip to the substrate. This is the classic failure mode called *purple plague* for dissimilar metals often used in microcircuits, such as aluminum wires bonded on gold pads. The activation energy for this failure mode has been published as 1.1 eV.[4] The manufacturer's warranty for this product is 8 years under normal-use conditions. How many units should be tested for 1 month at an elevated temperature to ensure that the MTBF is actually 8 years with 90% confidence?

Solution: Refer to Table 12.3, using the possible number of failures 0, 1, and 2 and for a 90% confidence. As can be seen in this table, the shortest test time necessary

TABLE 12.3
Unit-Years Required for Acceptance at 90% Confidence

Number of Failures	Test Factor	Test Time (Unit-Yr)
0	2.30	18
1	3.89	31
2	5.32	43

to prove that the MTBF of this package is 8 years to perform 18 unit-years without a failure. Although this appears long, testing 18 units for 1 year is equivalent to a test time for one unit of 18 years. For a test time of 1 month, we would therefore require 216 units to test. If none failed out of the 216 units after a month, the MTBF of 8 years is proven at least to a confidence level of 90%.

However, these units may be very expensive, and the supplier is willing to provide a maximum of 30 units for acceptance testing and does not want them tested any longer than 2 weeks. We can then use the Arrhenius model to reduce the number of units to test or reduce the length of time to test. At what temperatures should we test in order to perform this Product Reliability Acceptance Test (PRAT) with <30 units and within ≤2 weeks?

For the case of no failure allowed in order to pass the acceptance test, we need 468 units to test in 2 weeks (18 years × 26 2-week periods per year). With a maximum of 30 units, we need to accelerate the test by at least 15.6 times or rounded up to 20 times. With the acceleration factor equation and an activation energy of 1.1 eV, one can solve the temperature to achieve acceleration factors of 20, 40, and 60. This is shown in Figure 12.4 and Table 12.4.

TABLE 12.4
Temperatures in °C to Achieve Acceleration Factors of 20, 40, and 60

Temperature (°C)	Acceleration Factor
50	20
56.2	40
60	60

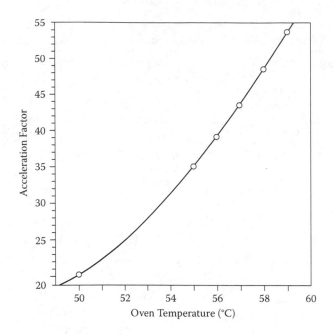

Figure 12.4 Plot of acceleration factors versus temperature based on activation energy of 1.1 eV.

Translating these acceleration factors into temperatures to test as well as the duration of the test time, one can obtain Table 12.5.

The acceptance criteria (in order of preference) for "proving" that the package will have an MTBF of 8 years are

1. Test 40 units for 2 weeks at 50°C, and accept if failure is ≤1.
2. Test 23 units for 2 weeks at 50°C, and accept if there is no failure.
3. Test 23 units for 1 week at 56°C, and accept if there is no failure.
4. Test 27 units for 1 week at 60°C, and accept if there is no more than one failure. (Note that there has been some caution published about exceeding a time acceleration factor of >50 with the use of the Arrhenius model for plastic pipes in the ISO standards,[3] which led the author to downgrade this scenario as the last choice.)

TABLE 12.5
Number of Units to Test for a Duration versus
Number of Failures Allowable

Temperature (°C)	Acceleration Factor	No Failure	One Failure
50	20	23 units/2 weeks	40 units/2 weeks
56	40	23 units/1 week	40 units/1 week
60	60	16 units/1 week	27 units/1 week

Censored Data

Observations whose exact failure times are unknown are called *censored observations* or *censored data*. The failure times are unknown because either the test was stopped prematurely or the system failed due to another unrelated component (for instance, there was a power failure or the monitoring equipment failed before the product did). Because censoring changes the proportion that failed relative to the ones that have not failed, the units censored need to be taken into account.

There are two types of censoring:

Type I: The batch of components is tested for a specified duration, and any survivor after this specified duration is a Type I censored observation.
Type II: The batch of components is tested until a specified number of failures are observed, and any survivor at that time is a Type II observation. Thus, the PRAT example just discussed is a Type II censored observation.

Nonparametric Reliability: The Kaplan–Meier Analysis

Oftentimes, one might not require a distribution in order to estimate the survival functions of a group of components whose cause of failures may be more than one. For example, we might want to determine the average life expectancy of a patient who has just been admitted into a hospital for colon cancer treatment. Within this population of patients, there might be subgroups, such as age <50 years or >50 years, male versus female, cigarette smoker versus nonsmoker, and so forth.

The Kaplan–Meier method is generally the method of choice for estimating the probability of surviving a time period interval i by calculating the probability $p_{(i)}$ such that

$$p_{(i)} = [(\text{Number at risk}) - (\text{Number Failed})]/(\text{Number at risk})$$

The reliability at time, I, is the product of the probabilities of operating through each of the previous time intervals before i. Thus,

$$R_{(i)} = \prod p_{(k)} \text{ for all } k < i$$

The following is an example of the lifetime of microelectronic packages tested per test scenario 1 above with censored observations.

Example

Suppose that it was decided to perform the performance reliability acceptance test (PRAT) per test scenario 1, which is

1. Test 40 units for 2 weeks (336 hours) at 50°C, and accept if failure is ≤1.

Supposed further that the actual failure data are collected as shown in Table 12.6, where 40 components were subjected to an environmental chamber of 50°C and continually functionally tested for time to failure.

The results show that the first failure occurred at 722.2 hours. No failure occurred before 2 weeks or 336 hours; therefore, the batch of components are accepted as meeting the requirement of MTBF ≥8 years. It can also

TABLE 12.6
Data for Time to Failure

Unit	Time (hours) to Failure	Censored?	Unit	Time (hours) to Failure	Censored?
1	722.2	No	21	6871	No
2	778.2	No	22	7091	Yes
3	1065	No	23	7624	No
4	1171	No	24	7667	No
5	1682	No	25	8150	No
6	3010	No	26	8298	No
7	3183	No	27	8385	No
8	3389	No	28	8636	No
9	3565	Yes	29	9859	Yes
10	3843	No	30	11015	No
11	4100	No	31	11123	Yes
12	4635	No	32	11271	No
13	4783	No	33	11740	No
14	4847	No	34	14631	No
15	5073	No	35	14842	No
16	5394	No	36	15824	Yes
17	5448	No	37	17786	No
18	5661	No	38	17945	No
19	5762	No	39	19768	No
20	5798	No	40	20035	No

be gleaned from the data above that the program manager decided to continue the test until all units failed in order to determine the actual life of the device. However, at certain intervals of the testing, some units were taken out (five in all) to determine and document the condition of the microcircuit at various ages. These packages are therefore censored (i.e., they were not failures at the times that testing was terminated). We will now see how these are estimated using the Kaplan–Meier method.

To compile Table 12.7, we need to compute from the number at risk and the number failed.

TABLE 12.7

Reliability and Unreliability of the Units Tested to Failure

Unit	Time (hours)	Number at Risk	Number Failed	$p_{(i)}$	$R_{(i)} = \Pi\, p_{(i)}$	$Q = 1 - R$	Censored? (No = 0, Yes = 1)
	0	40	0	1	1	0	0
1	722.2	40	1	0.975	0.975	0.025	0
2	778.2	39	1	0.974	0.95	0.05	0
3	1065	38	1	0.974	0.925	0.075	0
4	1171	37	1	0.973	0.9	0.1	0
5	1682	36	1	0.972	0.875	0.125	0
6	3010	35	1	0.971	0.85	0.15	0
7	3183	34	1	0.971	0.825	0.175	0
8	3389	33	1	0.97	0.8	0.2	0
9	3565	32	0	1	0.8	0.2	1
10	3843	31	1	0.968	0.774	0.226	0
11	4100	30	1	0.967	0.748	0.252	0
12	4635	29	1	0.966	0.723	0.277	0
13	4783	28	1	0.964	0.697	0.303	0
14	4847	27	1	0.963	0.671	0.329	0
15	5073	26	1	0.962	0.645	0.355	0
16	5394	25	1	0.96	0.619	0.381	0
17	5448	24	1	0.958	0.594	0.406	0
18	5661	23	1	0.957	0.568	0.432	0
19	5762	22	1	0.955	0.542	0.458	0
20	5798	21	1	0.952	0.516	0.484	0
21	6871	20	1	0.95	0.49	0.51	0
22	7091	19	0	1	0.49	0.51	1
23	7624	18	1	0.944	0.463	0.537	0
24	7667	17	1	0.941	0.436	0.564	0
25	8150	16	1	0.938	0.409	0.591	0
26	8298	15	1	0.933	0.381	0.619	0
27	8385	14	1	0.929	0.354	0.646	0
28	8636	13	1	0.923	0.327	0.673	0
29	9859	12	0	1	0.327	0.673	1

TABLE 12.7 *(Continued)*
Reliability and Unreliability of the Units Tested to Failure

Unit	Time (hours)	Number at Risk	Number Failed	$p_{(i)}$	$R_{(i)} = \Pi\, p_{(i)}$	$Q = 1 - R$	Censored? (No = 0, Yes = 1)
30	11015	11	1	0.909	0.297	0.703	0
31	11123	10	0	1	0.297	0.703	1
32	11271	9	1	0.889	0.264	0.736	0
33	11740	8	1	0.875	0.231	0.769	0
34	14631	7	1	0.857	0.198	0.802	0
35	14842	6	1	0.833	0.165	0.835	0
36	15824	5	0	1	0.165	0.835	1
37	17786	4	1	0.75	0.124	0.876	0
38	17945	3	1	0.667	0.083	0.917	0
39	19768	2	1	0.5	0.041	0.959	0
40	20035	1	1	0	0	1	0

$$p_{(i)} = [(\text{Number at risk}) - (\text{Number Failed})]/(\text{Number at risk})$$

is computed in the fifth column, and the reliability $R_{(i)}$ is computed in the sixth column:

$$R_{(i)} = \Pi\, p_{(k)} \text{ for all } k < i.$$

Table 12.7 shows this calculation for all 40 rows.

After performing this operation for all 40 units, one can plot the reliability (or survival) of all of the units tested as a function of test time. This is a Kaplan–Meier plot as shown in Figure 12.5. This plot shows that the MTBF at the 66.7% reliability (or 33.3% failure) is ~4850 hours, while the median failure life at the 50% reliability is ~6900 hours. The actual MTBF at room temperature is 4850 × 20 = 11.1 years, clearly exceeding the requirement of 8 years, because this was an accelerated test with an acceleration factor of 20 (see Table 12.5).

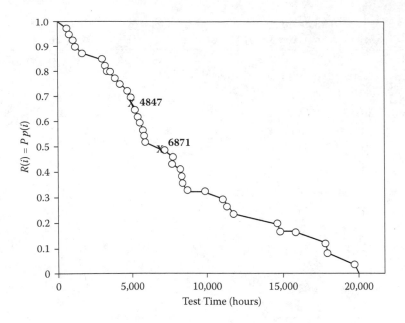

Figure 12.5 Kaplan–Meier plot for reliability of 40 electronic units.

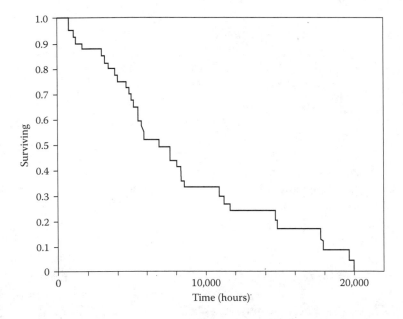

Figure 12.6 Survival or reliability plot from JMP®.

Calculating with Modern Statistical Software

Even with the relatively simple Kaplan–Meier method, the calculation of reliability is tedious; however, there are many sophisticated statistical software packages now commercially available to perform most statistics on reliability. One such software is the JMP® Discovery software from SAS Institute in Cary, North Carolina.[5]

The same result (Figure 12.6) and more are shown for the data in Figure 12.5 with this software. It computed the median time to failure as 6334 hours, which is in agreement with the hand-estimated one above.

Summary

Group	Number Failed	Number Censored	Mean	Standard Error
Combined	35	5	8486.6	964.067

Quantiles

Group	Median Time	Lower 95%	Upper 95%	25% Failures	75% Failures
Combined	6334.3	5073.5	8384.9	3971.4	11506

In addition, one can fit various distributions to the data. The fit for the Weibull distribution is shown in Figure 12.7.

It shows that the Weibull plot is a good fit to the data and that it predicted the median time to failure as 6334 hours.

The Effect of Censoring

It is not quite intuitively obvious what censoring actually does in the reliability plot if the five units censored were simply ignored. In this case, the result is shown in Case A below and shown in Figure 12.8. We simply recalculated the Kaplan–Meier plot without the censored data. This is easily done, of course, with

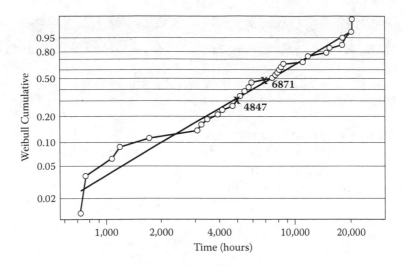

Figure 12.7 Weibull plot from JMP®.

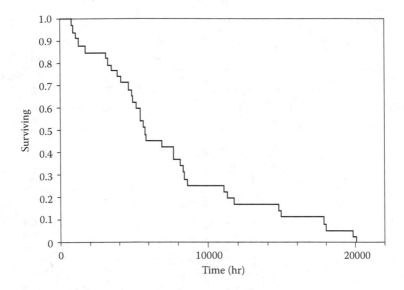

Figure 12.8 Survival or reliability plot for Case A.

the help of the statistical software, which makes this exercise more fun (and quicker to perform). In addition, we can find out if the number of censored units were ten instead of five as shown below in Case B.

Case A: No Censored Data

Summary

Group	Number Failed	Number Censored	Mean	Standard Error
Combined	35	0	7571.69	920.721

Quantiles

Group	Median Time	Lower 95%	Upper 95%	25% Failures	75% Failures
Combined	5711.2	4782.8	7667.2	3616	9825.8

In this case, the median life decreased from 6334 to 5711 hours. Thus, the effect of taking out the censored data appears to shift the entire plot toward the left-hand side of the reliability plot or decrease the lifetime of the tested devices. Would the opposite effect be observed if the numbers of censored data were increased to twice their number? Case B with Figure 12.9 will now be demonstrated.

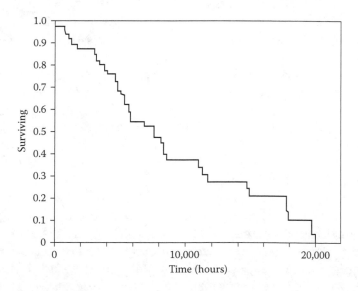

Figure 12.9 Survival or reliability plot for Case B.

Case B: Increase by Twice the Amount of Censored Data

Summary

Group	Number Failed	Number Censored	Mean	Standard Error
Combined	36	10	9080.87	980.255

Quantiles

Group	Median Time	Lower 95%	Upper 95%	25% Failures	75% Failures
Combined	7645.8	5393.7	9859.1	4367.5	13186

Case B shows that increasing the number of censored data by doubling each of them (notice that now we have ten as the number censored versus five beforehand) shifted the plot toward the right-hand side of the test time (i.e., the median lifetime has increased from 6334 to 7646 hours).

Thus, censoring shifts the timeline of the plot to the right toward a longer lifetime when more units are censored.

A Special Case: A Self-Healing Nano-Bio-Material Electronic Circuit

This book addresses many special cases because of the rapid advancement in the synthesis and applications of nanomaterials. Of interest for the estimation of reliability would be the special case for a nano-bio-electronic circuit that has the capability of curing itself. For instance, when an open occurs with an interconnection or via, the circuit would be capable of growing a shunt around this open connection. Conversely, if a short developed between two lines of nanowires, one or both shorted wires would cut off the shorted portion and form a shunt around the cut-off line. Assume again that this self-curing process can occur very quickly, say, in microseconds. For this special case, the hazard plot would be analogous to a square well (i.e., a very sharp vertical line at the start of use and ending at a low but constant failure rate λ followed

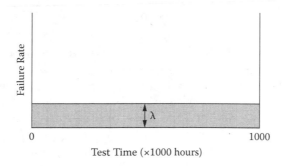

Figure 12.10 Square well hazard plot for self-healing devices.

again by a very sharp wear-out region). The end-of-life would be the time when the self-curing process is overcome by the failure rate. The MTBF of such a device would still be 1/λ. The lifetime of the circuit is the time when it encounters its wear-out mode and cannot cure itself anymore. This is shown in Figure 12.10. Thus, the lifetime of these types of devices will solely depend on their synthesis (or DNA) and the environmental stresses encountered. It is conceivable that this lifetime of the device can be very long. This will open up a period in our civilization when the lifetime of the device or system will be much longer than that of the human user's life or interest. This will become the period of the user-defined reliability (i.e., reliability as defined by the users' interest). For environmental considerations, such systems may contain a switch that will initiate a graceful "death" for the device to be turned on by its owner when it is no longer useful.

Summary

The concept of systems reliability derives from the basic rules of unit probabilities and the application of these basic rules into ever-increasing complexity. From these concepts, mean time between failures, reliability growth, accelerated testing, and data censoring can be developed. Finally, an actual case study of how to conduct performance reliability acceptance testing is developed for an application-specific integrated circuit (ASIC) with failure activation energy of 1.1 eV. Data from this case study were utilized to estimate the nonparametric reliability of the device using the Kaplan–Meier analysis technique coupled with modern statistical software. Finally, a special and hypothetical case of a self-healing nano-bio-material-based circuit is presented, and its reliability is discussed.

A substantial portion of this chapter can be attributed to statistics and reliability concepts learned during a 2-year master's-level technical training course called Management Problems Analysis while the author was with General Electric Company. Other materials were taken from experience while working with various programs at the Johns Hopkins University Applied Physics Laboratory. To both of these organizations, I owe most of what I learned about statistics and reliability.

Reliability through replication.

References

1. Duane, J.T., "Learning Curve Approach to Reliability Monitoring," *IEEE Trans. Aerospace*, 2(2):563–566, April 1964.
2. *NIST / SEMATECH e-Handbook of Statistical Methods*, http://www.itl.nist.gov/div898/handbook/, created June 1, 2003; last updated February 16, 2010.
3. ISO TC 138/SC 5, dated 01/14/2002.
4. Shirley, C.G., and Shell-DeGuzman, M. "Moisture-Induced Gold Ball Bond Degradation of Polyimide-Passivated Devices in Plastic Packages," in *31st Annual International Reliability Physics Symposium*, pp. 217–226, 1993.
5. JMP® Statistical Discovery Software, www.jmp.com, last updated Friday, October 15, 2010.

13

Test and Evaluation Techniques from Very-Large-Scale Integration (VLSI) to New Developments in Micro- and Nanoscale Technology (MNT)

William Paulsen

Preparing for self testing.

Contents

Introduction

Test and evaluation is the aspect of systems engineering that provides quantitative measurements of functionality, performance, and defects of a system or components that comprise a system.[1,2] The goal is to have a framework for systems engineers to make decisions on a system's or component's readiness at each stage in the systems engineering life cycle.

It is important to distinguish between the terms *test* and *evaluation*:

- *Test* is a procedure to collect quantitative data from the operation of a model, or a prototype, or a manufactured system or component.
- *Evaluation* is the analysis of test data, to enable objective assessment of design progress or the suitability of a system or component.

Test and evaluation procedures are often performed as part of the *verification* and *validation* of a system, where

- *Verification* is the process of determining that the documented requirements on a system or subsystem are indeed satisfied, "Is the system constructed correctly?"
- *Validation* is the process of determining that a completed system satisfies the needs of the user, "Is this the system that the customer wants?"

Systems Engineering Test and Evaluation

Chapter 2 of this book included a brief summary of Test and Evaluation in the context of the classic waterfall systems engineering methodology. Here, the test and evaluation stage resides between the *Implementation* stage and the *Integration and Deployment* stage. Major tasks in this stage are to

- Define test procedures to achieve verification of the system, including setup procedures, environment requirements, and data collection.
- Develop a collection of regression tests that must be performed after any design change.
- Perform the system test review (STR) during which the testing procedures, tests, results, and corrective actions are examined by a cross-functional team.
- Hold the system readiness review (SRR) before deployment or manufacturing begins to ensure that all open test- and evaluation-related issues are resolved.

The Role of Test and Evaluation Engineers

Test and evaluation engineers actively participate in all stages of the development process. They work with the engineering team to ensure that the system design will enable effective tests to be performed. For example, subsystem components will likely need to be unit tested before being integrated into the full system. These components must provide adequate accessibility to ensure that acceptance criteria are passed. During the early requirements and needs feasibility stages, test and evaluation engineers help to ensure that all specified requirements are eventually testable in a manufactured product. To support this, it is often necessary to perform testing and to evaluate results on early samples of subsystem components. Such test and evaluation results can provide valuable information to system designers to enable reductions in manufacturing costs. They also work with manufacturing and process engineers to specify how manufacturing test and evaluation will be performed and assist systems engineering in performing technology qualifications during trade studies by providing evaluations of new technologies that are based on quantitative analyses of test data.

As part of the systems engineering process, test and evaluation engineers develop test plans that describe exactly how the system will be tested, what data will be collected, and how the data will be evaluated and compared with the requirements on the system. Test plans typically include a list of tests, detailed conditions and test regimens, their objectives, and how each system requirement is covered by the tests. The format and content are often well defined for particular industries. For software development, the "IEEE Standard for Software System Test Documentation"[3] is used, and for electronic equipment for military use, "Reliability Test Methods, Plans, and Environments for Engineering Development, Qualification, and Production"[4] is used. Test and evaluation engineers are also responsible for ensuring that the test environment is acceptable, for example, that proper calibration of test and data acquisition equipment is achieved.

During early stages, the analyses are performed using reliability models of a system. In later stages, manufactured systems are subjected to above-normal environmental stress levels with the intent to create an artificially accelerated life. Once components of the system are available, test and evaluation engineers perform analyses and interpretations of test data. They provide information to systems engineers that permit them to accurately assess manufacturing yields;

define guard bands to ensure that specifications and margins are safely met; and assess system costs, schedule, and performance trade-offs. They perform tests and evaluations with the goal to ascertain the predicted reliability of a system.

Traditional/Classical Systems Engineering Challenges in Micro- and Nanoscale Technology Development

Several aspects of micro- and nanoscale technology (MNT) systems restrict them from fitting into the traditional, classical systems engineering test and evaluation approaches. The microtechnologies used in fabricating integrated circuits (ICs), and some Micro Electro Mechanical Systems (MEMS)[5] and nanoprocessors, are very mature technologies and have mature test and evaluation techniques and processes. However, many areas in the MNT realm are less mature; therefore, test and evaluation procedures are developing and evolving.

Aspects of MNT are an emerging, but currently immature, technology in which systems are designed using a capability-based, synthesis-driven systems engineering design methodology, as was discussed in Chapter 1 of this book. In MNT development, a system is synthesized from a collection of MNT subsystems that have known characteristics, and these subsystems will drive many of the system's requirements and definitions. In this scenario, test and evaluation data are used to assist in the derivation of the requirements. And as the manufacturing process matures, test and evaluation may discover that the measured characteristics between components are reduced. This information is then used to tighten the margin specifications on the system. Such capabilities-driven design implies an agile development methodology as presented in Chapter 4, in which development activities are iterative and progress spirals toward the target. An implication of capabilities-driven design is that test and evaluation must be an "integrated aspect of the Systems Engineering process from beginning to end"[6] (p. 60).

Immature MNT technologies and subsystems typically have inadequate models that can be analyzed or simulated as was discussed in Chapter 10. However, tests and evaluations of manufactured samples or components can be used to provide empirical data to assist in characterizing models and can provide continuing data to reduce discrepancies between MNT modeling and simulation and empirical data. It is also important to take into account that MNT systems can consist of a very large number of interacting subsystems that will typically display emergent properties that are difficult to comprehend fully at the early stages of design. Test and evaluation procedures of systems with emergent properties must be prepared to accommodate the discovery of unknown unknowns, which are failure modes that had not been expected.[7]

MNT technology will be used in autonomous, adaptable systems.[8] These systems are subject to uncertain or unknown external stimuli and exhibit complex response behaviors. Full system specifications, against which traditional test and evaluation is performed, will generally not be available. Nontraditional approaches for test and evaluation should be considered, such as run-time verification techniques that actively monitor for correct system behavior, and are discussed later in this chapter. MNT components will also be subject to manufacturing uncertainties in which small manufacturing tolerances that are introduced into a component can have significant changes in behavior or performance of the component. Probabilistic testing[9] is an approach that can be used to estimate the probability of faults, or defects, in a component. This technique associates the stimulus that is applied to a component with the likelihood of identifying defects. Repetitive stimuli are often used in order to identify transient, or nonpersistent, defects. Test and evaluation are used to derive a better understanding of uncertainties by obtaining quantitative measures of performance and defect variations in components and systems.

MNT subsystems may include built-in self-repair (BISR) capabilities that permit manufacturing defects to be located and replaced with "spare" parts. An MNT component can either repair the defect without external actions, or the component might require intentional testing for defects. When a defect is discovered during testing, additional analysis might be required to create a plan or a sequence of operations that will cause the MNT component to be manipulated to affect a repair.[10] Note that this is not the same as self-healing MNT sealants or coatings that typically have nanocontainers or additional layers that release particles to produce a repair. In contrast, this concept of BISR is to activate error-correcting algorithmic approaches to recovering from faults that occur in data that are streamed to a system. For example, the Reed-Solomon error correction is a scheme for appending information to data at the source such that if the data were to become corrupted, full recovery of the data would be possible up to a specified level of corruption.[11] BISR approaches have been implemented in Dynamic Random Access Memory (DRAM) integrated circuits (ICs) since their commercial inception. Because IC manufacturing yields are less than 100% and DRAM integrated circuits contain highly regular structures, it was determined that redundancy is cost effective.[12]

Development of Very-Large-Scale Integration (VLSI) Chips

Microcircuits or integrated circuits such as very-large-scale integration (VLSI) chips are small electronic circuits that are manufactured in the surface of thin semiconductor materials. Figure 13.1 shows a VLSI chip that is packaged and ready to be assembled onto a printed circuit board. The typical sizes of integrated circuit chips range from about 1 millimeter (mm) to 20 mm on each side and contain up to about 10 billion transistors. VLSI technology has seen continuous increase in permissible chip complexity since the 1960s and follows Moore's Law, a prediction that was made in 1965 that the number of transistors on an IC will double every 2 years.[13] This increase in complexity has necessitated improvements in systems engineering processes for VLSI. Because fields in the MNT realm are currently at the infant stage, and we expect to see continuous increases in capabilities, the path of MNT systems engineering is likely to be similar to VLSI. An excellent description of the VLSI design methodology is given in the classic book, *Introduction to VLSI Systems*.[14]

Figure 13.1 An example of a very-large-scale integration (VLSI) chip. (Courtesy of The Johns Hopkins University Applied Physics Laboratory.)

A VLSI chip consists of a collection of transistors and wires that serve to interconnect the transistors that are formed on the substrate during manufacturing. Some of the wires are designated to be accessible when a chip is mounted in a package to permit external communication with the chip. A package ranges in size from about 10 mm on a side up to about 50 mm. Packages are later inserted onto a printed circuit board for integration with the rest of the system.

The transistor was invented at Bell Labs during the 1940s with one transistor mounted inside a single package that was about 10 mm in size. Figure 13.2 shows individually packaged transistors. Starting in the early 1960s, it became possible to manufacture several transistors in one integrated circuit package, and, by the 1970s, it became possible to manufacture ICs with several thousand transistors. Figure 13.3 shows an enhanced three-dimensional (3D) view of a stardard part in an VLSI chip. The different shades of gray represent the different materials that are used in the manufacturing process. Figure 13.4a shows a representation of one standard cell as might be depicted by a computer-aided design (CAD) software program. The different colors represent materials that are to

Figure 13.2 Individually packaged transistors. (Image from http://en.wikipedia.org/wiki/File:Transistorer.jpg.)

be manufactured. Figure 13.4b shows the electrical view of the cell in which the physical characterstics have been abstracted away to enable design and analysis at the electrical signal level of detail. Figure 13.5 shows one VLSI chip being held in the tweezers and a wafer in the background that will be sliced into many chips.

Figure 13.6 shows a packaged integrated circuit with the top of the package removed. Note the thin wires that connect the IC to the package.

Some notable achievements in VLSI design and the associated systems engineering processes and methodologies include

- Development of a capability-driven design methodology in which continuous increases in chip complexity, reliability, and lower manufacturing costs resulted in new, unanticipated systems in which such VLSI chips are embedded. One example of this is the preponderance of VLSI usage in automobiles where microprocessors are used to monitor and control subsystems that had traditionally been mechanical or analog devices.[15]

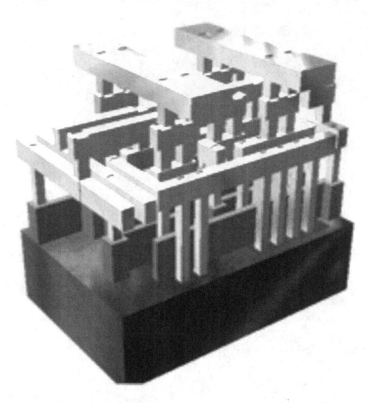

Figure 13.3 An enhanced image that represents the three-dimensional structure of several transistors that form a standard logic cell on an integrated circuit. (Image from http://en.wikipedia.org/wiki/File:Silicon_chip_3d.png.)

- Techniques for designing VLSI chips have continued to require increases in the level of abstraction so that regular structures are defined and assembled in a top-down methodology often automatically with the assistance of computer-aided design (CAD) tools. Without such a rigid design methodology, the complexity challenges of designing, testing, and manufacturing a VLSI chip (which, as previously stated, can currently contain up to about 10 billion transistors) would be enormous. For example, without a standard collection of structures that each consist of several up to hundreds of transistors, it would be necessary to design and test a VLSI chip at the circuit level using a software tool such as the Simulation Program with Integrated Circuit

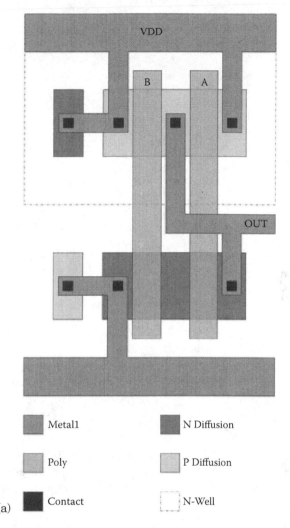

Figure 13.4 (a) A standard logic cell, as displayed by a computer-aided design software tool. (Image from http://en.wikipedia.org/wiki/File:CMOS_NAND_Layout.svg.) (b) The electrical view of very-large-scale integration (VLSI) standard logic cell. (Image from http://en.wikipedia.org/wiki/File:CMOS_NAND.svg.)

Emphasis (SPICE),[16] and it would take years of computer time for analysis before allowing a design to proceed to manufacturing.

- Test and evaluation of VLSI chips has also steadily improved, including acknowledgements that test and

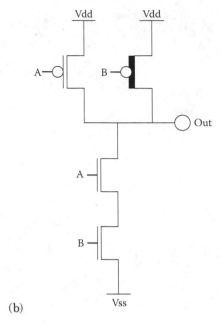

(b)

Figure 13.4 (Continued)

    evaluation must be considered very early in the design cycle and that special provisions must be included in designs to assist in the testing of each manufactured chip.

- A branch of VLSI design methodology of VLSI chips has been developed in which the functionality can be configured after manufacturing. Field-programmable gate array (FPGA) devices sacrifice price and performance for the flexibility to have their behavior defined at the system's power-up time.[17]

Evolution of VLSI Chip Design

The test and evaluation issues for VLSI chips are twofold: (1) during the design of a chip, it is essential to ascertain that the functionality is correct, and (2) during manufacturing, it is essential to test each chip to be sure that no defects were introduced. Smaller chips could easily be designed manually, and it was trivial to test the design of a chip before manufacturing

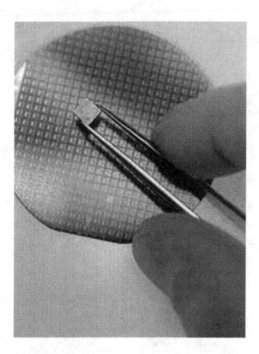

Figure 13.5 One very-large-scale integration (VLSI) chip held in the twee-zers, cut from the wafer in the background.

by using software simulation tools and by fabricating a sample chip. It was easy to devise a collection of tests for each manu-factured chip because of the limited complexity. However, as the chip manufacturing process matured and the complexity of chips increased, design and testing problems began to appear. It became too costly in time and labor to use a mechanical drafting team to manually draw the sizes and positions of all the small features that compose one transistor when there are thousands of transistors to be drawn.

CAD tools were introduced to allow chip designers to graphically lay out a chip and to save the chip design into com-puter files. Other CAD tools extracted information from these files to permit checking that design rules are satisfied and to allow simulation of the design and to test the functional-ity against specifications on the chip. Once a design has been deemed ready for manufacturing, the CAD files are processed into instructions for the tools that fabricate the chips. Masks, which are similar to blueprints in that they describe each of

Figure 13.6 A very-large-scale integration (VLSI) chip mounted in a package, with the top of the package removed for visibility. The size of the chip inside the package is about 10 mm on each side. (Image from JHU/APL, 1649PS00A.jp, if not OK then this image might be used: http://en.wikipedia.org/wiki/File:153056995_5ef8b01016_o.jpg.) (Courtesy of The Johns Hopkins University Applied Physics Laboratory.)

the many layers in a chip, are created from these CAD files. But again, as the manufacturing process continued to mature, the permitted complexity of chips grew. Manual layout, even with the assistance of early CAD tools that enabled graphical design of the transistors of a chip, became untenable. The solution was to introduce the concept of a standard cell design methodology. A standard cell is a collection of transistors and interconnect wires that perform a specific Boolean logic function. A standard cell library consists of a collection of cells that each performs a unique function. Each cell is designed and tested just once but can be replicated, or placed, many times on a chip. However, placements of library cells are sometimes parameterized or fine-tuned, but this step is typically not so complex and does not require extensive analysis.

Standard Libraries of Cells

Standard cells raised the abstraction level of the design process so that designers could focus their design process on activities that are closer to the system requirements. Productivity improved and testing became easier. Another aspect to using standard cells is the standardization of a methodology for placing the cells on a chip. Rather than permitting ad hoc placements in no particular pattern, standard cell methodology requires placement of cells in a regular pattern on the chip called *Manhattan style,* because the layout is similar to the streets in Manhattan, New York.

The standard cell methodology permitted CAD tools that could place and interconnect (or route) the cells automatically, almost in a "cookbook" approach. The optimal placement and routing algorithm is equivalent to the Steiner tree problem,[18] which has a goal to interconnect a collection of vertices in two-dimensional space with the minimum length of edges. The computation complexity of solving the Steiner tree problem is NP-Complete,[19] which means that the computation time grows exponentially with the size of the problem. For example, suppose an IC has n cells to be placed and interconnected. An algorithm that will find the best, or optimal, placement and interconnection solution will have a complexity of $O(2^n)$. In practice, electronic CAD tools use iterative algorithms that converge on acceptable solutions that might be completely optimal in reasonable times. These are called *heuristic*[20] or experienced-based problem-solving approaches and are commonly used in other domains to solve similarly complex problems. In addition, the CAD tools typically need to meet performance requirements and design rules, so an optimal result, while desirable, is not necessary.

More Complex VLSI Chip Designs

As the allowed complexity of manufactured VLSI chips increased, new design methodologies were introduced to help control the overall design efforts that such chips required. The system-on-chip[21] methodology allowed complex components to be easily incorporated into chip designs. Figure 13.7

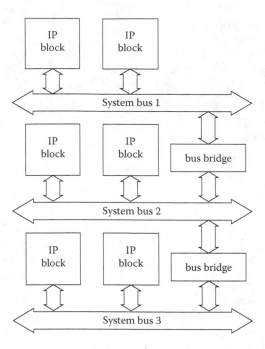

Figure 13.7 A collection of Intellectual Property (IP) blocks that are interconnected by System Buses.

shows a collection of Intellectual Property (IP) blocks that are interconnected by system buses. An IP block contains a well-defined, high-level functionality and is typically acquired from an outside source. A system bus, or other defined standard interface, provides the communication infrastructure between the IP blocks. The IP blocks have standard interfaces to the system buses which allow them to be easily used together and require no customization.

Ultimately, the VLSI design methodology allowed a large variety of commercial off-the-shelf (COTS) VLSI chips to be incorporated into complex, low-cost capability-based designs.[22] The enabling factors that permitted the growth of COTS VLSI microchips included

- Simplified methodologies that allowed designers who have extensive domain knowledge but limited VLSI microchip implementation experience to enter the COTS arena

- Evolution of standards for chip interface configurations and functional behaviors, including standards for logic design and simulation analysis
- Evolution of standards for development of models of COTS parts such that these models can easily be inserted into system simulations, thereby allowing test and evaluation of systems to be performed at early stages in the design process
- A steady drop in manufacturing costs of VLSI chips, thus encouraging further usage of COTS parts. Costs have also been reduced by the evolution of the foundry model in which the design of a chip is separated from the manufacturing of the chip, enabling designers to seek out the lowest-cost manufacturer.

Test and Evaluation Issues in VLSI Chip Design

At the design stage of a VLSI chip, test and evaluation involve verification that the chip will perform as required. Note that although advanced VLSI chip technology requires a capability-driven design methodology, as the technology matures the collection of capabilities begins to stabilize. Design tools are used that can synthesize a system from hardware description language code as the requirements and using standard cell library as the target implementation.

Several kinds of tools support test and evaluation at the design stage. A simulation of a system model can be performed in which a description of the design is used to represent the functionality. Stimulus can be applied to the simulator, and responses can be observed and evaluated. Unfortunately, simulation test and evaluation cover only as much as what is provided by the stimulus. In addition, the number of possible interactions among the components in a chip design grows exponentially with the number of these components, therefore making exhaustive simulation testing infeasible.

The incompleteness of simulation for verification has given rise to formal and semiformal approaches for verification. A mathematical model is derived from the VLSI design, and a set of formal specifications of the requirements on the design must be created. A tool can then perform an analysis based on mathematical logic proof techniques and can verify that the design conforms to the specifications. Unfortunately, formal analysis has two significant problems: (1) a full set of formal specifications on the VLSI design must be created, which is often too difficult to obtain, and (2) the mathematical analysis is often too computationally complex to perform. The problem has been shown to be equivalent to the Boolean satisfiability problem, which has a solution complexity that is NP-complete.[23]

Test and evaluation of VLSI chips have worked around the limitations of simulation and formal mathematical analyses with several strategies. Industry standards have been developed that permit the uniform application of testing methodologies for all components on a chip.[24] These methodologies encourage modeling and verification at higher levels of abstraction, such as at the transaction level,[25] to reduce the complexities of simulation. The methodologies also include semiformal methods to improve simulation coverage of design. Rather than requiring a full system specification, the standards include mechanisms to express hierarchically functional specifications at different locations in a design. Also, stimulus to a design can be expressed as constraints on values of the input stimulus to a chip design.[26] From these constraints, software tools will automatically generate test patterns and will also monitor the behavioral specifications.

In addition, the *Virtual Platforms*[27] technology has been developed allowing designers to work with a very-high-level system simulation model, which consists of a collection of "virtual" parts that represent subsystems. These virtual parts are intended to be available before the full system design is available but can provide an early simulation of behavior. This, in turn, is used to help guide the development of the system's requirements. Note that the *Virtual Platform* technique is applicable to any system where requirements are developing simultaneously with subsystem capabilities.

Test and Evaluation Issues in VLSI Chip Manufacturing

During chip manufacturing, samples of small-scale integrated circuits could easily be tested in laboratories at the desks of workers using traditional electronic test instruments. But as chip volumes increased, the integrated circuit automatic test equipment[28] (ATE) industry introduced machines in which chips could be mounted and then subjected to stimulus, and responses from chips could be analyzed. The chip manufacturing process produces a fairly large percentage of nonfunctioning chips with yields that range from 50% up to about 90%, requiring that each manufactured chip be tested.

Design for manufacturing (DFM)[29] is a collection of techniques used to improve testing of chips after manufacture. Software tools are typically used to automatically generate test patterns that are applied to each chip under test. However, complex chips require support in the design to allow access to areas that are deep in the design. The IEEE standard for boundary scan[30] defines a mechanism to open up access to the deeply nested internal parts of a chip for use by the ATE tester. Built-in self-test (BIST) embeds additional functionality in a design to enable self-diagnosis functionality at the power-up of a chip or to allow reporting of performance-related data to external testing. The primary goal is to understand and thereby improve the yield of manufactured parts.

Self-Monitoring VLSI and Electrical Systems

Even with full testing of electrical systems with embedded VLSI chips, it is possible for such systems to be unintentionally or deliberately damaged so that the systems do not perform correctly. For installations where correct behavior is expected at all times, the concept of self-monitoring and self-repair has evolved. A separate subsystem monitors, or observes, the system's behavior when the system is in active use. A description, or a specification, of correct behavior is defined, and the monitor checks the intended behavior against the actual observed system behavior. If incorrect behavior is observed, the monitor will calculate a recovery path and will attempt to signal the system to begin a repair action. Note that physical damage might prohibit recovery.

Lessons from VLSI Chip Design and Challenges for Micro- and Nanoscale Technology (MNT) Test and Evaluation

The discussion of the evolution of VLSI chip design and test can be applied to the evolution that is foreseen in MNT. The core MNT technology will continue to advance and will drive new system design methodologies. However, just as in the evolution in VLSI technology, eventually there will be solutions that allow these capabilities to be harnessed into systems that can be properly tested and evaluated. Abstractions of MNT subsystem properties are essential to allow methodical approaches to designs and tests. Models of MNT subsystem defects are needed. They will be used to guide the development of externally applied tests and, for the development of internal built-in self-tests, will guide any self-repair capabilities. The ability to allow observability and controllability to internal parts of VLSI chips has been extremely valuable for test and evaluation, and a similar capability will be necessary for MNT systems. It will also be necessary to understand the issues with interconnecting MNT subsystems with sufficiently accurate models of interfaces to ensure that the infrastructure that connects the MNT subsystems and to non-MNT subsystems can be designed and tested correctly. Finally, defect rates for MNT must improve to an acceptable level to enable systems engineering test and evaluation approaches to be used in a reliable and predictable way.

Test equipment is needed that can accommodate MNT systems to collect data to assist in characterization of the model, identify defective parts, and to ensure compliance with any risk constraints. For MNT subsystems that are to be embedded inside VLSI chip designs, traditional VSLI CAD tool models of the interface to the MNT subsystem will be needed. For test and evaluation, this includes simulation models to enable the MNT subsystems to be simulated with the remainder of the VLSI design.

Summary

Test and evaluation engineering practices play a critical role in all stages of any systems development. This chapter has presented the traditional features of test and evaluation in systems engineering and has discussed the challenges of an evolving MNT technology that has a bottom-up, capability-driven design methodology. A comparison was made with the evolution of VLSI microchip manufacturing, design, and test and evaluation strategies. The significant factors that enabled the success of VLSI microchip manufacturing include sufficient understanding of the manufacturing process such that accurate models can be developed to support analysis and verification of significant properties of a microchip at a wide range of levels of abstraction, understanding of manufacturing design principles to permit automation of the design process and understanding and characterizations of manufacturing failure modes, and to permit efficient testing of parts. For the successful integration of subsystems that are built with evolving MNT, we can expect a similar evolution in test and evaluation methodologies.

References

1. Kossiakoff, A., and Sweet, W., *Systems Engineering Principles and Practice*, John Wiley, New York, 2003.
2. *Test and Evaluation Management Guide*, Defense Acquisition University Press, Fort Belvoir, VA, January 2005.

3. "IEEE Standard for Software and System Test Documentation," *IEEE Standard 829-2008*, 1–118, July 18, 2008, doi: 10.1109/IEEESTD.2008.4578383.

4. MIL-HDBL-781, *Military Handbook, Reliability Test Methods, Plans, and Environments, for Engineering Development, Qualification, and Production*, 1987.

5. Osiander, R., and M.A.G. Darrin, editiors, *MEMS and Microstructures in Aerospace Applications*, CRC Press, Boca Raton, FL, 2005.

6. Bell, W.D., and Brown, C.D., "Systems Engineering and Test and Evaluation—The Integrated Process," *ITEA Journal* 2010; 31: 57–62.

7. McManus, H., and Hastings, D., "A Framework for Understanding Uncertainty and Its Mitigation and Exploitation in Complex Systems," *Proceedings of the Fifteenth Annual International Symposium of the International Council on Systems Engineering* (INCOSE), July 2005.

8. Silberberg, D.P., Mitzel, G.E., "Information Systems Engineering," *Johns Hopkins APL Technical Digest* 2005, Vol. 26, Part 4, 343–349.

9. Krishnaswamy, S., Markov, I.L., and Hayes, J.P., "Tracking Uncertainty with Probabilistic Logic Circuit Testing," *Design and Test of Computers, IEEE* 2007; 24(4): 312–321, doi: 10.1109/MDT.2007.146.

10. Nicolaidis, M., Achouri, N., and Anghel, L., "Memory Built-In Self-Repair for Nanotechnologies," *On-Line Testing Symposium, 2003. IOLTS 2003. Ninth IEEE,* pp. 94–98, July 7–9, 2003; doi: 10.1109/OLT.2003.1214373.

11. Reed, I.S., and Solomon, G., "Polynomial Codes over Certain Finite Fields," *Journal of the Society for Industrial and Applied Mathematics (SIAM)* 1960; 8(2): 300–304, doi:10.1137/0108018.

12. McConnell, R., and Rajsuman, R., "Test and Repair of Large Embedded DRAMs," *Test Conference, 2001. Proceedings. International*, 63–172, 2001.

13. Moore, G.E., "Cramming More Components onto Integrated Circuits," *Electronics Magazine* April 19, 1965; 38(8): 114–117.

14. Mead, C., and Conway, L., *Introduction to VLSI Systems*, Addison-Wesley, Reading, MA, 1979.

15. Turley, J., "Motoring with Microprocessors," *EE Times*, August 11, 2003.

16. Nagel, L., and Pederson, D., "SPICE (Simulation Program with Integrated Circuit Emphasis)," University of California, EECS Department, April 1973.

17. Brown, S., and Rose, J., "FPGA and CPLD Architectures: A Tutorial," *IEEE Design and Test of Computers* 1996; 13(2): 42–57, doi:10.1109/54.500200.

18. Hwang, F.K., Richards, D.S., and Winter, P., *The Steiner Tree Problem*. Elsevier, North-Holland, New York, 1992 (Annals of Discrete Mathematics, vol. 53).

19. Karp, R.M., "Reducibility Among Combinatorial Problems." In R.E. Miller and J.W. Thatcher (editors). *Complexity of Computer Computations*. New York: Plenum. 1972, 85–103.

20. Blum, C., and Roli, A., "Metaheuristics in Combinatorial Optimization: Overview and Conceptual Comparison." *ACM Computing Surveys*. 2003; 35: 268–308.

21. Tummala, R.R., and Madisetti, V.K., "System on chip or system on package?," *IEEE Design and Test of Computers* 1999; 16(2): 48–56, doi:10.1109/54.765203.

22. Boehm, B., and Abts, C., "COTS Integration: Plug and Pray?," *Computer* 1999; 32(1): 135–138, doi:10.1109/2.738311.

23. Cook, S.A., "The Complexity of Theorem Proving Procedures," in *Proceeding of the Third Annual ACM Symposium on Theory of Computing*, pp. 151–158, Association for Computing Machinery, 1971.

24. "Universal Verfication Methdology (UVM) 1.0 User's Guide," Accellera, May 2010.

25. Glasser, M., *Open Verification Methodology Cookbook*, Springer, New York, 2009.

26. Yuan, J., Pixley, C., and Aziz, A., *Constraint-Based Verification*, Springer, New York, 2006.

27. Popovici, K., and Jerraya, A., "Virtual Platforms in System-on-Chip Design," *Proceedings of the Design Automation Conference*, June 2010.

28. Kazamaki, T., "Milestones of New-Generation ATE," *IEEE Design and Test of Computers* 1985; 2(5): 83–89, doi:10.1109/MDT.1985.294821.

29. Carballo, J.-A., Zorian, Y., Camposano, R., Strojwas, A.J., Kibarian, J.K., Wassung, D., Alexanian, A., Wigley, S., and Kelly, N., "Guest Editors' Introduction: DFM Drives Changes in Design Flow," *IEEE Design and Test of Computers* 2005; 22(3): 200–205, doi:10.1109/MDT.2005.61.

30. Joint Test Action Group (JTAG) [IEEE Standard 1149.1, Standard Test Access Port and Boundary-Scan Architecture].

14

Developing and Implementing Robust Micro- and Nanoscale Technology Programs

Janet L. Barth

Developing cost effective
test and evaluation methods.

Contents

Introduction

Technology development programs are critical to increasing the capability of emerging and deployed systems and, in the case of private companies, to long-term growth and profits. The success of research and development programs has inherent uncertainty; therefore, potential return of these projects is riskier than for well-established programs. The program implementation plan must include a process to track and assess the risk of the new technology development throughout the life cycle. The risk of incorporating a new technology in a program is tracked by the program's systems engineering process to ensure that the realized benefits of the technology will not have an adverse impact on the program's budget and schedule. Risk management is particularly important for micro- and nanotechnologies (MNTs), which are complex systems. And the requirement for developing interfaces to adapt MNTs for human use adds further complexity in the development of MNTs and complicates assessing programmatic risk.

This chapter discusses characteristics of an organizational culture that not only fosters innovation but also employs appropriate, tailored systems engineering techniques to increase the likelihood that new technology developments will move beyond scientific curiosity to useful products. The implementation of development programs is reviewed with information specific to the development of new technologies. Topics include management activities such as planning, risk management, and assessment of progress. Quality management, specific to MNTs, is also reviewed. The chapter concludes with a discussion of best practices for technology development programs.

Successful Technology Development—Organizational Culture

To achieve robust technology development programs leading to the successful development of smart systems, it is paramount for the technology development (TD) manager to create a culture that fosters innovation, rapid development, and accelerated insertion into applications. This organizational culture is needed to keep pace with technology innovation. The success of a new technology development is the responsibility of the technology development manager. The TD manager not only sets the research vision for an organization but must also ensure that technology projects are completed accurately, within budget and on time.

It is a difficult challenge for advanced technologies to bridge the gap between the research laboratory and operational systems. The uniqueness, complexity, and uncertainty inherent in micro- and nanotechnology (MNT) development increase the risk that these new technologies will not reach a maturity level that is required to be embedded in existing or new applications or to be successful stand-alone commercial products.

According to an analysis performed by the National Research Council,[1] common characteristics of successful technology transition efforts include the following:

- The establishment of "Skunk Works-like" environment—these groups are committed to new technology transfer into internal programs and/or commercial markets
- Multidisciplinary teams led by champions who inspire and motivate their teams toward specific goals
- Team determination to make the technology succeed—which may include making the technology profitable and demonstrating to customers that they need the technology
- Use of expanded mechanisms of open and free communication—especially involving the ability to communicate an awareness of problems that will affect process goals
- Willingness of the champion to take personal risk—such leadership results in the willingness of the organization to take risks at the enterprise level

Additionally, success stories from commercial, sports, and defense industries suggest that the characteristics of such a culture include the following:

- Acceptance of risk, anticipation of failure, and plans for alternatives
- A flexible environment with the ability to accommodate change during the development process
- Open communication in all directions without regard to hierarchy
- Widespread sense of responsibility and commitment to success that exceed defined functional roles
- Valuing innovation over short-term economic efficiency
- A passionate focus on the end-user's needs

It is clear that innovation requires an organizational culture that accepts, defines, and manages risk. Chapter 6 discussed technology programs that did not fare well with a traditional top-down systems engineering approach or were managed without applying systems engineering processes. Successful technology development requires defining and managing risk throughout all phases of the development program by including rigorous systems engineering processes and having systems engineers (i.e., "product" systems engineers) embedded on technology development teams. This is especially critical for MNT developments due to their uniqueness and complexity, the immaturity of the processes and tools used in their development, and the rapid pace of their development.

Implementation of Technology Development Programs

A description of the life-cycle stages in development programs is shown in Table 14.1 (from Chapter 2). The advancement of the development program is controlled by decision gates where progress is reviewed against established, stakeholder coordinated metrics, and technical and schedule risks are assessed. Here we will delve further into systems engineering life-cycle stages as they apply to new technology development. Issues specific to MNT development will be discussed drawing on information presented in other chapters in this book.

The uniqueness and complexity of MNT development requires a defined process to review and evaluate these programs as they progress. The purpose of the process is to accept the uncertainty of the MNTs and manage risk to maximize the organization's return on research and development investments and to increase the likelihood of success for innovations. For a technology that is being developed to be an enabler for a larger system development, the process must also track the technology readiness level (TRL) (see Chapter 3).

For the TD manager, bridging the gap between the concept and production stages carries inherent uncertainty and requires a carefully thought-out program implementation plan. Risk is managed by setting *decision gates* where the options are to execute the next stage, continue in the current stage, return to a preceding stage, hold the project activity, or terminate the project. Often, technology development

TABLE 14.1
Description of Life-Cycle Stages

Life-Cycle Stages	Purpose	Decision Gates
Concept	Identify stakeholders' needs; explore concepts; propose viable solutions	*Decision options:* *– Execute next stage* *– Continue this stage*
Development	Refine system requirements; create solution description; build system; verify and validate system	*– Go to a preceding stage* *– Hold project activity* *– Terminate project*
Production	Produce systems; inspect and test (verify)	
Utilization	Operate system to satisfy users' needs	
Support	Provide sustained system capability	
Retirement	Store, archive, or dispose of the system	

Source: Adapted from Committee on Accelerating Technology Transition, National Research Council, *Accelerating Technology Transition: Bridging the Valley of Death for Materials and Processes in Defense Systems*, The National Academies Press, Washington, DC, 2004.

programs are managed using the same system engineering management plan (SEMP) as for traditional programs (e.g., the National Aeronautic and Space Administration's [NASA] 7120.5D[2]), frequently producing negative results for program advancement.[3] Decision gate criteria used in traditional SEMPs to determine whether the program can proceed to the next phase will likely guarantee that a new technology program will be canceled because the uncertainties and risks are difficult to define, poorly understood, or may be too large.

Cooper developed a modified process for tracking progress for technology development programs that consists of five phases and four decision points.[3] The phases or "stages" in Cooper's model are a set of best practice activities that reduce the unknowns in the development at each phase, thereby reducing risk from phase to phase. The decision points determine whether the technology development process will receive the resources to proceed to the next phase. Figure 14.1 (after

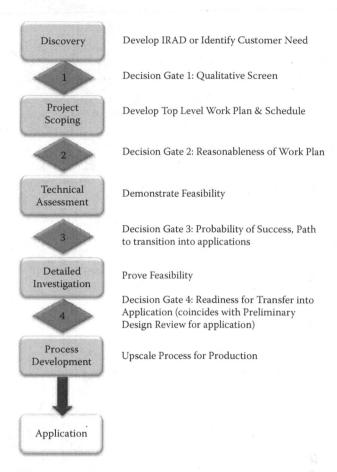

Figure 14.1 Technology development process for technology development programs. (Adapted from Cooper, Robert, Managing Technology Development Projects, *IEEE Engineering Management Review*, 35(1), 2007.)

Cooper) shows the phases and key decision points from discovery to process development. The next section presents a brief description of each phase of the process.

Phases of Technology Development Programs

The process starts with *discovery*. The idea can be the result of a competitive research initiative, a technology road-mapping exercise, strategic planning that is part of a regular business plan, a brainstorming session, or input from a customer with a specific need. Ideas for new technology development are next screened at the first decision gate. The criteria for passing from the first phase are largely qualitative and determine if the discovery warrants additional investment of research funds. Criteria for success can include likelihood of success, return on investment, fit to the organization's overall strategic plan, or criticality of the customer's need.

The second phase is *project scoping*, which defines the scope of the project and creates a top-level work plan and schedule. During this time, a literature search, a patent and intellectual property search, and resource gaps assessment are performed. Colleagues are engaged, and potential outside partners are identified for teaming. The output product of this phase is the *preliminary technical assessment* of the program that is presented at Decision Gate 2. The second decision gate is also mostly qualitative and determines if the top-level work plan is reasonable and the projected resources are adequate to support the next phase of the program.

Phase 3 is the *technical assessment* effort that must demonstrate the viability of the discovery. During this time, the ad hoc research team is formed to perform technological analyses and experiments to demonstrate feasibility and to address required interfaces to existing systems. The team also defines the resources required to complete the project including the level of effort, maturity of modeling and analysis tools, adequacy of laboratory facilities and equipment, and required research disciplines. At this point, the success of the development can be increased by indentifying a dedicated product systems engineer at the beginning of this phase. For a technology that will be embedded in a larger system, the systems engineering team for the larger program also must be involved.

At the end of this phase, the technical assessment is conducted at Decision Gate 3, the first rigorous screen for the program. This review is modeled after traditional systems engineering principles. The review assesses the level of development to measure design maturity, reviews technical risk, and determines whether to proceed to the next level of development. For MNT development, this review also evaluates emerging properties of the MNT and assesses the (positive and negative) impact on customer requirements. At this decision gate, the technical review addresses program risk and eases the transition to detailed investigation phase by

- Assessing the maturity of the design/development effort
- Assessing the maturity of modeling and simulation tools
- Clarifying design requirements for the end product
- Challenging the design and related processes
- Checking proposed design configuration against technical requirements, customer needs, and system requirements
- Evaluating the system configuration at different stages of the development
- Providing a forum for communication, coordination, and integration across all disciplines
- Establishing a common configuration baseline from which to proceed to the next level of design
- Recording design decision rationale in the decision database

These formal technical reviews are typically preceded by a series of technical interchange meetings where issues and problems and concerns surface and are addressed. The formal technical review is not the place for problem solving but to verify problem solving has been done; it is a process rather than an event.

Planning for technical reviews must be extensive and up front and early. Important considerations for planning include

- Timely and effective attention and visibility into the activities preparing for the review
- Identification and allocation of resources necessary to accomplish the total review effort
- Tailoring consistent with program risk levels
- Timing consistent with availability of required data and resources
- Establishing event-driven entry and exit criteria
- Where appropriate, conduct of incremental reviews

- Review of all system functions and the impact of new technology
- Confirmation that all system elements are integrated and balanced

Reviews should consider the testability, producibility, needed training, and supportability for the system, subsystem, or configuration item being addressed. The depth of the review is a function of the complexity of the system, subsystem, or configuration item being reviewed and the related maturity of the technology. Where the system is pushing state-of-the-art technology, the review will require greater depth and insight than if for a commercial off-the-shelf product. Items that are complex or an application of new technology will require more detailed scrutiny.[4]

Criteria for success are similar as for the first decision gate (likelihood of success, return on investment, fit to the organization's overall strategic plan, or criticality of the customer's need); however, the criteria will be more rigorous due to the increased maturity of the MNT innovation and because the next phase (detailed investigation) requires a large increase in the financial investment. The process will assess if the technology product continues to fit the overall strategic plan for the organization, has strategic leverage, has a high probability of technical success, and has a path to transition into applications. The technical screening will also assess whether the product has high potential for up-scaling of production. To increase the assurance that there is a feasible path to transition into operations, systems engineering must be engaged at this key decision point and the stages leading up to the decision.

The detailed investigation is performed in the fourth phase of the technology development program to prove technological feasibility and to assess the scope of the technology and its value to the organization. The program technical lead, the TD manager, and the product systems engineer must work closely together to set milestones and project reviews within this phase to ensure that the technology development stays on course and that the program is being managed to cost and schedule. In this phase, it is critical to engage tailored systems engineering processes to increase the likelihood of successful development. Sound systems engineering processes will track the technology product capability against requirements for targeted applications, develop and implement risk management strategies during development, and capture capabilities that are driven out in this investigation phase. The TD manager also begins to prepare a technology transition plan to present at the decision gate at the end of this detailed investigation phase.

Using modeling and simulation is critical early in this phase of development. Modeling and simulation provide virtual duplication of products and processes and represents those products or processes in readily available and operationally valid environments. Until an MNT product is available with full system specifications, the TD manager and product systems engineer will rely heavily on the results of modeling and simulation to control costs and manage risk in the MNT transition to the system that it enables. Chapter 2 discussed the importance of modeling and simulation in the systems engineering process, and Chapter 10 reviewed modeling and simulation capabilities for MNT scale systems. As high-fidelity prototypes become available near the end of the detailed investigation phase, test and evaluation must be conducted to ensure that all customer requirements are being met. However, in the case of MNT technology development, full system specifications may never be available for testing. In this case, successful new technology deployment will rely on a combination of highly developed simulations and nontraditional test and evaluation techniques such as those developed for the very-large-scale integration (VLSI) integrated circuits. Chapter 13 discussed the evolution of VLSI test and evaluation and techniques and outlines requirements for similar nontraditional techniques for MNTs.

The results of the detailed investigation phase are presented at the final decision gate, Decision Gate 4, which coincides with the preliminary design review (PDR) if the product is being inserted into a larger system. This decision gate determines if the technology product is ready for transition into production. The management team must also determine if the technology is ready for upscaling. Questions to ask are

- Is the technology well understood to the point where the application is known?
- Have performance requirements for the technology been established and include thresholds and goals?

If the systems engineering team is using the NASA definition of technology readiness levels (TRLs), the technology needs to be at TRL 6 (i.e., a high-fidelity system/component prototype that adequately addresses all critical scaling issues is built and operated in a relevant environment to demonstrate operations under critical environmental conditions).

Chapter 3 discussed the system engineering process as it is applied to technology maturation and the requirement that a new technology be at TRL 6 by the system preliminary design review. Figure 14.2 shows how a technology development plan fits into the overall flow of the systems engineering process for a target system application.[5] The need for a change during a system's life cycle can come from many sources and affects the configuration in infinite ways. Customer needs can increase, be upgraded, be deleted, or are in flux, and technology advancement may allow the system to perform better or less expensively.[4] MNTs are complex systems; therefore, they usually do not have stagnant configurations and may display emergent properties that may be beneficial or detrimental to customer requirements.

Technology Development Management

The TD manager[6] oversees the transition or optimization of technology products or capabilities within existing or new systems. In this role, he or she must act as a technical liaison with the technology applications manager, project managers, program systems engineers, and the technology development team's product system engineer.

The TD manager is responsible for directing the technical activities during the life cycle of the technology development program. Activities include

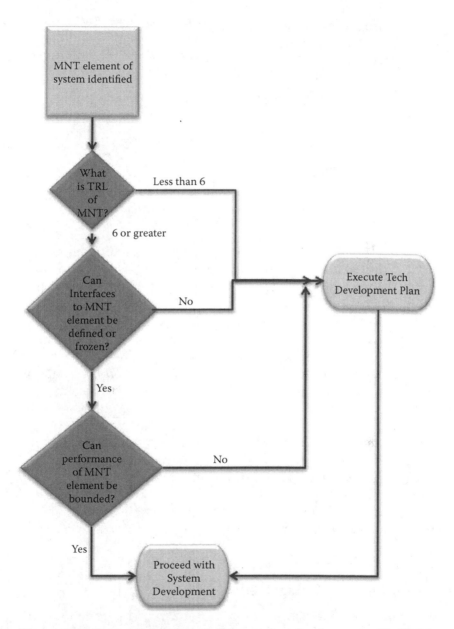

Figure 14.2 Notional decision tree for system development with Multi-Scale Systems (MuSS) (Kusnierkiewicz).

- Definition of program objectives, milestones, product capability, and cost
- Technical planning
- Requirements management
- Interface management
- Technical risk management
- Configuration management
- Technical data management
- Technical assessment
- Decision analysis
- Quality management
- Lessons learned

Technical Planning

Technical planning is a process for the identification, definition, and planning of the technical effort necessary to meet project objectives for each phase of the technology development program. The technology development manager must determine deliverable work products from technical efforts, technical reporting requirements, entry and success criteria for technical reviews or decision gates, product and process measures to be used, critical technical events, data management approach, technical risks to be addressed in the planning effort, and tools and engineering methods to be employed. The manager is also the team lead who is responsible for obtaining stakeholder commitments to the technical plans, planning the approach to acquire and maintain technical expertise needed throughout the program, and issuing authorized technical work directives to implement the technical work. The overriding system engineering management plan (SEMP) and other technical plans must be put in place to monitor and manage progress.

Requirements Management

The management of the technical requirements includes providing bidirectional traceability and change management for established requirement baselines over the life cycle of the system products. The TD manager is responsible for preparing or updating a strategy for requirements management, selecting an appropriate requirements management tool, training technical team members in established requirement management procedures, conducting expectation and requirements traceability, managing expectation and requirement changes, and communicating expectation and requirement change information. The need for a change during a system's life cycle can come from many sources and affects the configuration in infinite ways. Customer needs can increase, be upgraded, are different, or are in flux, and technologies may be developed that allow the system to perform better or less expensively.[4] A change in requirements creates challenges for any new technology development. The challenges are even greater for MNTs because they are complex systems (Chapter 1), there are scaling issues (Chapter 5), and they need to interface with macrosystems (Chapter 11). MNTs exhibit emerging behavior, so it is important for the product development systems engineer to track properties that can be enabling.

Interface Management

The central activity of the system dictates the design requirements of the system. Design decisions regarding interfaces should be made consistent with what directly or indirectly contributes to the central activity of the system. In order to efficiently manage the system interfaces, the central activity of the system must first be identified and always remain the top priority as each component of the system is broken down and planned and it is seen what can be measured or assessed.

Interface management is the establishment and use of formal processes to establish interface definition, details, and compliance among the end products and enabling products. This includes preparing interface management procedures, identifying interfaces, maintaining interface documentation, disseminating interface information, and conducting interface control. MNT developments pose additional challenges to the management of interfaces due to their complexity. Also, interfaces may be multitiered, and parameters at the interfaces are difficult to measure.

Figure 14.3 Interface elements of small, smart systems.

Embedded MNTs mean that heterogeneous macro-/meso-/ micro-/nano-elements must all be integrated together (see Figure 14.3), which requires careful tracking of interface requirements during the development. As discussed in Chapter 11, micro- and nanoscale interface design and control principles cannot be directly transcribed from fundamental systems engineering due to the dynamic nature of the interface at small size scales. Yet, fundamental systems engineering principles that dictate interface treatment can be built upon to devise a plan for the treatment of MNT system interfaces. The author of Chapter 11 notes that there are four basic practices from macro-interface control that can be adapted and applied to MNTs:

- Define system boundaries.
- Identify the internal and external interfaces.
- Map out the functional and physical allocations at each interface.
- Manage the interfaces and incorporate necessary feedback mechanisms.

Technical Risk Management

Technical risk management is paramount to successfully introducing a new technology into a program. It requires continuous examination of the risks of technical deviations from the plans and identification of potential technical problems before they occur so that risk-handling activities can be planned and invoked as needed across the life of the product. This includes developing the strategy for technical risk management, identifying technical risks, conducting technical risk assessment, developing technical risk mitigation processes, monitoring the status of each technical risk, and implementing technical risk mitigation and contingency action plans when applicable thresholds have been triggered. To leverage bottom-up nanotechnology to exploit self-healing properties, adaptability, and scalability, product systems engineering may need to create entirely new concepts for handling uncertainty not only in the environment but in the state of the system as it evolves.

Configuration Management

Configuration management is the process of documenting the configuration of the product at various points in time. This involves systematically controlling changes to the product configuration and preserving the integrity and traceability of the database throughout the full life cycle of the program. Activities include establishing configuration management strategies and policies, identifying baselines to be under configuration control, maintaining the status of configuration documentation, and conducting configuration audits. The use of configuration management early in the technology development process is critical for MNTs due to their complexity and emerging characteristics.

Technical Data Management

Technical data management is the process of identifying and controlling data requirements. This includes acquiring, accessing, and distributing data needed to develop, manage, operate, and support system products; managing the disposition of data as records; analyzing data use; obtaining technical data feedback for managing the contracted technical efforts; and assessing the collection of appropriate technical data and information. Technical data management strategies and policies must be formally established to maintain stored technical data, provide technical data to authorized parties, and collect and store required technical data.

Technical Assessment

Technical assessment involves monitoring the progress of the technical effort and providing statistical metrics to support system design, product capability and performance, and technical management efforts. This includes developing technical assessment strategies and policies, assessing technical work productivity, assessing product quality, and conducting technical reviews. As discussed throughout this book, nanotechnologies may have emerging properties. It is critical for the product systems engineer to understand and track these properties against the requirements matrix to (1) ensure that requirements are met and (2) to capture enabling properties that may enhance the capability of the system.

Technical Decision Analysis

Technical decision analysis provides for the evaluation of technical decision issues, technical alternatives, and their uncertainties to support decision making. This is done throughout technical management, system design, and product realization to evaluate the impact of decisions on performance, cost, schedule, and technical risk. It includes establishing guidelines for determining which technical issues are subject to formal analysis processes, defining the criteria for evaluating alternative solutions, identifying alternative solutions to address decision issues, selecting evaluation methods, selecting recommended solutions, and reporting the results and findings with recommendations, impacts, and corrective actions.

Quality Management

The uniqueness of MNTs requires a closer look at the reliability aspect of the quality management plan. Quality management is the planned and systematic activities necessary to provide adequate confidence that the product or service will meet the given requirements. The generic elements of a good quality management plan apply to the development and production of MNTs. The quality management plan must be carried out early in the formulation phase of the project and includes a broad range of activities: change control; procurement; receiving, processing, fabricating, assembly, test, and inspection control; contamination control; handling, packaging, packing, and storage controls; quality records; quality audits; process improvement; reliability; and safety. In many organizations, quality management plans are governed by ISO standards such as the AS9100:2001.[7]

Reliability is a critical element of quality management. Reliability is the ability of a system or component to perform its required functions under stated conditions for a specified period of time.[8] Chapter 10 discussed the importance of modeling and simulation and reliability, and Chapter 12 discussed reliability theory and gave an overview of reliability considerations for the technology development manager. The manager must first understand the failure mechanisms, material properties, processing and fabrication, and interfacing issues. All of these drive the strategies for life testing, accelerated stress testing, screening, reliability prediction, and so forth. Second, the additional challenges to quality assurance result from the lack of reliability and failure data. For new technologies, it is nearly impossible to quantify reliability with rigorous statistics. This challenge is addressed using the "physics of failure" approach as discussed in Chapter 12. Finally, there is a need to educate statisticians and reliability engineers about the micro- and nanoscale world with regard to qualification testing and application-specific environmental testing protocols being developed for MNTs.

Another unique aspect to consider for MNTs is the ability to ensure reliability through built-in self-repair and exploitation of their small scale by employing massive redundancy. The implications of these qualities will be discussed in greater detail in Part 4 of this book.

Micro Electro Mechanical Systems (MEMS) Reliability

A critical part of understanding the reliability of any system comes from understanding the possible ways in which the system may fail. In MEMS, there are several failure mechanisms that have been found to be the primary sources of failure within devices. In comparison to electronic circuits, these failure mechanisms are neither well understood nor easy to accelerate for life testing. In any discussion of failures, the definition of failure mechanisms, or causes of failure, often overlaps with the definition of failure modes, or observable failure events.[9] Table 14.2 gives several known failure modes for MEMS devices.

TABLE 14.2

Micro Electro Mechanical Systems (MEMS) Failure Modes

MEMS Failure Mode	Description	Impact
Mechanical fracture	Breaking of a uniform material into two separate sections	Catastrophic failure of device
Stiction	Surfaces join together due to strong primary bonds	In most MEMS devices, failure of device
Wear	Removal of material from a solid surface as the result of mechanical action	Fatigue cracks, flaking, increase in voltage required to drive the device
Delamination	A materials interface loses its adhesive bond	Shorting or mechanical impedance, loss of mass alters the mechanical characteristics of the structure
Environmentally induced failure mechanisms	Vibration, shock, humidity effects, particulates, temperature changes, electrostatic discharge	Wide range of effects specific to the environment[a]
Stray stresses	Stresses in thin films that are present in the absence of external forces	Small stresses—noise in sensor outputs; large stresses—lead to mechanical deformation
Parasitic capacitance	Unwanted capacitive effect in a device either between the device and substrate or within the device	Unwanted electrical and mechanical behavior
Dampening effects	Increased dampening due to system degradation	Decrease in resonant frequency yields increase in sensitivity; large structural dampening yields changes in resonance that will alter output, which can be a long-term reliability concern

[a]   For more detail, see Jeng, S.-L., Lu, J.-C., and Wang, K., *IEEE Transactions on Reliability*, 56(3), 2007.

Nanotechnology Reliability

Compared to MEMS technologies, the understanding of reliability for nanoscale technologies is in its infancy stages. Research in reliability is at the initial stages of applying the concepts of reliability to the nanoscale. Progress in understanding the physical failure mechanisms during fabrication and operation are critical to the advancement of systems enabled by nanoscale technologies. Central to this understanding is our ability to model the degradation, reliability functions, and failure rates of nanoscale systems. The emerging field of research on nanotechnology reliability addresses four main tasks:

- Introduction of concepts and technical terms of reliability to nanotechnology in an early state
- Identification of physical failure mechanisms of nanostructured materials and devices during fabrication process and operation
- Determination of quality parameters of nanodevices, failure modes, and failure analysis, including reliability testing procedures, and instrumentation to localize nanodefects
- Modeling of reliability functions and failure rates of nanosystems[10]

The directive to conduct basic research on issues related to the development and manufacture of nanotechnology, including metrology, reliability, and quality assurance; processes control; and manufacturing best practices is in the charter of the U.S. National Institute of Standards and Technology (NIST).[11] NIST supports the U.S. nanotechnology enterprise from discovery through production via its Center for Nanoscale Science and Technology (CNST), established in 2007. The CNST is building the infrastructure to assess the reliability of nanotechnology by providing rapid access to a world-class, shared-use nanofabrication facility (the NanoFab) and creating the next generation of nanoscale measurement instruments, made available through collaboration. NIST's Technology Innovation Program is investing in the development of transformational technologies necessary to advance the large-scale manufacture of nanomaterials. In coordination with the private sector, NIST provides the technical foundation and leadership to the development of international nanotechnology reliability standards that are key to opening new markets and facilitating trade and commerce.[12]

Moreover, the concept of exploiting self-healing properties, adaptability, and scalability of nanotechnologies to increase their reliability and the reliability of systems that they support is a developing capability that is being explored for third- and fourth-generation nanotechnologies. This is analogous to the microelectronics industry that used the small scale of devices to build in self-testing and redundancy. These concepts are developed in greater detail in Part 4 of this book.

Systems Engineering Approaches for Micro- and Nanotechnology Development

A study by the National Research Council[1] recommended "best practices" for technology development, which are especially critical for MNT development. The first is developing a "viral" process for technology development, meaning that the process is infectious and self-propagating. This process entails quick, iterative development cycles and prototyping of materials and products and must be done in conjunction with potential costumers. As already discussed in Chapter 10, viral development for MNTs is critically dependent on effective modeling of materials and processes that accelerate the iterative process by using predictive models to redesign the development processes.

The second best practice is increasing reliance on functional requirements rather than on specifications. One of the key limitations to the rapid insertion or development of new technology is the lack of information given to vendors about the relevant functional and technological needs. Instead, strict adherence to detailed but incomplete specifications is expected. Specifications are essential for ensuring that a technology product will have an extremely low probability of failure. Previously, we saw that applying overly restrictive project management practices or systems engineering processes on new technology developments increases the chances that promising developments will be stopped in early phases. In this case, overdependence on specification will decelerate the rate of transition.

The third best practice is developing a mechanism for creating successful teams in a sustainable way. Successful teams consist of committed and multidisciplinary individuals who implement iterative prototyping and work to function rather than to specification. Overriding all of the program responsibilities of the TD is the need to develop staff who not only excel in their field of research but also have a passion for seeing their technology products inserted into application areas. For MNT developments, success depends on nurturing multidisciplinary teams (physicists, chemists, biologists, engineers). In the future, multidisciplinary employees, including the product systems engineer, will be highly valued employees in organizations working on MNT-scale developments.

In addition to these best practices, engagement of systems engineering expertise early in the process of new technology development is critical to the success of MNT development programs. Including a product systems engineer who is cross-trained between MNT and systems engineering on the technology development team, directly supporting the TD manager, will greatly increase the success of technology transition from research into products and operations. Table 14.3 lists the responsibilities of product systems engineering for each development activity.

As MNTs mature from the early generations, the product systems engineer will play a critical role in deciding which systems engineering methodology to apply to the technology development process. As noted in Chapter 3, while risks and benefits of reliance on an immature MNT technology must be evaluated on a case-by-case basis, general guidelines may benefit the TD manager, the MNT technologist, and system engineers for deciding when to apply the various risk mitigation techniques. These guidelines are provided in Table 14.4 (from Chapter 3). The first consideration is to determine the most appropriate systems engineering methodology (see Chapter 4) to apply to the development activity. As nanotechnology development matures from generation

TABLE 14.3
Responsibilities of Product Systems Engineering

Development Activity	Product Systems Engineering Responsibility
Technical planning	Develop or update planning strategies for common technical processes; collect information for technical planning; define technical work to be done; schedule, organize, and cost the technical work; directly develop formal technical plans; obtain stakeholder agreements with the technical plans; develop technical work directives; directly capture work products and related information from technical planning activities
Requirements management	Develop strategies for requirements management; ensure that requirements are documented in proper format, baseline is validated, out-of-tolerance technical parameters are identified; approve changes to out-of-tolerance technical parameters; track between baselines; develop and maintain compliance matrices; review Engineering Change Proposals (ECPs) and provide recommendations; implement procedures; disseminate approved changes; capture work products from requirements management activities
Interface management	Develop procedures for interface management; direct interface management during system design; direct interface management during product integration; direct interface control activities; direct capture of work products from interface management activities
Technical risk management	Develop strategies to conduct technical risk management; identify risk; coordinate stakeholders; direct risk analysis; select risks for mitigation; develop risk mitigation/contingency action plans; plan implementation; capture work products from technical risk management activities
Configuration management	Develop strategies to conduct configuration management; identify items to be place under configuration control; establish baseline; contribute to configuration change control; be able to identify content of configuration control; direct systems engineer participation in configuration audits; capture work products from configuration management activities

TABLE 14.3 *(Continued)*
Responsibilities of Product Systems Engineering

Development Activity	Product Systems Engineering Responsibility
Technical data management	Develop strategies to conduct technical data management; direct data for storage; develop lessons learned; ensure measures to protect technical data; develop procedures to access technical data
Technical assessment	Develop strategies to conduct technical assessments; identify process measures; monitor progress against plans; determine degree to which product satisfies requirements; determine product performance variances; select corrective actions; identify type and when a technical review is needed; direct review material preparation; direct action item identification and resolution; chair technical review boards (e.g., Preliminary Design Review (PDR), Critical Design Review (CDR), Test Readiness Review (TRR)); capture work products from technical assessment activities
Decision analysis	Develop for when to use formal decision making and who will make decisions; establish criteria definition for types and range and rank criteria; select evaluation method and solution; identify and evaluate alternatives; capture work products from decision analysis activities

two into generation three, the application of the traditional waterfall methodology is less likely to result in a successful technology development. This topic is discussed in greater detail in Chapter 15.

Summary

The technology development manager of MNTs is responsible for developing and leading multidisciplinary teams of professional researchers and engineers. In the development environment, the manager must maintain a careful balance between an organizational culture that champions risk yet at

TABLE 14.4
Guidelines for Evaluating Alternate Approaches for Incorporating an Immature Micro- and Nanotechnology (MNT) System Component

Consideration	Guideline
1. What is the preferred development methodology? (waterfall versus agile)	If agile is preferred, make sure the technology development iterations are consistent with the system development iterations; otherwise, consider a waterfall or hybrid model with a dedicated technology development phase that precedes the system development.
2. What is the technology readiness level (TRL) of the MNT?	For TRL < 6, develop technology development plan. TRL 1 to 3 may necessitate a technology development phase that precedes the main development effort. In some cases, additional mitigations may be needed (e.g., identify an alternate design to replace the immature MNT, and identify impacts on system and programmatic resources).
3. Can the interfaces between the MNT and the system components be defined and "frozen"?	If they can, and if the risk is low of timely MNT maturation, then it may be low risk to proceed concurrently with system development. If not, a precursor risk-reduction technology development phase may be warranted.
4. Can the performance of the MNT component be bounded?	If not, a precursor technology development phase may be warranted.

the same time ensures that progress in development efforts is monitored with appropriate oversight that includes carefully defined *decision gates*. To successfully bridge the "valley of death" between research and operations (whether the new product is intended for insertion in a larger system or for commercialization), it is essential to include systems engineering expertise on the development team. As noted in Chapter 2, systems engineers must be well versed in systems engineering methodologies and also have in-depth knowledge of the technology area. Product systems engineers who are also trained in disciplines that are at the core to MNT development such as physics and chemistry will be key to successful developments in the next generations of micro- and nanoscale technologies.

Defining customer requirements.

References

1. Committee on Accelerating Technology Transition, National Research Council, *Accelerating Technology Transition: Bridging the Valley of Death for Materials and Processes in Defense Systems*, The National Academies Press, Washington, DC, 2004.
2. http://nodis3.gsfc.nasa.gov/displayDir.cfm?Internal_ID=N_ PR_7120_005D, Effective Date: March 06, 2007, Expiration Date: March 06, 2012.
3. Cooper, R., Managing Technology Development Projects, *IEEE Engineering Management Review*, 35(1), 2007.
4. *System Engineering Fundamentals*, Defense Acquisition University Press, Fort Belvoir, VA, January 2001.
5. Kusnierkiewicz, D., The Johns Hopkins Applied Physics Laboratory, Private Communication, March 2010.

6. "NASA Project Management and Systems Engineering Competencies," www.nasa.gov/offices/oce/appel/pm-development/pm_se_competencies.html, Last Updated: September 18, 2009.

7. "Quality Management Systems—Requirements for Aviation, Space and Defense Organizations, AS9100:2001," http://standards.sae.org/as9100c/, Last updated: January 8, 2011.

8. *IEEE Standard Computer Dictionary: A Compilation of IEEE Standard Computer Glossaries*. Institute of Electrical and Electronics Engineers, New York, 1990.

9. "MEMS Reliability Assurance Guidelines for Space Applications," B. Stark, ed., Jet Propulsion Laboratory, Pasadena, CA, JPL Publication 99-1, January 1999.

10. Jeng, S.-L., J.-C. Lu, and K. Wang, "A Review of Reliability Research on Nanotechnology," *IEEE Transactions on Reliability*, 56(3), September 2007.

11. "Department Organization Orders: National Institute of Standards and Technology," www.osec.doc.gov/omo/dmp/doos/doo30_2a.html, U.S. Department of Commerce, Office of Management and Organization, DOO 30-2A, Effective Date: November 19, 2008, Last Updated: February 18, 2010.

12. "Nanotechnology at NIST Overview," www.nist.gov/nano_overview.cfm, Date created: March 10, 2010, Last updated: October 5, 2010.

PART 4

Systems Engineering Applications— Toward the Future

Part 4 delves into the next generations of micro- and nanoscale systems and discusses future concepts that are integral to moving from the second to third and fourth generations of nanotechnology. The reader is reminded that these technologies are complex, and because they do not have stagnant configurations, systems engineering must account for the risk in their development (Chart IV.1). Details of the challenges of evolving micro- and nanoscale systems toward future generations of micro- and nanotechnology are reviewed to provide systems-oriented engineers a high level of familiarity and competence with the engineering and scientific details of these systems. As we explore the development of biomedical systems, the design of complex nanosystems, and the idea that smaller technologies with high information processing rates typically result in a savings or beneficial reallocation of human metabolic energy or technological energy, we are reminded of the importance of applying systems engineering principles to technology development from the outset (i.e., understanding requirements) (Chart IV.2). Predicting the way forward offers alternatives futures and high- or low-growth prediction (Chart IV.3).

CHART IV.1
Systems Engineering and Complex Systems

Complex systems do not usually have stagnant configurations. A need for a
change during a system's life cycle can come from many sources and affect
the configuration in infinite ways. The problem with these changes is that,
in most cases, it is difficult, if not impossible, to predict the nature and
timing of these changes at the beginning of system development. Accordingly,
strategies or design approaches have been developed to reduce the risk
associated with predicted and unknown changes. Well-thought-out
improvement strategies can help to control difficult engineering problems
related to

- Requirements that are not completely understood at program start
- Technology development that will take longer than the majority of the
  system development
- Customer needs (such as the need to combat a new military threat)
  that have increased, been upgraded, are different, or are in flux
- Requirements change due to modified policy, operational philosophy,
  logistics, support philosophy, or other planning or practices from the
  eight primary life-cycle function groups
- Technology availability that allows the system to perform better or less
  expensively
- Potential reliability and maintainability upgrades that make it less
  expensive to use, maintain, or support, including development of safety
  issues requiring replacement of unsafe components
- Service life extension programs that refurbish and upgrade systems to
  increase their service life

Operational distribution or deployment: Where will the system be used?

Mission profile or scenario: How will the system accomplish its mission objective?

Performance and related parameters: What are the critical system parameters to accomplish the mission?

Utilization environments: How are the various system components to be used?

Effectiveness requirements: How effective or efficient must the system be in performing its mission?

Operational life cycle: How long will the system be in use by the user?

Environment: What environments will the system be expected to operate in an effective manner?

Chart IV.2 Operational Requirements

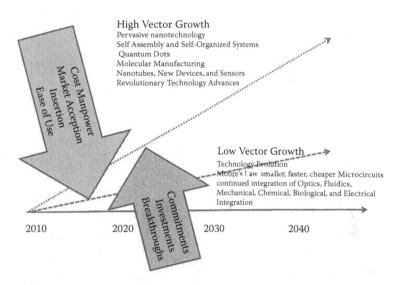

Chart IV.3 Predicting Two Alternative Futures for Nanoscale Technologies

15

Future Generations of Nanotechnology

M. Ann Garrison Darrin
Janet L. Barth

Generational issues?

Contents

Introduction

This chapter provides a baseline or calibration point for looking at the future of systems engineering and micro- and nanoscale technologies (MNTs). MNTs have been discussed in the context of systems engineering of multiscale systems. MNT covers the scales of 10^{-6} to 10^{-9} meters for those who want a linear reference point. There are several key assumptions that one will note in the chapters that have been gathered for this book. The first is that we are dealing with multiscale systems, with innate complexity. Earlier chapters introduced MNTs and the concepts of top-down fabrication and assembly versus bottom-up self-assembly. One finds alignment of the top-down approaches across the macro-, micro-, and nanoscales. Bottom-up or self-assembled technologies present less of an alignment and therefore offer opportunities to explore less traditional systems engineering concepts. This is our second assumption, that nanotechnologies demonstrated by the bottom-up manufacturing techniques have less in common than macro-, micro-, and nanoscale (top-down) systems. This unique region is explored in terms of translating classical systems engineering into the nano realm. A common set of definitions and taxonomy is supplied along with an in-depth discussion of complex systems with a focus on uncertainties and emergent properties. This chapter begins with the drivers for exploring systems engineering in the micro- and nano-world as the field of nanotechnologies continues to emerge. The final assumption is that the nano-world of technologies is not mature; therefore, we are emphasizing exploring systems engineering technologies in the research and development phases as emphasized in Chapter 3. As explained in the next section, the emerging field of nanotechnologies is just entering the third generation. It is interesting to note the lack of articles describing this third generation. The fourth generation is actually covered very well by futurists such as Kurzweil and Drexler, who are discussed in Chapter 18. The first forays into the third generation of nanotechnology are being seen with lab-on-a-chip technologies, such as bioassay, driven by the biomedical engineering world. Chapter 16 offers us a current state of the emerging third generation that includes concepts in self-assembly.

Looking to the Future of Systems Engineering and Micro- and Nanotechnologies

The International Council on Systems Engineering (INCOSE) developed a vision for systems engineering in 2020. This study found that technology innovations are the primary drivers that influence the capabilities of system products, as well as the practice of systems engineering. Key drivers will be the continuing evolution of information technology, with associated applications to both system implementations (both large and small, including microsystems) and to model-based techniques for systems engineering. This vision for the future is already seen in emerging conceptual and technological areas, such as complexity theory, nanotechnology, and genetic engineering that stretch the validity of present systems engineering processes.[1] Further, the same study noted technology trends today that will affect future systems. These trends include

- Increased miniaturization, including nanotechnologies
- Increased use of biotechnology
- Increased connectivity and interoperability
- Integrated process technology within the system

These are the trends that have shaped our chapter selections. In the conventional systems engineering approach, the project is recursively broken into subparts. The parts are then put together, with the task of the systems engineer to select and coordinate the subprojects appropriately.[2] The deconstructionism methodology does not, however, translate to the self-assembly seen in bottom-up nanofabrication techniques, so a constructionist approach must be sought. This synthesizing approach is driven by the complexity of the systems. Although the complexity of engineering projects has been increasing, it is important to recognize that complexity is not new. Engineers and managers are aware of the complexity of these projects and have developed systematic techniques to address them. There are several strategies commonly used, including modularity, abstraction, hierarchy, and layering. Modularity is a well-recognized way to separate a large system into parts that can be individually designed and modified. However, modularity incorrectly assumes that a complex system behavior can be reduced to the sum of its parts. As systems become more complex, the design of interfaces between parts becomes increasingly coupled, and eventually the process breaks down.[3] The decomposition of elements to their respective subsystems is in the functional review steps.

Drivers for Micro- and Nanotechnologies (MNTs) Systems Engineering

It is hard to imagine any technologist today who has not heard the "buzz" and promises of nanotechnologies and nanoscience in the twenty-first century. A simple Internet search combining *nanotechnology* and *promise* will yield close to a million hits in contrast to combining *microtechnology* with *promise* or a less-used term *macrotechnology* with *promise*. There is either a great deal of potential in nanotechnology or a sizable amount of hype. The truth lies somewhere in between.

The U.S. National Nanotechnology Initiative lists that among the potential nanoscale research and developments expected by 2015,

- Half of the newly designed advanced materials and manufacturing processes will be built using control at the nanoscale.
- Converging science and engineering from the nanoscale will establish a mainstream pattern for applying and integrating nanotechnology with biology, electronics, medicine, learning, and other fields.
- Life-cycle sustainability and biocompatibility will be pursued in the development of new products.[4]

Although all three of the above predictions are significant and relative to this work, the second trend drives the requirement to converge science and engineering at micro- and nanoscales. The systems engineering process at this juncture becomes an enabler for the future.

The combination of a need for product quality at the molecular scale with the economic necessity that feedback control systems utilize macroscopic manipulated variables motivates the creation of methods for the simulation, design, and control of *multiscale systems*. This incorporation of models that couple molecular- through macroscopic-length scales within systems engineering tools enables a systematic approach to the simultaneous optimization of all of the length scales of the process.[5]

The Swiss federally funded organization (Nano-Tera.CH) for nanoscale technologies emphasizes the increasing complexity of systems incorporating nanoscale technology compared with today's systems, demonstrating the true drivers for system process/integration approaches to the micro- and nanoscale (multiscale). "Innovative breakthrough ideas that enable true tera-scale system integration will play a central role in Nano-Tera.CH, opening up the possibility of achieving system complexities that are two-to-three orders of magnitude higher than today's state-of-the-art."[6] Nowhere has the trend toward multiscale systems been more evident than in the microelectronics field, where multiscale simulation has been successfully applied for several decades. The nanosciences world, in contrast, is still in an emerging phase for systems engineering approaches and tools.

Engineering complex systems is global in importance and impact. There are economic and technical changes sweeping the world that elevate the critical importance of systems engineering to the emerging and industrialized nations and to their peoples. There is also a deadlock, an impasse, which the art of systems engineering faces and which presently limits its contributions. It is time to revisit classical science and engineering on closed systems. The current expansion of system engineering into the open systems realm of multiscale systems prepares the community of the future. Classical science and engineering concentrate on closed systems. Physics and the second law of thermodynamics would have us believe that entropy, the degree of disorder, is increasing with time in a closed system. But if the systems we see and interact with daily are open systems, that knowledge is not useful. Could this be why classical science and engineering are out of step with the times?[7]

Engineering Perspective: Four Generations of Nanotechnology Applications

As noted in the introduction, commercial nanotechnology is not as mature as microtechnology. Here the four generations of nanotechnology products as described by Roco[8] are discussed in more detail. We previously referred to these four generations in Chapters 1 and 7.

The capabilities of nanotechnology for systematic control and manufacture at the nanoscale are envisioned to evolve in four overlapping generations of new nanotechnology products with different areas of research and development focus. Each generation of products is marked by the creation of commercial prototypes with systematic control of the respective phenomena and manufacturing processing.

1. *First generation of products (2001–): passive nanostructures*, illustrated by nanostructure coatings, dispersion of nanoparticles, and bulk materials—nanostructure

metals, polymers, and ceramics. The primary research focus is on nanostructured materials and tools for measurement and control of nanoscale processes. Examples are research on nanobiomaterials, nanomechanics, nanoparticle synthesis and processing, nanolayers and nanocoatings, various catalysts, nanomanufacturing of advanced materials, and interdisciplinary simulation and experimental tools.

2. *Second generation of products (2005–): active nano-structures*, illustrated by transistors, amplifiers, targeted drugs and chemicals, actuators, and adaptive structures. An increased research focus will be on novel devices and device system architectures.

3. *Third generation (2010–): three-dimensional (3D) nano-systems and systems of nanosystems* with various syntheses and assembling techniques, such as bioassembling; networking at the nanoscale and multiscale architectures.

4. *Fourth generation (2015–): heterogeneous molecular nanosystems*, where each molecule in the nanosystem has a specific structure and plays a different role. Molecules will be used as devices, and from their engineered structures and architectures will emerge fundamentally new functions.[9]

This time line is represented in Figure 15.1.

Migrating from the Machine Age to the Systems Age and Beyond

The concepts Russell Ackhoff put forward in 1981[10] define the difference between the Machine Age and the Systems Age. Table 15.1 compares these two approaches. In general, these differences correlate well with the macrotraditional systems engineering approaches versus the nanoscience-based synthesis approach. Science and engineering have many foundation principles based on the concept of reductionism. Reductionism is

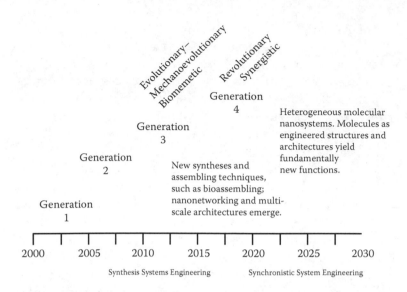

Figure 15.1 Time line for the four generations of nanotechnology.

defined as the analysis of complex things, data, and so forth, into less complex constituents or as any theory or method that holds that a complex idea, system, and so forth, can be completely understood in terms of its simpler parts or components.[11] This was based on Descartes in 1637 whose principles are stated as

- Accept only that which is clear and distinct as true.
- Divide each difficulty into as many parts as possible.
- Start with the simplest elements and move by an orderly procedure to the more complex.
- Make complete enumerations and reviews to make certain that nothing was omitted.

This reductionist path has served the systems engineering community well.

In 1981 Russell Ackhoff promoted the concept that although the reductionist approach has served the "Machine Age system thinking," there is a need to move into system-age thinking.[10] In applying systems engineering processes to the micro- and nanorealm, we will take a systems-age focus on functionality or capability. In an idealized systems engineering process, a set of customer-defined requirements derived from customer

TABLE 15.1
Comparison of Machine-Age Thinking and System-Age Thinking

Machine-Age Thinking	System-Age Thinking	Machine-Age Analysis	Systems-Age Analysis
Procedure	Procedure	Analysis focuses on structure; it reveals how things work	Synthesis focuses on function; it reveals why things operate as they do
Decompose that which is to be explained	Identify a containing system of which the thing to be explained is part	Analysis yields knowledge	Synthesis yields understanding
Explain the behavior or properties of the contained parts separately	Explain the behavior of the properties containing the whole	Analysis enables description	Synthesis enables explanation
Aggregate these explanations into an explanation of the whole (synthesis)	Then explain the behavior of the thing in terms of its roles and functions within its containing whole	Analysis looks into things	Synthesis looks out of things

agreements form the base inputs. This requirements-driven approach has led to the use of deconstruction (decomposition, reductionist) techniques. In the function or capability approach, the emphasis is on the synthesis or a constructionist approach. This synthesis approach is required considering both the complexity and uncertainty of these systems.

The Systems Engineer and Complex Systems with Technological Uncertainty

As we enter the third and fourth generation of the maturing field of MNT, we will need to rethink our approach to management. A summary of the characteristics of complex systems with technology uncertainty shows that change in management style for both the systems engineer and the systems management team will be required.

- *Technology maturity level*: Ranges from existing technologies that may be integrated in new and untried ways to key technologies that are not developed or are under development.
- *Research and development*: Considerable to extensive development is required along with concept and feasibility demonstrations.
- *Development style*: Distributed (even organizationally) multiple paths; evolutionary.
- *Test and analysis*: Prototypes are necessary; extensive development of technologies may be required; hybrid solutions using a mix of mature technologies may truncate test time.
- *System requirements*: Extensive interface with the sponsor includes the expectation of requirement changes and iterations.
- *Functional allocation/system resources*: Dynamic, complex, multimodal, and interface that are hard to define.
- *Design loop*: Cyclic or spiral, numerous iterations required; increased uncertainty pushes the point of design freeze to later in the cycle.
- *Systems engineer*: Emphasis is placed on adaptability of the system to mitigate risk rather than rigid discipline of the system for risk mitigation. Anticipate change and balance or juggle design and requirements trade space.
- *Management style*: Balance of formal and informal to mix interdisciplinary teams who form and reform in an ad hoc rather than hierarchical manner. Extensive ongoing interaction is required.

From the Third to the Fourth Generation

The promise of nano- and microtechnologies for the future will be enabled by investigations into applying system engineering

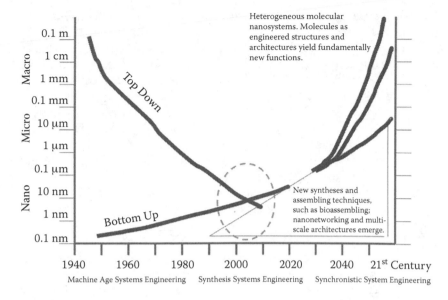

Figure 15.2 The convergence of capability coupled with the emerging third generation of nanotechnologies.

techniques. Figure 15.2 demonstrates the convergence of our fabrication techniques from bottom-up to top-down on a linear scale (y-axis) versus our time frame (x-axis) of entering the third generation. The next chapters just begin to touch on an area that is ripe for future development and exploitation and perhaps the new field of nanosystems engineering or nanosynthesis engineering. The true textbook of nanosystems (synthesis) engineering will need to reflect this activity as it relates to the third generation that is in development phase and the fourth generation that has yet to come to fruition.

References

1. INCOSE Systems Engineering Vision 2020 INCOSE-TP-2004-004-02, September, 2007.
2. Bar-Yam, Y., *About Engineering Complex Systems: Multiscale Analysis and Evolutionary Engineering.* New England Complex Systems Institute, Cambridge, MA.
3. Ibid.

4. Roco, M.C., Nanoscale Science and Engineering: Unifying and Transforming Tools. National Science Foundation, Arlington, VA, and Nanoscale Science, Engineering and Technology (NSET) Subcommittee, U.S. National Science and Technology Council, doi: 10.1002/aic.10087. Published online at Wiley InterScience (www.interscience.wiley.com).

5. Braatz, R.D., Alkire, R.C., Seebauer, E.G., Drews, T.O., Rusli, E., Karulkar, M., Xue, F., Qin, Y., Michael Jung Y,L., Gunawan, R.A., Multiscale systems approach to microelectronic processes, University of Illinois at Urbana-Champaign. Available online 14 July 2006.

6. www.nano-tera.ch/program/research/electronics.html.

7. Hitchins, D.K., *Putting Systems to Work*. Chichester, England: Wiley & Sons, 1992.

8. Roco, M.C., Nanoscale Science and Engineering: Unifying and Transforming Tools. National Science Foundation, Arlington, VA, and Nanoscale Science, Engineering, and Technology (NSET) Subcommittee, U.S. National Science and Technology Council, doi: 10.1002/aic.10087, published online at Wiley InterScience, New York, (www.interscience.wiley.com).

9. Roco, M.C., Nanoscale Science and Engineering: Unifying and Transforming Tools, National Science Foundation, Arlington, VA, and Nanoscale Science, Engineering and Technology (NSET) Subcommittee, U.S. National Science and Technology Council DOI 10.1002/aic.10087 Published online in Wiley InterScience (www.interscience.wiley.com).

10. Ackhoff, R., 1981, *Creating the Corporate Future: Plan or be Planned For.* New York: John Wiley & Sons.

11. Reductionism. *Dictionary.com. Collins English Dictionary— Complete and Unabridged, 10th Edition.* HarperCollins. http://dictionary.reference.com/browse/reductionism (accessed: December 30, 2010).

16

Biomedical Microsystems

Brian Jamieson
Jennette Mateo

Getting the specifications just right.

Contents

Introduction

The goal of this chapter is to provide an introduction to the design and development of biomedical microsystems, paying particular attention to unifying themes and challenges held in common by many of these systems. In the simplest terms, systems engineering can be seen as a formal approach to organizing and executing complex engineering tasks so that their chances of success are maximized. Systems engineering as a discipline was developed in the context of complex defense and aerospace projects, and a generous sprinkling of good engineering sense was "baked into" the field by practitioners after years of hands-on experience in those fields. In order for a new generation of systems-oriented engineers to similarly rise to the challenges facing the development of biomedical microsystems, a high level of familiarity and competence with the engineering details of these systems will be necessary. This chapter is intended to inspire first steps in that direction.

Definition of a Biomedical Microsystem

As described in previous chapters, MEMS (Micro Electro Mechanical Systems), microsystems, and nanotechnology are interrelated and overlapping fields with somewhat hazy boundaries and definitions. For the purposes of this discussion, a *biomedical microsystem* will be defined as a self-contained and autonomous system that performs a specific biomedical function or task, generally measuring or sensing a biological parameter of interest, processing the acquired data, communicating findings to the outside world, or acting upon the gathered data by actuating its environment. In general, a biomedical microsystem has some or all of the following parts:

- An internal source of power
- A sensor to interface with the biological world
- Circuitry for signal transduction and communication
- A communications system
- An actuator

Some examples of biomedical microsystems, according to this definition, are pressure sensors that measure blood pressure in a stent and neural probes that record the electrical activity of neurons (Figure 16.1a,b). Other examples are implantable neurostimulators, capsule endoscopes, subcutaneous glucose monitors, cochlear implants, and implantable hearing aids [1,2].

A nonexhaustive list of companies developing and marketing biomedical microsystems is given in Table 16.1. Several of these systems will be discussed in more detail in following sections. Note that *biomedical microsystem* is not synonymous with *implantable device*. For example, an externally worn sensor module measuring joint position and angle is a biomedical microsystem as it fulfills most of the criteria described above, whereas an artificial hip joint is implantable but is not a biomedical microsystem because it lacks basic autonomy and "intelligence." Still, it is not surprising that implantable devices provide some of the most compelling examples

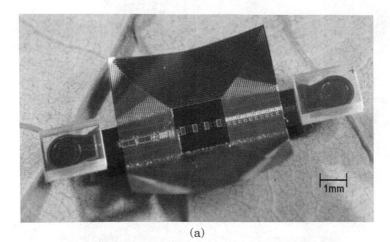

(a)

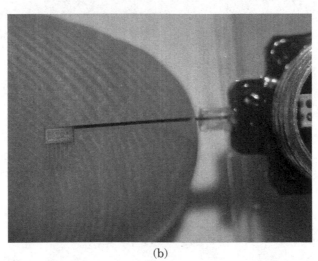

(b)

Figure 16.1 (a) Capacitive pressure sensor with integrated readout and wireless transmission circuitry and a deformable antenna structure for insertion in a stent. (b) Silicon neural probe with micromachined silicon-penetrating shanks, 177 um^2 iridium oxide recording sites, integrated circuitry for signal conditioning, and a monolithic silicon ribbon cable for connection to the hermetic skull cap connector. (Reprinted with permission from DeHennis, A.D., and Wise, K.D., *Journal of Microelectromechanical Systems,* 15, 678–685, 2006; and Wise, K.D., Anderson, D.J., Hetke, J.F., Kipke, D.R., and Najafi, K., *Proceedings of the IEEE,* 92, 76–97, 2004.)

of biomedical microsystems, because the need for small and autonomous systems in such applications is obvious, and implantable microsystems will be a major focus of this chapter.

TABLE 16.1
Some Companies Developing or Marketing Biomedical Microsystems

Company	Device	Medical Application	Webpage
Apogee	Microneedles	Transdermal drug delivery	www.apogeemems.com
CardioMEMS	Pressure sensor	Aortic and aneurism stent graft pressure monitoring	www.cardiomems.com
Second Sight	Retinal implant	Visual prosthesis	www.2-sight.com
Cyberkinetics	Penetrating microelectrode array	Recording from motor cortex for neuroprosthesis	www.cyberkinetics.com
Neuronexus	Penetrating electrode arrays	Recording and stimulation of central nervous system for research and clinical applications	www.neuronexustech.com
Issys	Pressure sensors, densitometers	In vivo pressure measurement, IV drug delivery monitoring	www.mems-issys.com
Microchips Chip Rx	Controlled-release fluidic reservoir	In vivo drug delivery	www.mchips.com www.chiprx.com
Sensors for Medicine and Science, Inc. (SMSI)	Implantable biosensor	Continuous implantable glucose monitoring system	www.s4ms.com

Finally, it is worth noting some devices that are *not* defined to be biomedical microsystems for the purposes of this chapter. Microfluidics is an active and growing field and is very well described in the literature with several good review papers and books detailing the state-of-the-art. Such systems offer almost limitless potential for the miniaturization of laboratory analytical instruments and clinical diagnostics. These systems could open the door for rapid analysis of samples outside of the laboratory setting, for example, at the point of patient care. Current systems featuring microfluidic chips tend to be relatively large benchtop instruments because the supporting electronics, fluidic subsystems (pumps and valves), and optics are typically external to the analytical microchip. The challenges associated with the system-level development and integration of microfluidic systems tend to be somewhat different from those driving the development of biomedical microsystems. The two greatest drivers for microfluidic systems are the desire to increase measurement throughput for drug discovery and to decrease the cost of the systems for point-of-care diagnostic applications. These drivers result in a whole different set of system considerations and design trade-offs than those under consideration here, and the market pull is not necessarily toward increased integration and miniaturization as with biomedical microsystems.

Design Challenges for Biomedical Microsystems

There are several considerations that are unique to the design and development of biomedical microsystems that present challenges to the systems engineer. Among them are the following broad categories:

- Environmental
- Power
- Communications
- Fabrication and manufacturing

Environmental Challenges

Microsystems intended for operation in realistic biological conditions almost always come in contact with physiological saline because warm saltwater is the basis for most life. Therefore, these systems must be designed not only to withstand this harsh environment without degradation, but in many cases, also to maintain proper functioning in light of the body's natural defenses against foreign invaders.

Packaging and Hermeticity

For the large number of biomedical microsystems that are intended to operate in vivo, the warm saline environment of the body is a substantial challenge for the long-term survival of the device. This is especially true since deployment of these systems is often carried out in a costly and invasive procedure that can represent risk to the patient. These systems are designed for working lifetimes of many decades, because replacement or upgrades can often only be carried out with another operation.

Historically, devices such as pacemakers and implantable cardiac defibrillators (ICDs) have been encased in sealed hermetic canisters made of titanium or other inert metals [5] (Figure 16.2) (tantalum, niobium, or stainless steel). Electrical interconnect between the electronics inside the sealed case and the electrical pacing leads is made through glass-fritted hermetic feed-throughs.

The paradigm of sealing a system in a preformed hermetic metal case is more difficult to implement at the microscale where metal shaping and laser or e-beam welding reach the end of their practical fabrication size scales. In addition, many systems have more complicated sensing and actuation interfaces to the body than simple pacing leads, and it is necessary in many cases that a part of the device be in direct contact with the biological system being tested. Therefore, a reliable method is required for encapsulating portions of an implanted microsystem that need to be hermetic (e.g., circuitry, electrical leads, and power sources) while providing for the long-term viability of the sensor portion that must necessarily be exposed to the

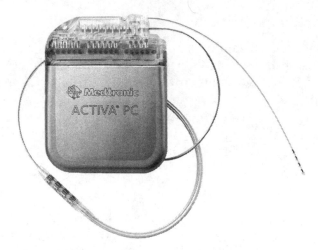

Figure 16.2 Hermetic titanium housing containing an implantable pulse generator (IPG) connected to a four-electrode pacing lead that is implanted in the brain for the treatment of Parkinson's disease. (Courtesy of Medtronic, Inc.)

body. In addition, from a fabrication point of view, it is desirable that hermetic packaging processes be carried out at the *wafer level*. By this, it is meant that the devices are encapsulated or sealed using a method that is not carried out on individual devices, but rather one that captures the inherent efficiency of processing many devices simultaneously on the same wafer.

Several approaches to solving these challenges have been developed and, in general, can be divided into wafer-bonding approaches and thin film encapsulation. Hermetic wafer-level packages in which circuits and other devices are sealed within a cavity between bonded silicon and pyrex wafers have been shown to remain hermetic in vivo for several years [6], and accelerated lifetime testing shows that they should remain so for many decades at normal body temperature (Figure 16.3a,b). Similar approaches utilizing fusion bonding of silicon wafers, gold eutectic bonding, and bonding with various intermediate layers have been shown to be effective methods of wafer-level hermetic encapsulation for implants. Some of this work [7] has illustrated the inclusion of electrical feed-throughs integrated into the device structure allowing the passage of signal leads into and out of the hermetic package.

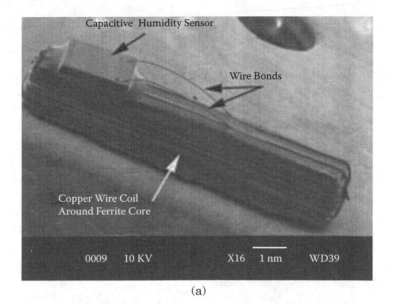

(a)

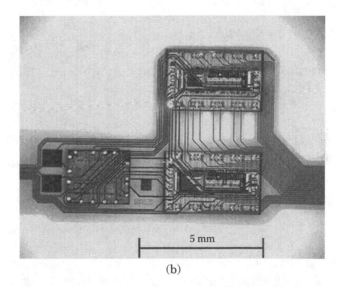

(b)

Figure 16.3 Examples of hermetic packages assembled using microfabrication techniques: (a) Anodically bonded pyrex/silicon capsule. (b) Thin-film encapsulation and integration of circuitry, sensors, and interconnect using polyimide. (Reprinted with permission from Stieglitz, T., Schuetter, M., and Koch, K.P., *Engineering in Medicine and Biology Magazine, IEEE,* 24, 58–65, 2005; and Harpster, T.J., Hauvespre, S., Dokmeci, M.R., and Najafi, K., *Journal of Microelectromechanical Systems,* 11, 61–67, 2002.)

A different approach to wafer-level biomedical microsystem packaging has been the use of thin film encapsulation [8]. Devices have been fabricated and utilized in vivo using encapsulation with spin-coated polyimide, plasma-deposited parylene (Figure 16.3a,b), chemical vapor deposited (CVD) polycrystalline diamond, Benzocyclobutene (BCB), electroplated gold, and a stress-compensated stack of silicon dioxide and silicon nitride. Despite this promising work, published data or verifiable commercial experience with these materials and conclusions about the long-term suitability of such hermetic, thin film packaging is relatively scarce and anecdotal. One of the largest and most extensive databases of thin film encapsulants tested under accelerated lifetime conditions was carried out under an National Insitutes of Health (NIH) contract (N01-NS-2-2347) by David Edell and his group at the Massachusetts Institute of Technology (MIT) and continued commercially by InnerSea Technologies (Bedford, MA) [9] (Figure 16.4). These tests showed excellent long-term survivability for silicone encapsulants and for stacks of thermally grown silicon dioxide and chemical vapor deposited (CVD) nitride, but reported more spotty results with thin film encapsulants such as parylene, liquid crystal polymer (LCP), and silicon carbide.

It should be noted that great attention is being paid to thin film encapsulation on fully implantable retinal stimulation systems for restoring sight to the blind by several academic and commercial groups [11–13].

Biofouling

Biofouling is a loosely defined term often used by biomedical engineers to describe the loss of function or sensitivity of a sensor or other device that is implanted in the body (See Figure 16.5). The body's response to the introduction of a foreign body involves several physical mechanisms operating on different time scales. A reaction begins almost immediately with an inflammation response (experienced as irritation, redness, and itching) associated with a chemical cascade that results in the arrival of neutrophils at the site. Later (within days), macrophages arrive as the body continues its attempts

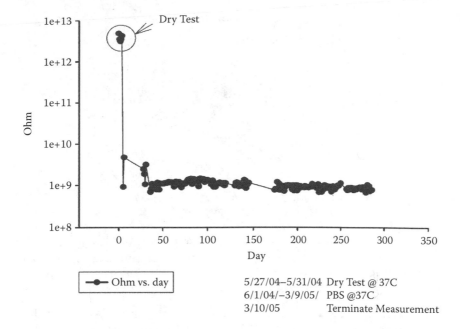

Figure 16.4 Impedance record for a test device encapsulated with a silicon dioxide/silicon nitride/silicon dioxide stack and soaked continuously in body-temperature phosphate-buffered saline for slightly less than 1 year. (Reprinted with permission from Edell, D., *Insulating Biomaterials*, Bedford, MA: NIH, 2002.)

to break down the foreign body. If there are recognizable molecular antigens on the foreign body, they will also trigger an immune response mediated by the macrophages and the various immunological cells within the body such as T cells. The acute, near-term inflammation response is paralleled by a longer time scale process in which tissue is deposited around the foreign body. Almost immediately upon implantation, protein buildup begins on the exposed surfaces of the device. In its extreme, this process of tissue growth eventually leads to a thick capsule around the foreign body that walls it off from the body and isolates the object, a condition referred to as fibrosis.

Surface properties, and in particular roughness, have been shown to have a major impact on the degree of fibrosis [11,14], and attention has been given to producing favorable cellular tissue responses through the intentional modification or engineering of the surface in contact with the tissue. It is important

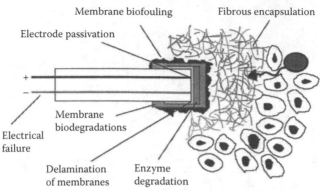

Figure 16.5 Failure of an implant resulting from responses including protein adsorpton (labeled *biofouling*) and fibrous encapsulation. (Reprinted with permission from Wisniewski, N., Moussy, F., and Reichert, W.M., *Fresenius' Journal of Analytical Chemistry*, 366, 611–621, 2000.)

to note that even if the extreme response of fibrosis is absent, as is any measurable degree of cytotoxicity (cell death) induced by the device, sensor function can be impaired by protein buildup on the sensor. In other words, demonstration of the basic "biocompatibility" of an in vivo biosensor is necessary but far from sufficient to ensure long-term sensor viability.

The separate phenomena of acute cell-mediated inflammation and the deposition of tissue are often lumped together in the catch-all term *biofouling*. Subcutaneous glucose sensors, for example, typically experience a loss of sensitivity of between 30% and 80% in the first 72 hours that they are implanted [15]. It is widely accepted that it is due at least in part to the deposition of material on the sensor, which is present in vivo but not in vitro, because this degradation begins almost immediately, does not occur in vitro, and is reversible with explantation and cleaning [15]. However, there is some controversy about exactly what to implicate in this process. Experiments that show the restoration of sensitivity with explantation make no real distinction between the cellular response and nonspecific deposition of proteins and tissue. Experiments using proteomics techniques on explanted sensors [16], however, have implicated proteins and, in particular, biomolecule fragments less than 15 kD for most of the material deposited on the sensor.

This evidence seems to point definitively at protein fouling. However, it has been argued [17] that the cellular response (as distinct from fibrosis) has been given insufficient attention and that cells such as macrophages can, in fact, disturb the microenvironment of the sensor (e.g., pH and O_2 concentration) and thus affect sensor accuracy. In short, it is clear that designing a stable in vivo biosensor presents serious challenges related to the dynamic responses on the part of the body in response to a foreign body. In the case of glucose monitoring, the early sensor sensitivity loss (and its unpredictable magnitude) leads to the need for frequent calibration (as much as four times a day) using finger sticks and an external meter. This is annoying to patients and, until this problem can be solved, precludes the use of the glucose biosensor as a continuous monitor for driving a closed-loop insulin delivery system.

Several approaches to dealing with biosensor sensitivity loss have been investigated. First, attention has been focused on removing the sensing element from direct contact with the body. Many biosensors, including glucose oxidase–based enzyme sensors for subcutaneous glucose monitoring have incorporated a semipermeable polymer membrane. Such membranes actually perform multiple functions. First, they encapsulate the biosensing molecule and fix it to the sensor. In addition, membranes are designed to be selectively permeable to the analyte of interest while excluding potentially interfering analytes. Most relevant to the current discussion, polymer membranes are intended to present a less appealing surface for the deposition of proteins and thus ameliorate the problems of biofouling. A similar approach is to coat the biosensor surface with self-assembling monolayers (SAMs), certain of which have been shown to reduce protein adsorption [18]. A somewhat different approach to preventing biofouling is microdialysis. In this approach, a very fine (and in some cases microfabricated) needle with a semipermeable membrane tip is inserted in the tissue. A buffering solution is circulated through the needle, and this fluid is brought into near-equilibrium with the tissue fluid. The fluid is then circulated to the biosensor for analysis. Using this approach, Ricci et al. demonstrated improved measurement stability [19]. However, the same study found other factors in addition to biofouling that contribute to sensitivity loss in the first few days of implantation, including a 15% decrease in sensitivity attributed to the inactivation of the glucose oxidase enzyme layer, a further 9% decrease in sensitivity attributed to electromediator leaching, and a 6% decrease attributed to background noise. Finally, periodic electrical stimulation has been shown to "clean" metal and polymer electrode recording sites from the buildup of adsorbed proteins [20]. This process may explain why deep-brain-stimulating electrodes, for example, can operate successfully in vivo for many years without an appreciable increase in site impedance. It is useful to consider whether this principle of periodic site cleaning through electrical stimulation could be generalized for use in other in vivo microsystems.

Power

Most current in vivo biomedical devices are powered either through an inductive link [2], batteries, or a combination of both (e.g., rechargeable subcutaneous batteries that are recharged via an inductive link). In the evolution of biomedical microsystems to smaller and more widely capable and useful devices, the ability to provide power to small systems in a wide range of anatomically useful areas (e.g., deep in the body) is one of the most severe problems and one for which no immediate solution appears to be on the horizon. Inductive links work well when power must be coupled to just below the surface of the skin (e.g., from an external charger to an implantable pulse generator implanted just below the skin). For deep implants or devices that are moving through the body (as in gastrointestinal imaging or sensing devices), an inductive link is simply not efficient or practical.

Battery-Powered Systems

It is useful to refer to some simplified design examples in order to understand the magnitude of the power constraints that must be dealt with in developing a typical in vivo autonomous biomedical microsystem of even moderate size. To begin with, just how big is a biomedical microsystem in the current generation of commercially available devices? Boston Scientific's BION system, described by the company as the world's smallest neuromodulator, is approximately 0.2 cm^3 in total volume.

The Bravo system for monitoring esophageal reflux (this product line was purchased from Medtronic by and Israeli startup, Given Imaging, in 2008) is just under 0.9 cm³. All of the capsule endoscopes currently on the market for small bowel imaging (including Given Imaging's Pillcam, Olympus' Endocapsule, and similar products from Mirocam and Omom) are substantially larger in volume, at approximately 2.5 cm³. For purposes of this illustration, we will consider battery-powered microsystems within this approximate range, from 0.1 to 2.5 cm³. The volume available for batteries would, of course, depend on the system implementation, but 30% is probably a reasonable upper limit for the purposes of our order-of-magnitude illustration and is consistent (for example) with the total volume occupied by the two silver oxide watch batteries found in Given Imaging's small bowel capsule endoscope. The volumetric power densities vary among different commonly used battery technologies for medical applications [21]; however, silver oxide is a common choice. From these parameters, we can derive an estimated device lifetime (or time between recharging) as a function of average continuous power consumption as shown in Figure 16.6.

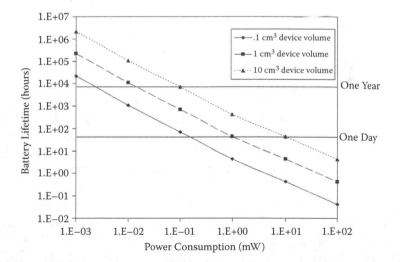

Figure 16.6 Design example showing how much lifetime can be expected from battery-operated microsystems of various sizes and average power consumption, with assumptions about battery technology utilized.

To place these numbers in context, we can return to some of our previous examples. For a typical small bowel capsule endoscope with a required lifetime of 10 hours, we can see from the figure that the maximum power consumption falls in the range of approximately 10 mW. This power budget places a severe limit on the performance of the current generation of capsule endoscopes which, as a result, operate at very low data rates (two to three image frames of only 90,000 pixels each).

Using the same power/lifetimes curve, we can consider the design of a hypothetical second-generation (wireless) deep brain stimulator. In this application, the depth of implantation prohibits charging through an inductive link and explantation for a battery change would involve brain surgery and is, therefore, out of the question. We can see that for a required lifetime of three decades, the maximum power consumption for a battery-operated system of this type would be limited to the sub-microwatt range. Given that the single-cell stimulation current threshold is on the order of 10 microamps, this is not a practical requirement. Either an improvement in battery power density on the order of one or more orders of magnitude will be necessary for such an application, the stimulation threshold for cell activation must be lowered (which is possible at least in part by miniaturizing electrodes and bringing them into more intimate contact with the target cells), or the device will have to be powered in a different way.

Energy Scavenging

Energy scavenging, the conversion of energy from ambient sources such as kinetic energy, light, or in situ chemical reactions, is a promising paradigm for powering biomedical microsystems. The concept of energy scavenging is familiar from self-winding watches in which the motion of the wearer's arm is utilized to move a proof mass and thus tension a spring through a gear train. Theoretical calculations show that (depending on the method of energy harvesting) 100 mW per cm^3 of continuously generated power may be available in implantable applications utilizing the intrinsic vertical motion of the human body to transfer energy to a proof mass. Several methods of power conversion from this proof mass motion have been proposed and tested including piezoelectric, electromagnetic, and electrostatic conversion [22]. In addition, scavenging concepts based on thermoelectric conversion, the creation of an electrochemical battery using bodily fluids or a fuel cell using glucose as fuel, have all been proposed. Substantial work on reliability and practical methods of implementation, fabrication, and integration remains to be done before any of these promising technologies are likely to be found in clinical use for biomedical microsystems.

Other Sources of Power

A substantial amount of research was carried out in the 1970s on the use of radioactive elements as a power source for implantable applications. The most promising research [23,24] was carried out on the use of ^{238}Pu with several systems implanted successfully in both dogs and humans. Energy was converted from the radioactive isotope either through the thermoelectric effect or by semiconductor detectors that allowed excitation of carriers following the capture of emitted beta particles (electrons) from the decaying isotope. Research in this area seems to have come to an abrupt halt with the rise of public opinion against radioactive devices in general, and was probably precluded as well by the availability of reliable rechargeable lithium iodine batteries for pacemakers beginning in the early 1970s.

Finally, some progress has been made on the use of externally coupled optical or ultrasonic signals as a means to provide power to an implant. Remon Medical has patented and developed an approach to using an external ultrasonic source to power an implanted pressure sensor for abdominal aortic aneurysm monitoring. The use of low power (red) lasers transmitted through the transparent medium of the eye [25–27] has been reported as a practical means to deliver energy to a retinal implant.

Communications

The most common method of choice for communicating with biomedical microsystems is through a radio-frequency (RF) electromagnetic or inductive link. Examples include telemetry on pacemakers and defibrillators, capsule endoscopes, pH and temperature measurement and drug delivery capsules, neuro-stimulators, stents, and joint monitoring pressure sensors and strain gauges [28]. The highest (known) data rate biomedical microsystem currently in use is the capsule endoscope in which sequentially captured images are sent from the capsule to RF receivers worn on the patient's torso. Given Imaging's capsule utilizes the ZL7102, a low-power application-specific integrated circuit (ASIC) manufactured by Zarlink operating in the Industrial Scientific, and Medical (ISM) band (433 to 434 MHz). According to the manufacturer's specifications, the chip has a data rate of up to 800 kilobits per second (kbps) and consumes roughly 7.5 to 15 mW of power (approximately one third to two thirds of the system's total power budget). Note that at this data rate, it is only possible to send one uncompressed 800 kilopixel image (with only one bit of color) per second. As illustrated by this example, the attenuation of RF in the lossy medium of the body and the increase in attenuation with increasing frequency introduces a difficult trade-off between power and data rate with current implantable systems.

Other methods of data transfer include an infrared link [29,30] (which is good only in cases of relatively shallow implantation and low data rate) and acoustic communication [31]. Acoustic-based communication systems offer the benefit that the absorption of acoustic signals are somewhat less lossy with increasing frequency as compared to RF, which in some circumstances can relax the inherent constraints between data rate and power consumption that plagues RF biotelemetry links. For an acoustic data link, a small piece of piezoelectric ceramic (e.g., lead zirconate titanate, or PZT) is used for the transmitting element on one end and for the receiver on the other. The piezoelectric element can be quite small and thus is suitable for miniaturization, and can operate over a wide range of frequencies both within and outside of the resonant frequency of the crystal.

Fabrication

Even though MEMS and microsystems technology have been accurately described as a powerful new paradigm for medical device manufacturing, there are some specifics regarding this approach to device fabrication that present challenges (Figure 16.7a,b). Specifically, the cost of a semiconductor fabrication facility is prohibitive for all but the largest research and development (R&D) budgets. Many small-scale commercial foundries have arisen to fill this need, and entry-level MEMS processes are in many cases relatively reasonable. An order-of-magnitude estimate for prototyping runs on 4-inch silicon wafers is in the neighborhood of $500 per wafer per photolithography step. This means, for example, that a modestly complex run of eight masking steps, with a batch of six to eight wafers, iterated twice in a 9-month period can fit within the budget of a Phase I Small Business Innovative Research (SBIR) proposal. Needless to say, there is a tremendous amount of variability in this cost estimate based on the details and complexity of the process under consideration, and these estimates apply only to a few iterations of a relatively simple and well-understood process.

When considering fabrication, it is important to keep in mind that MEMS differs from integrated circuit design development in that the fabrication process (i.e., the detailed semiconductor process steps used) is often unique to the device being fabricated. There are some standard MEMS processes [32,33] (e.g., Summit, Mumps), but they most often are simply not flexible enough to support the range of devices under consideration, and most new devices require custom process design. This means that the design process becomes tightly coupled to the commercial foundry partner who should be involved in the design process from the earliest stages. It is worth noting that medical microsystems applications, even in cases where the expected impact can be quite high, tend to be relatively low in volume. It is important to work with (and budget for) a fabrication partner with the willingness and ability to invest in effective and efficient nonrecurrent engineering at the beginning, with reasonable expectations about the production volume in later manufacturing stages. This is, in many cases, easier said than done, because most commercial foundries expect to make most of their revenue from the volume production of commercial parts. In addition, the stringent regulatory requirements for implantable devices place an additional burden on a manufacturing partner, and many MEMS foundries are not equipped to handle such demands.

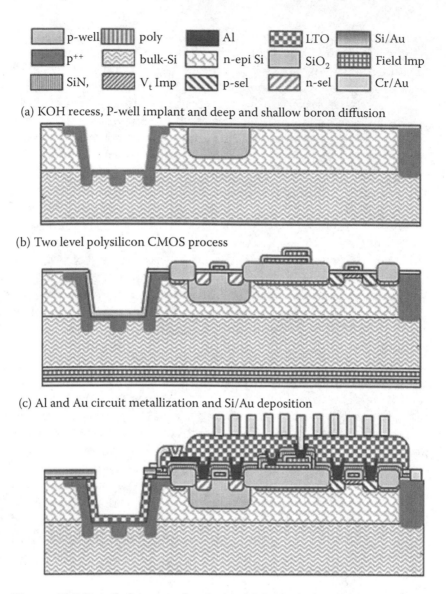

(a) KOH recess, P-well implant and deep and shallow boron diffusion

(b) Two level polysilicon CMOS process

(c) Al and Au circuit metallization and Si/Au deposition

Figure 16.7 Detailed process flow in (a)–(g) for a wireless microsystem for measuring in vivo pressure, illustrating the potential complexity of microsystem fabrication. This process was used to fabricate the device illustrated in Figure 16.1 and consists of a standard complementary metal-oxide semiconductor (CMOS) integrated circuits process combined with a Micro Electro Mechanical Systems (MEMS) process for creating the capacitive pressure sensor and the integrated antenna. (Reprinted with permission from DeHennis, A.D., and Wise, K.D., *Journal of Microelectromechanical Systems,* 15, 678–685, 2006.)

(d) Backside patterning and DRIE wafer thinning

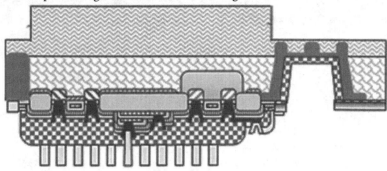

(e) Glass recess and metallization

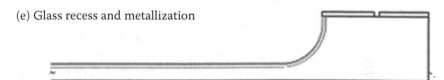

(f) Anodic bonding of the glass/Si substrates and glass thinning

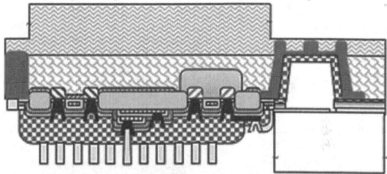

(g) Silicon etch to release sensor diaphragm and thin antenna area

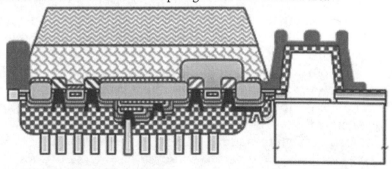

Figure 16.7 (Continued)

Example Biomedical Microsystems

Neural Interfaces

One of the earliest applications of microtechnology to a biomedical system was the development of microfabricated extracellular neural recording electrodes by Wise, Angell, and Starr at Stanford in the late 1960s [37–40] and early 1970s [34,35]. Using photolithographically defined gold interconnect, silicon bulk etching, and a modified version of the Bell Labs "beam lead" for transistor isolation, they were able to produce penetrating electrodes that were used to record single-unit action potentials from the auditory cortex of an anesthetized cat. Wise took the seeds of this technology with him to the University of Michigan where over the course of several decades it was developed to the point where large numbers of recording sites could be integrated onto two- and three-dimensional silicon arrays with penetrating shanks that displaced a very small volume of tissue and could record continuously for months in awake and behaving animals [36]. The devices, which became known as *neural probes* or *Michigan probes*, were integrated monolithically with complementary metal-oxide semiconductor (CMOS) circuitry to amplify, multiplex, digitize, and transmit the neural signals. In some cases, the probes were integrated with channels for microfluidic drug delivery. Through a contract with the National Institutes of Health (NIH), thousands of these neural probes were distributed to neurophysiology researchers throughout the world, resulting in a deep body of knowledge regarding the acute and chronic implantation of these devices in the central and peripheral nervous systems of rodents, primates, and other animals [37–40]. Subsequently, Neuronexus Technologies (www.neuronexustech.com) was spun out of the University of Michigan, making these devices available commercially for research purposes (Figure 16.8a,b).

The long-range motivation for neural probe technology development has been the promise of neuroprosthesis, the interfacing of man-made electrical devices with the nervous system for the purposes of replacing lost neural function. In these systems, the communication direction can be in either

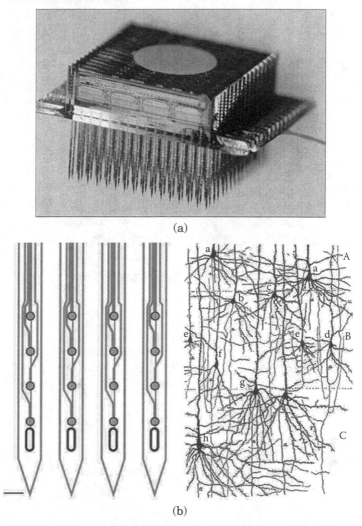

(a)

(b)

Figure 16.8 (a) Three-dimensional neural stimulation probe, with monolithically integrated circuitry for stimulus generation and current steering,
designed to electrically stimulate and record from the neurons in a three-
dimensional area of neural tissue in the cortex of the brain. (b) The scale
and spacing of stimulating sites compared to the spacing and scale of the
target cortical neurons (right panel). This type of device is used in studies of
basic neurophysiology and is also proposed for use in neuroprosthetic applications in which the device would stimulate target neurons in the visual
cortex to produce visual information, or would record from the motor cortex
to produce control signals in a severely paralyzed patient. (Reprinted with
permission from Wise, K.D., Anderson, D.J., Hetke, J.F., Kipke, D.R., and
Najafi, K., *Proceedings of the IEEE,* 92, 76–97, 2004.)

direction (i.e., the transmission of information to the nervous system by electrically stimulating neurons to fire action potentials or the recording of neural activity in the form of action potentials to receive information from the nervous system). For example, in a cochlear implant, the stimulation of the hair cells of the cochlea to produce hearing in the deaf is affected by a multichannel stimulating electrode controlled by an external speech processor and implantable pulse generator (Figure 16.9a,b). Similarly, stimulating electrode arrays chronically implanted in the retina (or in the visual cortex) are being explored as a possible method for restoring sight in some visually impaired people [42]. Going in the other direction, cortical recording electrode arrays are being explored as a means for providing the direct cortical control of external

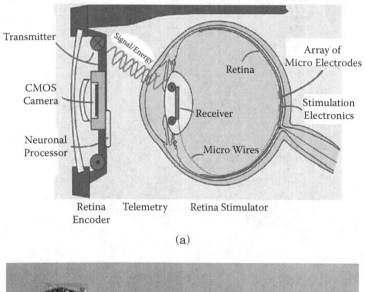

(a)

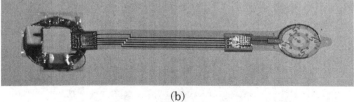

(b)

Figure 16.9 (a) Retinal implant for visual prosthesis. (b) Prototype retinal implant microsystem. (Reprinted with permission from W. Mokwa, *Measurement Science and Technology*, 18, R47, 2007.)

prostheses. For this last application, a silicon-based electrode array was developed through the 1990s at the University of Utah by Richard Normann and his collaborators. Using this device, Cyberkinetics Neurotechnologies was granted an investigational device exemption (IDE) by the U.S. Food and Drug Administration (FDA) in 2008 for limited trials of the motor control of prostheses in humans, and the results have been published extensively [43–45]. This effort continues in the form of Braingate, a noncommercial consortium consisting of researchers from Brown University and other institutions.

In 2002, Medtronic received approval from the FDA for the Activa system, a therapy for Parkinson's disease that works by stimulating cells in the deep subthalamic nucleus affected by this disease. Activa therapy has now been received by over 70,000 patients, and research is underway in both commercial and academic settings to produce a second-generation system that will utilize microsystems concepts to improve the functionality and limit the side effects of these systems (Figure 16.10).

A variety of other neural stimulators and neuromuscular stimulators are either already on the market or are under development, including bladder and sacral nerve stimulators

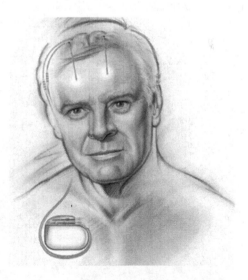

Figure 16.10 Deep brain stimulator (DBS) placement. A large number of groups worldwide are working on making DBS electrode systems less bulky and more highly integrated. (Courtesy of Medtronic, Inc.)

[46,47] to treat urinary and fecal incontinence, respectively; vagal nerve stimulators [48,49] to treat depression; and functional electrical stimulation (FES) systems [50] to stimulate skeletal muscle in paralyzed patients. For all of these systems, the goal of developing more highly integrated, smaller, lower-power, and more targeted therapies will hinge on the ability of micro- and nanoengineering to deliver on its promise. Specifically, an ideal neural interface would be fully implantable with no external leads and would have a large number of very small and precisely controlled stimulation or recording sites that could be used to deliver current to a very small number of cells without stimulating others. The ideal system would also be very small, in order to minimize cell damage during implantation, would be hermetic for many decades, and would have recording and stimulation sites that stayed free of protein buildup for its lifetime.

Capsule Endoscopy and Gastrointestinal (GI) Tract Sensing

Another interesting case study in the development of biomedical microsystems is capsule endoscopy (Figure 16.11). Because it could not be reached by traditional endoscopes, for many years, the only way to image the small bowel was through a radiographic series that provided useful diagnostic information but lagged far behind the quality of imaging possible in the esophagus, stomach, and colon through endoscopy. In 2001,

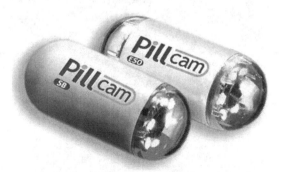

Figure 16.11 Given Imaging's capsule endoscope for small bowel (left) and esophageal (right) imaging. (Courtesy of Given Imaging.)

the first application of double-balloon-based enteroscopy (DBE) was reported [51,52]. It was invented by Dr. H. Yamamoto, supported by Fujinon Inc., which introduced a dedicated system for DBE in 2003. This made it possible to perform endoscopy of the small bowel and opened up this organ to direct observation by the gastroenterologist. Unfortunately, the DBE procedure is time consuming and tedious, and gastroenterologists were slow to adopt it. At about the same time, the employment of wireless capsule endoscopy technique was first published [53]. A year later, Given Imaging received FDA approval to market the PillCam, a swallowed capsule known as a small bowel capsule endoscope that captures and transmits a sequence of images from the small bowel. These images are transmitted to receivers on the patient's body and are stored on a data recorder from which they are later retrieved for review by the physician. Over a million of these procedures have now been performed worldwide, and Olympus and several other companies have introduced similar products to the marketplace. Given Imaging is currently working on a version of the capsule that would image the colon, offering an adjunct therapy or even a possible replacement for traditional colonoscopy.

The PillCam capsule is an elegant piece of engineering and packaging. The optical train is composed of a transparent dome with an optical prescription that forms the top one third of the capsule in conjunction with a low-aperture wide-angle lens held in an injection-molded plastic lens holder. The lens holder is baffled, so the flash from the four white light-emitting diodes (LED) mounted next to the lens holder does not pass directly onto the imager. Electrical interconnect from the printed circuit board (PCB) holding these LEDs is made through small, spring-loaded "pogo pins" similar to those found in cell phones. The imager is a charge-coupled device (CCD) containing about 90,000 pixels. The control circuitry is an application-specific integrated circuit (ASIC) of unknown manufacturer, and the RF communications chip (as described previously) is a chip manufactured by Zarlink. An antenna of approximately 7 windings and 3.5 mm diameter is included in the distal capsule end (opposite the transparent dome). The capsule is shipped to the physician in packaging that includes a permanent magnet and is turned on when a microfabricated magnetic reed switch is activated as the capsule is removed from the package. All of the electronic components are mounted on "rigid flex" boards—a stack of printed circuit boards connected by integrated polyimide ribbon cables that can be folded into a compact cylindrical shape. Two 1.5 V silver oxide batteries provide power to the device.

The story of Given Imaging is interesting on several fronts. The company was funded in its early days largely through the Israeli government, which appears to have provided the continuity and patience necessary for the ambitious development path that was necessary to turn this clinical need into a commercial reality. In addition, the company decided to fully commercialize and market the device (notwithstanding a licensing agreement with Ethicon Endo for an esophageal capsule) rather than selling the company to an established medical device company with existing distribution channels. Finally, it is interesting that the inventor of the capsule and former chief technology officer of Given Imaging, Dr. Gavriel Idan, was originally a missile engineer who worked for the RAFAEL Armament Development Company. It is fitting to note in this book about systems engineering that arguably one of the most successful examples of a biomedical microsystem was developed by a systems engineer.

The field of biomedical microsystems for gastrointestinal applications is expanding rapidly. In 2008, Philips Medical introduced the i-pill [54], an electronic drug delivery capsule that uses a microdiaphragm pump to deliver medication to a targeted region of the small bowel. The privately held company SmartPill markets a pH and temperature monitoring capsule that is used in clinical studies of delayed gastric emptying and for other purposes related to GI tract motility. And encouragingly, the research literature is full of projects describing improved techniques for imaging the GI tract and, in some cases, for delivering therapeutic agents or performing procedures such as biopsy [55,56].

In Situ Blood Pressure Monitoring

Small implantable systems for measuring blood pressure in vivo have been developed and tested since at least the 1960s.

MacKay built a small cuff-type electrode [57] through which an artery could be passed with a tethered subcutaneous transmitter sending the pressure data out of the body. In the 1980s, with the advent of MEMS, compact capacitive and piezoelectric silicon pressure sensors with impressive performance characteristics were developed largely for automotive applications. As these sensors became more widespread and accepted, the exploration of their use in biomedical applications began, and references to in vivo pressure sensing became numerous [58–62].

In 1999, CardioMEMS was formed to commercialize a technology for wireless passive pressure sensing based on a capacitive pressure sensor that was licensed from Georgia Tech where it was invented by Mark Allen. The CardioMEMS wireless pressure sensor has been used in a large human clinical trial where it was implanted in a branch of the pulmonary artery to monitor the recovery of patients who had suffered heart failure. It was shown in this trial that the ability to monitor pulmonary artery pressure, and thus adjust the medication of patients to regulate blood pressure following heart failure, decreased 6-month rehospitalization rates by 30% as compared to a control group [63].

The CardioMEMS device is a capacitive, micromachined pressure sensor formed from a pair of bonded fused silica (silicon dioxide) wafers. The external wires, which form the mechanical connection to the blood vessel wall, are nickel/titanium coated with polytetrafluoroethylene (PTFE) (Teflon®). The device is powered by passive telemetry, meaning that RF power is coupled into the device by an external transmitter. The capacitive sensor, in combination with an on-board inductor, forms a resonating tank circuit. As the capacitance of the sensor changes, the resonant frequency of the tank shifts, which can be measured as a change in reflected power at the external transmitter. Power is not a concern in this application because of the relatively low bandwidth of the pressure signal and the ultralow power needed for the passive telemetry approach. There were no indications of substantial tissue encapsulation during the clinical trials, a problem that has dogged implanted stent applications. A company representative offered the explanation that the devices are implanted with relatively little trauma into healthy tissue. This is in contrast to stents in which the vessel is already damaged (occluded), and the implantation and deployment process causes further trauma to the vessel, which is expected to initiate an inflammatory response.

Implantable pressure sensors are not yet FDA approved for this or any other medical application; however, the demonstration of safety and efficacy by CardioMEMS seems to be a big step in that direction. The research literature is full of references to implantable pressure sensors for applications such as stent and hydrocephalus monitoring, and there are several companies in addition to CardioMEMS developing products in this space.

Conclusions

The novel fabrication and packaging techniques enabled by MEMS have created an opportunity to engineer and develop advanced microsystems capable of having a huge impact on biology and medicine. Examples of existing biomedical microsystems include neural interfacing systems, capsule endoscopes, wireless pressure monitors, implantable glucose sensors, and drug delivery systems. The ability of this field to continue to mature will depend on the ability to address substantial challenges at the systems level, including power management, management of environmental conditions, the development of efficient communications, and the introduction of practical and agile manufacturing channels. These challenges are similar in concept to the challenges that were successfully addressed in the defense and aerospace industries in the 1960s and 1970s, largely through the application of system engineering principles. The practical example of pioneers in biomedical microsystems such as Gavriel Idan, Robert Fischell, and many others who came from an aerospace background and were well-versed in systems engineering principles should give great encouragement to the current generation of systems engineers who wish to have an impact in the important and growing field of biomedical microsystems.

References

1. Arshak, A., K. Arshak, D. Waldron, D. Morris, O. Korostynska, E. Jafer, and G. Lyons, "Review of the potential of a wireless MEMS and TFT microsystems for the measurement of pressure in the GI tract," *Medical Engineering and Physics,* vol. 27, 347–356, June 2005.

2. Liu, W., P. Singh, C. DeMarco, R. Bashirullah, M. Humayun, and J. Weiland, "Semiconductor-based implantable microsystems," in *Handbook of Neuroprosthetic Methods.* Boca Raton, FL: CRC Press, 2002.

3. DeHennis, A.D. and K.D. Wise, "A fully integrated multisite pressure sensor for wireless arterial flow characterization," *Journal of Microelectromechanical Systems,* vol. 15, 678–685, 2006.

4. Wise, DK.D., .J. Anderson, J.F. Hetke, D.R. Kipke, and K. Najafi, "Wireless implantable microsystems: high-density electronic interfaces to the nervous system," *Proceedings of the IEEE,* vol. 92, 76–97, 2004.

5. Jiang, G. and D. Zhou, "Technology advances and challenges in hermetic packaging for implantable medical devices," in *Implantable Neural Prosthesis 2—Techniques and Engineering Approaches for Biological and Medical Physics,* D. Zhou and E. Greenbaum, Eds.: New York: Springer, 2009.

6. Najafi, K., "Packaging of Implantable Microsystems," in *Sensors, 2007 IEEE,* 2007, 58–63.

7. Stark, B.H. and K. Najafi, "A low-temperature thin-film electroplated metal vacuum package," *Journal of Microelectromechanical Systems,* vol. 13, 147–157, 2004.

8. Stieglitz, T., M. Schuetter, and K.P. Koch, "Implantable biomedical microsystems for neural prostheses," *Engineering in Medicine and Biology Magazine, IEEE,* vol. 24, 58–65, 2005.

9. Edell, D., "Insulating biomaterials," Bedford, MA: NIH, 2002.

10. Harpster, T.J., S. Hauvespre, M.R. Dokmeci, and K. Najafi, "A passive humidity monitoring system for in situ remote wireless testing of micropackages," *Journal of Microelectromechanical Systems,* vol. 11, 61–67, 2002.

11. Meyer, J.-U., "Retina implant—a bioMEMS challenge," *Sensors and Actuators A: Physical,* vol. 97–98, 1–9, 2002.

12. Seo, J.-M., S.J. Kim, H. Chung, E.T. Kim, H.G. Yu, and Y.S. Yu, "Biocompatibility of polyimide microelectrode array for retinal stimulation," *Materials Science and Engineering: C,* vol. 24, 185–189, 2004.

13. Po-Jui, C., D.C. Rodger, S. Saati, M.S. Humayun, and T. Yu-Chong, "Microfabricated implantable parylene-based wireless passive intraocular pressure sensors," *Journal of Microelectromechanical Systems,* vol. 17, 1342–1351, 2008.

14. Sharkawy, A.A., B. Klitzman, G.A. Truskey, and W. Reichert, "Engineering the tissue which encapsulates subcutaneous implants. I. Diffusion properties," *Journal of Biomedical Materials Research,* vol. 37, 401–412, 1997.

15. Wisniewski, N., F. Moussy, and W.M. Reichert, "Characterization of implantable biosensor membrane biofouling," *Fresenius' Journal of Analytical Chemistry,* vol. 366, 611–621, 2000.

16. Gifford, R., J.J. Kehoe, S.L. Barnes, B.A. Kornilayev, M.A. Alterman, and G.S. Wilson, "Protein interactions with subcutaneously implanted biosensors," *Biomaterials,* vol. 27, 2587–2598, 2006.

17. Wilson, G. and Y. Hu, "Enzyme-based biosensors for in vivo measurements," *Chemical Reviews,* vol. 100, 2693–2704, 2000.

18. Sharma, S., R.W. Johnson, and T.A. Desai, "Poly(ethylene glycol) interfaces for the control of biofouling in silicon-based microsystems," in *Second Annual International IEEE–EMB Special Topic Conference on Microtechnologies in Medicine and Biology,* 2002, 41–45.

19. Ricci, F., F. Caprio, A. Poscia, F. Valgimigli, D. Messeri, E. Lepori, G. Dall'Oglio, G. Palleschi, and D. Moscone, "Toward continuous glucose monitoring with planar modified biosensors and microdialysis: study of temperature, oxygen dependence and in vivo experiment," *Biosensors and Bioelectronics,* vol. 22, 2032–2039, 2007.

20. Kipke, D.R., "Implantable neural probe systems for cortical neuroprostheses," in *Engineering in Medicine and Biology Society, 2004. IEMBS '04. 26th Annual International Conference of the IEEE,* 2004, 5344–5347.

21. Takeuchi, K.J., R.A. Leising, M.J. Palazzo, A.C. Marschilok, and E.S. Takeuchi, "Advanced lithium batteries for implantable medical devices: mechanistic study of SVO cathode synthesis," *Journal of Power Sources,* vol. 119–121, 973–978, 2003.

22. Romero, E., R.O. Warrington, and M.R. Neuman, "Energy scavenging sources for biomedical sensors," *Physiological Measurement,* vol. 30, R35, 2009.

23. Laurens, P., "Nuclear-powered pacemakers: an eight-year clinical experience," *Pacing and Clinical Electrophysiology,* vol. 2, 356–360, 1979.

24. Parsonnet, V., L. Gilbert, I.R. Zucker, R. Werres, T. Atherley, M. Manhardt, and J. Gort, "A decade of nuclear pacing," *Pacing and Clinical Electrophysiology,* vol. 7, 90–95, 1984.

25. Doron, E.K.Y. and A. Penner, "Implantable medical device with integrated acoustic transducer," Caesarea, IL: Remon Medical Technologies, Ltd., 2009.

26. Eyal, M., M. Mauricio, D.W. James, J.G. Robert, Y.F. Gildo, T. Gustavo, V.P. Duke, M.O.H. Thomas, L. Wentai, L. Gianluca, D. Gislin, A.S. Dean, J. Eugene de, and S.H. Mark, "Retinal prosthesis for the blind," *Survey of Ophthalmology,* vol. 47, 335–356, 2002.

27. Zhou, D.D. and R.J. Greenberg, "Microsensors and microbiosensors for retinal implants," *Front Bioscience,* vol. 10, 166–179, January 1, 2005.

28. Ziaie, B., "Implantable wireless microsystems," in *BioMEMS and Biomedical Nanotechnology*, M. Ferrari, R. Bashir, and S. Wereley, Eds.: New York: Springer, 2007, 205–221.

29. Arneson, R.M., R.W. Bandy, J.K. Powell, and B.R.A. Jamieson, "Ingestible low power sensor device and system for communication with same," Columbia, MD: Innurvation, Inc., 2008.

30. Bandy, W.R.G., B.G. Jamieson, K.J. Powell, K.E. Salsman, R.C. Schober, J. Weitzner, and M.R. Arneson, "Ingestible endoscopic optical scanning device," Columbia, MD: Innurvation, Inc., 2010.

31. Penner, A.T.A. and E. Doron, "Systems and method for communicating with implantable devices," Caesarea, IL: Remon Medical Technologies, Ltd., 2009.

32. Allen, J., "MEMS fabrication," in *MEMS and Microstructures in Aerospace Applications*. Boca Raton, FL: CRC Press, 2005.

33. Fang, W., J. Hsieh, and H.-Y. Lin, "Towards the integration of nano/micro devices using MEMS technology," in *Nanomechanics of Materials and Structures*, T.J. Chuang, P.M. Anderson, M.K. Wu, and S. Hsieh, Eds.: Dordrecht: Springer Netherlands, 2006, 151–159.

34. Wise, K.D., J.B. Angell, and A. Starr, "An integrated-circuit approach to extracellular microelectrodes," *Biomedical Engineering, IEEE Transactions on,* vol. BME-17, 238–247, 1970.

35. Wise, K.D. and J.B. Angell, "A low-capacitance multielectrode probe for use in extracellular neurophysiology," *Biomedical Engineering, IEEE Transactions on,* vol. BME-22, 212–219, 1975.

36. Csicsvari, J., D.A. Henze, B. Jamieson, K.D. Harris, A. Sirota, P. Bartho, K.D. Wise, and G. Buzsaki, "Massively parallel recording of unit and local field potentials with silicon-based electrodes," *Journal of Neurophysiology,* vol. 90, 1314–1323, August 2003.

37. Ahrens, K.F. and W.J. Freeman, "Response dynamics of ento-rhinal cortex in awake, anesthetized, and bulbotomized rats," *Brain Research,* vol. 911, 193–202, 2001.

38. Csicsvari, J., H. Hirase, A. Czurko, and G. Buzsáki, "Reliability and state dependence of pyramidal cell-interneuron synapses in the hippocampus: an ensemble approach in the behaving rat," *Neuron,* vol. 21, 179–189, 1998.

39. Csicsvari, J., B. Jamieson, K.D. Wise, and G. Buzsáki, "Mechanisms of gamma oscillations in the hippocampus of the behaving rat," *Neuron,* vol. 37, 311–322, 2003.

40. Heetderks, W.J. and E.M. Schmidt, "Chronic, multi-unit recording of neural activity with micromachined silicon microelectrode," in *Rehabilitation Engineering and Assistive Technology Society of North America*, 1995, 659.

41. Mokwa, W., "Medical implants based on microsystems," *Measurement Science and Technology,* vol. 18, R47, 2007.

42. "Second Sight," www.2-sight.com/.

43. Donoghue, J.P., A. Nurmikko, M. Black, and L.R. Hochberg, "Assistive technology and robotic control using motor cortex ensemble-based neural interface systems in humans with tetraplegia," *Journal of Physiology,* vol. 579, 603–611, March 1, 2007.

44. Donoghue, J.P., A. Nurmikko, G. Friehs, and M. Black, "Chapter 63: Development of neuromotor prostheses for humans," in *Supplements to Clinical Neurophysiology*. vol. 57, M. Hallett, L.H. Phillips, D.L. Schomer, and K.M. Massey, Eds.: New York: Elsevier, 2004, 592–606.

45. Hochberg, L., M. Serruya, G. Friehs, J. Mukand, M. Saleh, A. Caplan, A. Branner, D. Chen, R. Penn, and J. Donoghue, "Neuronal ensemble control of prosthetic devices by a human with tetraplegia," *Nature,* vol. 442, 164–171, 2006.

46. Braun, P.M., C. Seif, C. van der Horst, and K.-P. Jünemann, "Neuromodulation—Sacral, peripheral and central: current status, indications, results and new developments," *EAU Update Series,* vol. 2, 187–194, 2004.

47. Fall, M., K. Ahlstrom, C.-A. Carlsson, A. Ek, B.-E. Erlandson, S. Frankenberg, and A. Mattiasson, "Contelle: Pelvic floor stimulator for female stress-urge incontinence A multicenter study," *Urology,* vol. 27, 282–287, 1986.

48. "Cyberonics," http://us.cyberonics.com/en.

49. Schwartz, P.J. and G.M. De Ferrari, "Vagal stimulation for heart failure: background and first in-man study," *Heart Rhythm,* vol. 6, S76–S81, 2009.

50. Stein, R.B., "Functional electrical stimulation," in *Encyclopedia of Neuroscience*, R.S. Larry, Ed. Oxford: Academic Press, 2009, 399–407.

51. Yamamoto, H., Y. Sekine, Y. Sato, T. Higashizawa, T. Miyata, S. Iino, K. Ido, and K. Sugano, "Total enteroscopy with a non-surgical steerable double-balloon method," *Gastrointestinal Endoscopy*, vol. 53, 216–220, 2001.

52. Sunada, K. and H. Yamamoto, "Double-balloon endoscopy: Past, present, and future," *Journal of Gastroenterology*, vol. 44, 1–12, 2009.

53. Iddan, G., G. Meron, A. Glukhovsky, and P. Swain, "Wireless capsule endoscopy," *Nature*, vol. 405, 417–417, 2000.

54. "IntelliCap," in *Philips Research*: www.research.philips.com/initiatives/intellicap.

55. Moglia, A., A. Menciassi, P. Dario, and A. Cuschieri, "Capsule endoscopy: progress update and challenges ahead," *Nature Reviews: Gastroenterology and Hepatology*, vol. 6, 353–362, June 2009. Wiley, NY.

56. Thomas, D.W. and D. Jacques Van, "Optical biopsy: a new frontier in endoscopic detection and diagnosis," *Clinical Gastroenterology and Hepatology: The Official Clinical Practice Journal of the American Gastroenterological Association*, vol. 2, 744–753, 2004.

57. MacKay, R., *Sensing and Transmitting Biological Information from Animals and Man*, 1998, Wiley, NY.

58. Peng, C., W.H. Ko, and D.J. Young, "Wireless batteryless implantable blood pressure monitoring microsystem for small laboratory animals," *Sensors Journal, IEEE*, vol. 10, 243–254, 2010.

59. Allen, M.G., "Micromachined endovascularly-implantable wireless aneurysm pressure sensors: from concept to clinic," in *Digest of Technical Papers. TRANSDUCERS '05. The 13th International Conference on Solid-State Sensors, Actuators and Microsystems*, 2005, 275–278 vol. 1.

60. DeHennis, A. and K.D. Wise, "A double-sided single-chip wireless pressure sensor," in *The Fifteenth IEEE International Conference on Micro Electro Mechanical Systems*, 2002, pp. 252–255.

61. Frischholz, M., L. Sarmento, M. Wenzel, K. Aquilina, R. Edwards, and H.B. Coakham, "Telemetric implantable pressure sensor for short- and long-term monitoring of intracranial pressure," in *EMBS 2007. 29th Annual International Conference of the IEEE Engineering in Medicine and Biology Society*, 2007, p. 514.

62. Po-Jui, C., S. Saati, R. Varma, M.S. Humayun, and T. Yu-Chong, "Wireless intraocular pressure sensing using microfabricated minimally invasive flexible-coiled LC sensor implant," *Journal of Microelectromechanical Systems,* vol. 19, 721–734, 2010.

63. "CardioMEMS completes CHAMPION clinical trial study," Atlanta: CardioMEMS, Inc., 2010.

17

Stability and Uncertainty in Self-Assembled Systems

I.K. Ashok Sivakumar

"We might have done better if we'd
emphasized adaptation over redundancy."

Contents

Introduction

In this chapter, exploiting the uncertainty inherent at the molecular level to achieve stability is explored instead of investigating means of extricating unknown phenomenology. Control and feedback are demonstrated as essential to any hybrid solution of nanotechnology integration. Quantitative methods in terms of static and dynamic self-assembly (SA) are also considered. *Self-assembly* is a term used to describe processes in which a disordered system of preexisting components forms an organized structure or pattern as a consequence of specific, local interactions among the components, and without external direction. SA in the classic sense can be defined as the spontaneous and reversible organization of molecular units into ordered structures by noncovalent interactions. The first property of a self-assembled system that this definition suggests is the spontaneity of the SA process: the interactions responsible for the formation of the self-assembled system act on a strictly local level—in other words, the nanostructure builds itself. Chapter 9 discussed in depth the concepts of SA in nanofabrication. Self-assembly can be classified as either static or dynamic. In the static case, the ordered state forms as a system approaches equilibrium, reducing its free energy. In contrast, thermodynamic equilibrium may fluctuate in dynamic SA, where specific local interactions allow components to self-organize (SO).

In order to quantify, exploit, and control the uncertainty in a self-organizing system, it is necessary to determine whether a system is moving toward or away from equilibrium and whether its nonlinearities are impacting its stability. Lyapunov stability theory is applied to assist in these determinations. It is evident that the integration of molecular components may require more manual oversight or control through robust and flexible design to achieve system homeostasis.

As introduced in earlier chapters and shown in Figure 17.1, Mike Roco of the U.S. National Nanotechnology Initiative described four generations of nanotechnology[1] and depicts the current time period as the intersection between top-down and bottom-up breakthroughs. Thus, we are at an epic junction between eras of nanotechnology development and the requirements of integration.

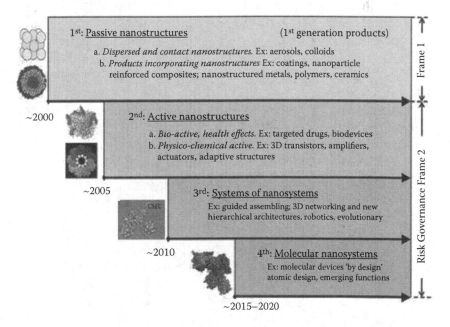

Figure 17.1 The four potential generations of nanotechnology development: (1st) passive nanostructures, materials designed to perform one task; (2nd) active nanostructures for multitasking (e.g., actuators, drug delivery devices, and sensors); (3rd) nanosystems with thousands of interacting components; and (4th) integrated nanosystems, functioning like a mammalian cell with hierarchical systems within systems.

Although the timeline may be different now than when originally proposed, the overall chronology of events has followed quite closely. Breakthroughs have synthetically produced passive and active nanostructures with multiple applications from space technology to medical devices.

Embracing the Uncertainty

Throughout this book, it has been demonstrated that nanotechnology is not simply about the final product but encompasses the underlying process intrinsic to any manufacturing or fabrication design. Fundamental principles at the macroscale are not accurate at the nanoscale. Even though traditional approaches of systems engineering described in Chapter 2 have successfully been demonstrated to build reliable systems by extricating uncertainty (reduce variability, eliminate failure modes, etc.) from individual components, this method may prove futile as random processes and entropy are fundamental laws at molecular scales. An amended systems engineering process would be to exploit such inherent uncertainty at various steps and dynamically adapt to attain final products.

A systems engineer in the nano-world must be cognizant that there will be uncertainty amidst reactions and components and must coerce the system's location in terms of other quantitative metrics. The optimal strategy in guiding the system to a (perceived) state of stability or equilibrium for SA is to use control and feedback mechanisms rather than dictate a strict and rigid path a priori in the process.

Systems engineering has had several examples of dealing with such unknowns and several agile systems engineering techniques are discussed in Chapter 4. Probabilities and estimates of output variables are often calculated and integrated to mitigate uncertainties.[2] Other algorithms including multiattribute utility theory, analytic hierarchy process,[3] or Bayesian team support[4] have been proposed but contain approximations that are difficult to simulate at the molecular scale. Because the methods of dealing with uncertainties fall into the purview of statistics, it is not surprising that these references contain many statistical concepts.

Feedback and Control

It follows that some level of control is an essential requirement to any large system using nanotechnology. Feedback and control have long been used in linear time-invariant (LTI) systems and have also been shown in simple, nonlinear, dynamic systems. In systems with feedback, a controller manipulates inputs and measured outputs of the system to obtain the desired output and system stability[5] (Figure 17.2). Unfortunately, this simple model of control becomes quickly intractable with nonlinear systems under uncertainties. However, the theory and process can still be applied in a more discrete manner at the molecular level. It may prove infeasible to regulate and respond to each reaction at the molecular scale, but processes should be put in place to ensure the system is progressing toward a well-defined end state, suitable for SA and capable of spawning future generations.

This process is akin to "Dead Reckoning," often used by seamen for years on unchartered waters. Briefly stated, the process involves estimating one's position (or state) based on a previously known position (or state) using measured variables from the environment and across time. In terms of

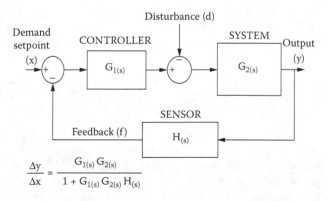

$$\frac{\Delta y}{\Delta x} = \frac{G_{1(s)} G_{2(s)}}{1 + G_{1(s)} G_{2(s)} H_{(s)}}$$

Figure 17.2 A control diagram portraying the concept of negative feedback.

nanotechnology processes, the metric could be a threshold or offset from a local equilibrium or a quantifiable measure of the systems' stability as described in more detail in subsequent sections. It is noted that this process causes errors to accumulate over time; however, this simply underscores the point that error is not to be avoided under such a dynamic system, as long as control mechanisms are successful in guiding the integrated system to achieve a higher level of stability.

Bartosz Grzybowski's group at Northwestern University, Evanston, Illinois, provided an interdisciplinary approach to understanding SA and SO in both equilibrium and nonequilibrium ensembles at various length scales. His group has developed design rules that allow "synthesis" of self-assembling systems from various types of interactions or phenomena. This work has focused on fundamental statistical-mechanical questions[6] concerning the role of energy dissipation in such systems, extension of thermodynamic concepts (e.g., temperature) to nonequilibrium ensembles, the influence of system's geometry on the emerging structures/patterns,[7] and the ability to "reverse engineer" a desired self-assembled structure into the properties of its constituents. Top-down and bottom-up approaches converge in biology-oriented work on the SA of cell components responsible for cell motility and metastasis.[8] Using special substrates (prepared by top-down reaction-diffusion processes) for controlled cell spreading and motility, they have been able to deconstruct (bottom-up) the spatial and temporal aspects of SA of microtubules and actin stress fibers. The knowledge of how these components assemble and interact with one another is important to the future ability to exploit self-assembled and self-organized systems.

Quantifying Nonlinearities in Self-Assembly

Understanding SA and SO are imperative to take advantage of fabrication processes at the nanoscale. Although nonlinearities may be prevalent at this size, stable systems can still be highly regulated and controlled. For example, the use of Coulombic interactions can control the supramolecular synthesis of finite, well-defined nanostructures.[9] The energy associated with the separation of ion pairs was demonstrated to regulate precisely guanosine self-assembly into discrete G-quadruplexes.

In order to build a formal foundation in monitoring and controlling SA for fabrication, it is prudent to incorporate metrics that evaluate the system's stability. Algorithms to accomplish this take into account how the self-assembled system responds to small perturbations to its state under nonlinear stochastic conditions. One proposed method is based on Lyapunov exponents.[10]

This approach can characterize the rate of separation of infinitesimally close trajectories and determine if a system is globally asymptotically stable *without* needing to solve the associated differential equations and finding such trajectories. The metric in a SA system would be to find a Lyapunov function and ensure that it satisfies the required properties for stability, divergence, and so forth.[11] A common technique is to develop a Lyapunov candidate function where the form and parameters are chosen in advance. Then, it remains to find the values of these parameters to show that the stability hypothesis holds.

Because a full explanation of Lyapunov theory is out of the scope of this text, a brief summary is provided here.[12]

Given the differential equation

$$x = f(x), x(0) = x_0 \tag{17.1}$$

the Lyapunov exponents, $\{\lambda_1, \lambda_2, \dots \lambda_n\}$ (for an n-dimensional phase space) are a characterization of the asymptotic properties of the solution $x(t, x_0)$ via analysis of the linear problem (for ease of notation, the dependence of the solution on x_0 is suppressed)

$$y' = f_x(x(t))y \tag{17.2}$$

Formally, these exponents associated to Equation 17.2 are defined as follows. Let $\{p_i\}$ be the columns of an initial conditions (full rank) matrix Y_0, and define the numbers λ_i, $i = 1, \dots, n$,

$$\lambda_i(p_i) = \lim_{t \to \infty} \sup \frac{1}{t} \log \left\| Y(t) p_i \right\| \qquad (17.3)$$

where $Y(t)$ is the solution of

$$Y' = f_x(x(t))Y, \quad Y(0) = Y_0 \qquad (17.4)$$

When the sum of these numbers is minimized as we vary over all possible initial conditions Y_0, the numbers are called *Lyapunov exponents* of the system.[13] It follows that the set of Lyapunov exponents will be the same for almost all starting points of an ergodic component of the dynamical system.

It is shown that a dynamic system $\dot{x} = f(x)$ is *Lyapunov stable* about an equilibrium point x_{eq} if state trajectories are confined to a bounded region whenever the initial condition x_0 is chosen sufficiently close to x_{eq}.[14] This definition is somewhat difficult to use when measuring system stability. Therefore, the Lyapunov direct (or second) method in finding a candidate function with certain properties is a common technique. Assume we find such a candidate function V, such that $V: \mathbf{R}^n \to \mathbf{R}$, and V has continuous first partial derivatives and V is positive-definite in a region surrounding the origin. Assume it is known that the origin is an equilibrium point. Lyapunov theory states that if the function V' *is negative-semidefinite* in the same region, then the origin is a *stable* equilibrium point. Furthermore, if V is positive definite and V' is negative-semidefinite throughout the state-space, the origin must be globally stable.

This condition is only sufficient and not necessary, but it is still a powerful tool when evaluating nonlinear systems. Stated somewhat differently, Zak's summary of applying the continuous matrix Lyapunov equation to check stability of a system is an excellent presentation of the practicality regarding the theory.[15] He shows that the Lyapunov matrix equation is

$$A^T P + PA = -Q \qquad (17.5)$$

It then follows that the Lyapunov derivative of the candidate function is

$$V' = \frac{d}{dt}V = -x^T Q x, \quad \text{s.t. } Q = Q^T > 0 \qquad (17.6)$$

Then, the real matrix A, is asymptotically stable (i.e., all eigenvalues of A have negative real parts *iff* for $Q = Q^T > 0$, the solution P of Equation 17.5 is (symmetric) positive definite).

Given a molecular nanoscale system, we can thus proceed through the following steps to determine whether the system is asymptotically stable:

1. Given A, Select any Q that is symmetric positive definite (e.g., the Identity matrix I_n).
2. Solve Lyapunov Equation 17.5 such that $P = P^T$.
3. If P *is positive definite*, $\rightarrow A$ is asymptotically stable, else $\rightarrow A$ is NOT asymptotically stable.

As discussed in Chapter 18, information content and entropy play a large role in determining the overall system's characteristics. The Lyapunov spectrum can be used to give an estimate of the rate of entropy production and of the fractal dimension of the considered dynamical system. One hypothesis is that the sum of a system's negative Lyapunov exponents can indicate a system's propensity to dynamically self-assemble, whereas the positive exponents may provide quantitative information on the rate of such SA.[16] Further investigation of this hypothesis is required to determine whether this holds for SA in general systems.

In summary, when examining nanotechnology as a complex nonlinear system, this method is useful in determining stability. Because system fabrication at the molecular level has many complex components, it may be more critical to characterize the system behavior in terms of equilibrium and stability during SA rather than solve the differential equations and obtain an exact solution. Table 17.1 summarizes the relationship between the Lyapunov characteristic (first) exponent and system stability.

TABLE 17.1
Relationship between Characteristic (or Dominant) Lyapunov Exponent and System Stability[1]

λ	System Description
>0	Likely chaotic or unstable divergence; physical example can be found in Brownian motion
=0	Lyapunov stable, system in some steady-state mode; volume of space is constant along trajectory; neutral fixed point
<0	Globally asymptotically stable; dissipative and system achieves stability or converges toward fixed points

Source: Karcz-Dule.ba, I., Chaos detection with Lyapunov exponents in dynamical system generated by evolutionary process ICAISC 2006, LNAI 4029, 380-389. Berlin: Springer-Verlag.

Engineering Self-Assembly under Stochastic Conditions

In addition to the nonlinearities of the system, many uncertainties are constantly prevalent at the molecular scale. One such example is the impact of thermodynamic equilibrium on SA of nanomaterials. Much of the current work in SA has focused on static SA, implying the presence of thermodynamic equilibrium. However, in dynamic self-assembly, there is continuous association and disassociation of products. At any instant in time, noncovalent interactions ensure the lowest-energy products are present in the greatest proportion.[18] Dynamic SA is still an area of active research and may be important when dealing with larger systems of higher complexity. As novel structures are created, self-organization far from equilibrium may actually amplify fluctuations into coherent oscillations, leading to fabrication of structures otherwise impossible in static SA.[19] Moreover, such processes occurring outside of thermodynamic equilibrium underlie many forms of adaptive and intelligent behaviors in natural systems.[20]

As science tries to catch up, systems that incorporate nanotechnology are already being developed. Thus, the need for a holistic methodology is imminent. Recent research in the medical field portrays this concern. Here, it is important that the overall system dynamics are accurately monitored and one does not neglect the "human in the loop" as part of the system. Specifically, it has been shown possible to control electrical activity in neurons by an externally applied magnetic field using temperature-sensitive ion channels and magnetic nanoparticles attached to cell membranes.[21] As larger systems evolve with such breakthroughs, it follows that the following heuristics or externalities can aid in achieving stable designs at the molecular level of self-assembly under stochastic conditions:[22]

1. Identifying suitable interactions
2. Balancing interactions and potentials
3. Choosing a common reference scale
4. Implementing effective feedback and control
5. Defining and tracking synthesis and integration metrics

In essence, the facets of systems engineering must be more relaxed during nanoscale integration and are counterintuitive because each component has larger uncertainties of possible failure. An ill-devised strategy would be to codify and strictly regulate the means by which interactions occur at this level. Rather, it behooves the system engineer to peer through both top-down and bottom-up lenses and initiate control mechanisms at well-defined intervals. Tools available for controlling the system may include increasing or decreasing the external energy (magnetic field) during synthesis, catalyzing known reactions, and creating new reactions with system by-products.

Nanosystems engineers can also take cues from research done in uncertainty management for engineering planning and design. The traditional engineering task of optimizing technology for a set of criteria must be adapted because of uncertainties. Specifically, strategies such as flexibility and robustness must be interwoven through the design process under such a scenario. A robust system will allow functionality in spite of unforeseen circumstances; the architecture of the system should permit alternate trajectories toward stability in order to circumvent disabled areas. This type of statistical quality control is being used in industrial programs and is referred to as *robust engineering processes*, optimized by design of experiments, or in programs referred to as *Six Sigma*. In terms of flexibility, the system should be able to use control and feedback as specified earlier to alter atom configuration or other structural processes as certain optimal stability or equilibrium metrics change. Specifically, systems should have "options" built in where an option is *not* defined as a synonym for "alternative" but is, rather, a deliberate choice made possible during the system planning phase. Systems engineering principles that address uncertainties should thus consider the tails of the probability distribution of likely outcomes and create options to deal with unstable courses of action. Similar to the financial industry, these "options" should be "exercised" depending upon circumstances and the state of the system.[23]

It would be naïve to believe that such system engineering principles will design themselves and that trial and error is a viable strategy. While academia, industry, and government convene on larger quantities of nanoproducts, an integration scheme will slowly become a never-ending nightmare if early standards are not adopted. Furthermore, the number of interacting particles and subsystem interfaces that make up even the most basic nanosystems are exactly the environment where systems engineering principles are the most needed.

Molecular and Computational Symbiosis

Nondeterministic self-assembly has deep roots in computer science, going as far back as the 1950s and the self-organizing patterns initially proposed by Alan Turing.[24] His seminal work on the computational theory of the mind includes findings between "the state of the system" and overall homogeneity and even symmetry. Recent work in examining DNA structure and SA are analogous to a bottoms-up fabrication approach, and results have shown the ability to create periodic patterns on the molecular scale.[25] At the cellular level, biological components such as fatty acids, nucleic acids, and amino acids continuously work cohesively together to regulate processes to achieve stability. Biological examples abound, and one that is clearly illuminating is the assembly of a bacterial flagellum.

Amazingly, the flagellum is a motor organelle and a protein export and assembly device. Six of the components of the export apparatus are now believed to be located within the flagellar basal body.[26] While it appears similar to the man-made outboard motor, this flagellum assembles spontaneously with high degrees of regulation from neighboring proteins and nucleic acid instructions. As shown in Figure 17.3,[27] the assembly occurs as a linear process in a highly specified order of formation: base, hook, and filament. A key critical process that occurs in this "system" assembly is that once the base is built, the remaining parts are assembled from a variety of proteins exported through the base's center. This type of instructional-based flow that utilizes feedback to orchestrate a multitude of components with robust flexibility is a shining example from nature of successful systems engineering at the molecular scale.[28]

New developments in technology have aided in determining such critical links between the parts and processes within the cell. Current research using photo-activated localization microscopy (PALM) has allowed scientists to map the cellular locations of three proteins central to bacterial chemotaxis (the Tar receptor, CheY, and CheW) with a precision of 15 nm. These maps support the notion that stochastic self-assembly can create and maintain approximately periodic structures in biological membranes without direct cytoskeletal involvement or active transport.[29] Practically, SA offers the possibility to organize molecules in a given architecture through a subtle interplay between different noncovalent interactions. Thus, models of SA are not simply biological but have deep connections at the chemical and physical layers.

In time, artificial intelligence and man-made computation algorithms have started bridging the gap between cellular SA and larger and larger systems. Further advancements will be made when the components, the regulation and feedback, and atomic structure are aligned to achieve stability or an optimal path toward stability and cyclic SA. The ability to use protein molecules as fundamental simple machines has allowed

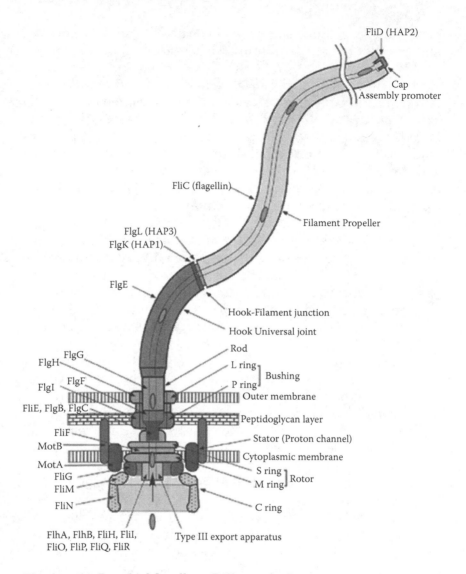

FliD (HAP2)

Cap
Assembly promoter

FliC (flagellin)

Filament Propeller

FlgL (HAP3)
FlgK (HAP1)

FlgE

Hook-Filament junction

Hook Universal joint

Rod

FlgG
FlgH
L ring
FlgF
FlgI
P ring
Bushing
Outer membrane
FliE, FlgB, FlgC
Peptidoglycan layer
FliF
Stator (Proton channel)
MotB
Cytoplasmic membrane
MotA
S ring
FliG
Rotor
FliM
M ring
FliN
C ring

FlhA, FlhB, FliH, FliI,
FliO, FliP, FliQ, FliR
Type III export apparatus

Figure 17.3 Bacterial flagellum. *Different shading* represent different protein components.[27]

genetic and biological engineers to create systems and will soon pave the way to develop a system of systems. As Drexler forecasted, "Advances in the technologies of medicine, space, computation and production—and warfare—all depend on our ability to arrange atoms."[30] In other words, demonstrating stochastic SA as a viable means of organizing complex entities into stable structures is a critical result to move from generation 3 to generation 4 in Figure 17.1.

Computation and artificial intelligence are only in their infancy when compared to the evolution of even a simple bacterial cell. Analogous to Minsky's claim of a mind being able to function from many "mindless" parts and agents,[31] a similar mode of thought is required to move nanotechnology forward. In this light, machine learning or mechanisms by which computational techniques can predict trends using supervised and unsupervised learning will be the primary means by which *systems of molecular systems* will be developed over time. Thus, the concepts of uncertainty and feedback are interwoven throughout the technological framework of molecular systems. These features are not simply the guiding forces toward systems engineering optimizations and system stability but are ironically the characteristics of the system that define its powerful capabilities.

Summary

Exploiting molecular uncertainty through flexible control and robust feedback, system engineers are poised to utilize stability and equilibrium heuristics throughout the self-assembly process. It is recognized that other countries besides the United States have active, well-funded programs to develop nanoscale technologies. These include Japan, China, Australia, Singapore, Russia, Israel, Brazil, India, and many nations within the European Union. The president of India, himself a nuclear physicist, has gone so far as to say that advanced nanotechnology must be developed by his country's scientists because, among other things, it "would revolutionize the total concepts of future warfare."[32]

In the end, policy makers and industry executives have their own motivations for such prognostications. However, from a scientific and engineering standpoint, the worlds of top-down and bottom-up fabrication are colliding at the nanoscale, which inherently has molecular and system uncertainties. A system approach requires accepting which individual uncertainties are worth accepting, which are too critically close to system failure, and what design decisions can be made before, during, and after any self-assembly process to still achieve success.

References

1. Roco, M. Nanoscale Science and Engineering: Unifying and Transforming Tools. *AIChE Journal*, 50: 890–897 (2004),
2. Felix A., Standard Approach to Trade Studies: A Process Improvement Model that Enables Systems Engineers to Provide Information to the Project Manager by Going beyond the Summary Matrix.
3. NASA, *NASA Systems Engineering Handbook*, Section 5.1 Trade Studies, NASA SP-610S 1995.
4. Ullman, D., and Spiegel, B. Trade studies with Uncertain Information; Sixteenth Annual International Symposium of the International Council On Systems Engineering (INCOSE) 8–14 July 2006
5. Boland, T., and Fowler, A. A systems perspective of performance management in public sector organisations, *International Journal of Public Sector Management*, 13(5): 419–446 (2000).
6. Fialkowski, M., Bishop, K.J.M., Klajn, R., Smoukov, S.K., Campbell, C.J., and Grzybowski, B.A. Principles and implementations of dissipative (dynamic) self-assembly, *J. Phys. Chem. B*, 110: 2482 (2006).
7. Bishop, K.J.M., and Grzybowski, B.A. Localized chemical wave emission and mode switching in a patterned excitable medium, *Phys. Rev. Lett.* 97: 128702 (2006).
8. Grzybowski, B.A., Bishop, K.J.M., Campbell, C.J., Fialkowski, M., and Smoukov, S.K. Micro– and nanotechnology via reaction-diffusion, *Soft. Mater.* 1: 114 (2005).
9. Gonzales-Rodriguez, D. et al., G-quadruplex self-assembly regulated by Coulombic interactions, *Nature Chemistry*, 1(2): 151: 2009.
10. Khalil, H. *Nonlinear Systems* (3rd edition). Upper Saddle River, NJ: Prentice-Hall, 2002.
11. www.stanford.edu/class/ee363/lectures/lyap.pdf.
12. Dieci, L. Jacobian Free Computation of Lyapunov Exponents. *Journal of Dynamics and Differential Equations*, 14(3), July 2002.
13. ibid.

14. http://academic.csuohio.edu/richter_h/courses/mce647/mce647_8_hand.pdf.
15. http://cobweb.ecn.purdue.edu/~zak/ECE_675/Lyapunov_tutorial.pdf.
16. Fialkowski, M. et al., Principles and implementations of dissipative (dynamic) self-assembly, *J. Phys. Chem. B*, 110: 248, 2 (2006).
17. Karcz-Duleba, I., Chaos detection with Lyapunov exponents in dynamical system generated by evolutionary process, ICAISC 2006, LNAI 4029, 380-389. Springer-Verlag.
18. Pfeil, A. et al. Helicate self-organisation: Positive cooperativity in the self-assembly of double-helical metal complexes. *J. Chem. Soc., Chem. Communications*. 1992, 838–840.
19. Geblinger, N. et al. "Self-organized nanotube serpentines." *Nature Nanotechnology* 3: 195–200 (2008).
20. Fialkowski, M. et al. Principles and implementations of dissipative (dynamic) self-assembly, *J. Phys. Chem. B*, 110: 2482 (2006).
21. Knöpfel, T., and Akemann, W. Nanobiotechnology: Remote control of cells, *Nature Nanotechnology* 5, 560–561 (2010) doi:10.1038/nnano.2010.163.
22. Fialkowski, M. et al. Principles and implementations of dissipative (dynamic) self-assembly, *J. Phys. Chem. B*, 110: 2482 (2006).
23. de Neufville, R. et al. Uncertainty management for engineering systems planning and design. Engineering Systems Monograph. MIT Engineering Systems Symposium, March 2004.
24. Turing, A.M. The chemical basis of morphogenesis. *Philosophical Transactions of The Royal Society of London, series B* 237: 37–72, 1952.
25. Winfree, E., Liu, F., Wenzler, L.A., and Seeman, N.C. Design and self-assembly of two-dimensional DNA crystals, *Nature*. 394, 539–544: (1998).
26. Macnab, R.M. How bacteria assemble flagella. *Annu. Rev. Microbiol*. 57: 77–100 (2003).
27. Yonekura, K. et al., Growth mechanism of the bacterial flagellar filament. *Research in Microbiology* 153: 191–197 (2002).
28. "Self-assembly of the bacterial flagellum: no intelligence required." www.biologos.org/blog/self-assembly-of-the-bacterial-flagellum-no-intelligence-required/.
29. Greenfield, D., McEvoy, A.L., Shroff, H., Crooks, G.E., Wingreen, N.S., et al. Self-organization of the *Escherichia coli* chemotaxis network imaged with super-resolution light microscopy. *PLoS Biol* 7(6): e1000137: (2009). doi:10.1371/journal.pbio.1000137.

30. Drexler, K.E. *Engines of Creation*, New York: Anchor Books. 1986.
31. Minsky, M. *The Society of Mind*. New York: Simon and Schuster. 1988.
32. Treder, M., *Turn on the Nanotech High Beams*. Future Brief 2005.

18

The Role of Mechanoevolution in Predicting the Future of Micro- and Nanoscale Technologies

Bradley Layton

"Tell me about the 4th generation of nanotechnology."

Contents

Introduction

In this chapter, we take a "Kurzweilian" approach [1–3] to interpreting some of the technologies discussed thus far and cast these into a framework of *mechanoevolution*, a term recently introduced to describe how machines are selected by and form symbioses with humans (e.g., [4,5]). Although the prediction of specific technologies that may appear in the future is typically unreliable, if we place bounds upon the limits of sophistication that a technology is capable of reaching with a combination of Gould's Left-Wall Hypothesis [6], Shannon's Information Theory [7], and Carnot's mathematical presentation of the Second Law of Thermodynamics [8], some headway might be gained. To do so, we enlist the metric, α, Table 18.1, which relates the physical entropy generation rate of a specific technology to its mathematical information generation rate. In general, the manufacturing of both small technologies and very large technologies consumes more energy per unit mass to produce than human-scale technologies and thus generates more environmental entropy per unit mass of product during a production cycle. Unexpectedly, however, during their operating life cycle, both smaller technologies and larger technologies have superior

TABLE 18.1
Nomenclature

Symbol	Meaning	Units
b	Bits	1
I	Information	b
$\dot{I}$	Information processing rate	b s^{-1}
E	Energy	J
k_B	Boltzmann's constant	1.38 E-23 J K^{-1}
mEPP	Minimum entropy production principle	—
MEPP	Maximum entropy production principle	—
N	Gravitational-to-electrostatic-force ratio	1
p	Probability	0 – 1
P	Power	W, J s^{-1}
S	Entropy	J K^{-1}
$\dot{S}$	Entropy generation rate	J K^{-1} s^{-1}
$\ddot{S}$	Entropy acceleration rate	J K^{-1} s^{-2}
S_U	Entropy of the universe	J K^{-1}
T_H	High temperature	K
T_L	Low temperature	K
t_L	Lifetime	s
α	Entropy-information coefficient or "sustainability ratio"	J b^{-1} K^{-1}
α_m	Mass-specific entropy-information coefficient	J b^{-1} kg^{-1} K^{-1}
ϕ	Specific power throughput	W kg^{-1}
η	Efficiency	1
σ	Entropy generation rate	J K^{-1} s^{-1}
Ω	Energy probability weight	0 – 1

Note: The symbol, H, with dimensions of bits, which has conventionally been used to denote Shannon entropy has been avoided in order to make the clear distinction between physical entropy, S, with dimensions of joules per kelvin and the abstract amount of information, I, with dimensions of bits required to fully describe the entropy of a system.

information-to-entropy ratios than mesoscale technologies. The result is that the information throughput density of microdevices and likely nanodevices typically "pay off" on an energy-invested to quality-of-life basis as compared to their predecessors. This is because smaller technologies with high information processing rates typically result in a large savings or beneficial reallocation of human metabolic energy or technological energy. This chapter begins with a historical perspective on the beginnings of technology, discusses key stepping stones, and concludes with a series of vignettes on where our technologies may lead us.

Confluence

We have arrived at confluence of our own species' evolution and the "evolution" of our technologies. The confluence of these two streams, one biological and the other technological, is becoming increasingly intermixed and, in many cases, inextricably so. This confluence has manifested as an irrepressibly large number of human as well as ecological symbioses with our machines. Just as the downstream waters of two previously unjoined streams become indistinguishable through the embedded micro- and nanotechnologies discussed in Chapter 16, the boundaries between our technologies and ourselves are disappearing.

This syncytium with nature that we have woven with our numerous technologies began with our accidental and fortuitous discovery of fire. The subsequent mastery of a variety of thermal energy sources for supplementing body heat, cooking food, and warding off predators essentially pushed us over a precipice whereby our ability to funnel energy, and to a limited extent, entropy gave us a distinct advantage over all other species on the planet. What particularly distinguishes man's use of fire to create order and predictability in his immediate environment is that it represents our species' ability to acquire energy without eating it. In fact, the mastery of fire was perhaps the key event that laid the first few stones upon the path that has enabled us not only to live more comfortably but also to transform matter through various phases, and, indeed, to aggressively and somewhat indiscriminately transform matter into pure thermal energy, via nuclear weaponry.

Fire was the beginning of mankind's use of external energy sources to produce order; a sequence that began with campfires and gas stoves is now converted to other forms of energy via secondary technologies via the power grid that drives additional technologies such as laptop computers and personal digital assistants (PDAs). Indeed, combustion is responsible for approximately 85% of all technological energy consumed [9].

Nearly all technologies, unlike all biologies, are "born of fire." With very few exceptions, every human on the planet relies upon the technological harnessing of energy for maintaining a high quality of life. The campfire and the gas stove serve the purpose of supplying thermal energy to our "corporeal selves" and thus enhance the probability of propagating our "genetic selves." The laptop and the PDA, on the other hand, consume energy to maintain our "extracorporeal selves" and thus serve the purpose of maintaining and distributing our "memetic selves" [10].

Entropy Partitioning

The central thesis of this chapter is that the fundamental purpose of our technologies, be they micro, nano, or macro, is to partition entropy. Specifically, the sole purpose of all our technologies is to "deentropicize" our corporeal selves and, in some instances, our immediate environment in order to increase our probability of survival. In some cases, our technologies are used to map the external world, essentially converting that which was previously unknown into a useful, abstract, portable set of bits. This recording and internalization serve the same purpose of direct deentropization. In either case, energy must be converted, resulting in a net production of physical entropy. It is my contention that the most successful and sustainable technologies will be those that create the greatest amount of information while keeping entropy production to a minimum.

Imagine that we were to be suddenly stripped of all technologies. In a sense, we would cease to be fully human. Imagine Bill Gates without the microprocessor, Jeff Gordon without the internal combustion engine, or John Glenn without the rocket. Each would still be a Homo sapiens, but the human potential of each of these men, specifically his ability to rise to prominence within the ecosystem, the technosystem, the media, and the economic infrastructure, would be diminished to the point that each of them would likely be less remarkable. In the absence of their respective technologies, Bill Gates might be leading an effort to move from dirt scrawlings toward the abacus, Jeff Gordon might be domesticating horses, and John Glenn might be leading expeditions into uncharted terrestrial destinations. However, without their advanced technologies, the amplification of their innate human abilities would be greatly dampened. The identity, power, and influence of each of these three men, and to a lesser extent the average person, are inextricably tied to their abilities to harness the power of their machines.

As described by Kurzweil [3], Chaisson [11,12], and Coren [13,14], the trend of increasing complexity of technology across all scales from nano to macro will continue to advance in a "Moore's Law fashion" [15], and the next few generations of humans will likely include members with abilities beyond those alive today. For example, medical technology has already enabled people to have better visual acuity than their genetics prescribed [16,17]. Recently, it was also proposed that a not-so-distant singular genetic mutation event led to our enlarged brains, and thus mental capacity compared to chimpanzees [18]. Presumably, in a simplistic and fundamental way, this led to the accelerating sophistication of our technologies.

What specific symbioses will be formed between the emerging technologies discussed in the preceding chapters of this book and the next generation of humans? Whatever form they ultimately take, these symbioses will define our fate as a species on this planet and potentially beyond.

This same seemingly irreversible trend toward increased human–machine symbiosis permeates our species even with very simple technologies. I have chosen to call them irreversible, because to abandon our technologies would be an abandonment from our path of competition with other species with the natural environment and, indeed, with mortality. For the technologist to do so, would be to make himself "less fit," so we press on with increasingly sophisticated technologies, essentially creating new worlds of high-tech "guns, germs and steel," which by design, or by default, suppress societies without them. As we will see shortly, there are four fundamental paths or modes in which a society may choose to proceed. In the author's opinion, the one with the lowest overall $S:I$ ratio, or "sustainability ratio," α, a variable to be defined shortly, is the most likely to persist.

For example, Industrial Revolution technologies such as eyeglasses, shovels, spinning wheels, and needles all contribute to their own selection by consistently performing their respective tasks. Each of these technologies, of course, also contributes to the "selectability" of their respective users. If we briefly examine each of these technologies in a manner similar to that taken by Henry Petroski [19], we easily see that eyeglasses enhance the survival potential of their users by allowing information to stream into the user's brain at a greater rate, $\dot{I}$, and to a greater degree of accuracy than would be possible with the user's imperfect ocular lens. The shovel, while it cannot amplify mechanical power, can amplify mechanical stress from foot to soil by a factor of 10^3 to 10^6, enabling the shoveler to gain access to water or mineral resources more quickly than would be possible with more primitive technologies such as sharp rocks or sticks. Even though water and mineral resources are difficult to equate in units of either mechanical power, P, or information, b, we will see shortly that enhanced access to both enables the user to effectively "bend down" his or her own entropy curve (i.e., $\ddot{S} < 0$), per Ziegler's minimum entropy production principle (mEPP) [20]. (Note that a single dot above a symbol denotes the first derivative with respect to time, and two dots denote the second derivative with respect to time.) Shovels and shovel-like technologies also give us the ability to sculpt and contour the land in a manner that lessens the energy required to navigate the terrain, thus allowing for faster access to additional material or information resources. More sophisticated in many ways than both eyeglasses and shovels, if only by the presence of moving parts, the spinning wheel enables the organization of natural fibers into long, continuous one-dimensional structures to be subsequently arranged into membranous structures for maintaining warmth, avoiding the sun's rays, or in the case of ornamental dress, elevating the perception of one's social status or sexual attractiveness and thus the probability of reproduction. The humble needle, a favorite of Petroski, enables the production of more sophisticated arrangements of woven fiber-based clothes; two-dimensional membranes can now be shaped into intersecting tubes to conform to bodies, domesticated animals, or technologies. Each of these technologies represents a specific arrangement of matter that was "selected for" based upon its specific ability to interact with light and matter; eyeglasses bend light, shovels move soil, spinning wheels funnel fibers, and needles guide thread. In all four cases, the specific funneling of light and matter benefits the human user in ways that enhance the practitioner's Darwinian chances for survival. It is the case with each of these technologies and with all technologies that their ability to control the movement or bending of light and matter make the immediate human world less entropic and thus more predictable for the human user. A more predictable, less entropic world is, of course, one with either less randomness or less perceived entropy at the human scale. However, since the second law of thermodynamics is never violated, the reduction in entropy caused by each of these technologies is channeled to the molecular scale where it is absorbed by the environment at a commensurate rate. Thus, it is one of the primary theses of this chapter that $\dot{S}_{humanity} > \dot{S}_{universe}$ and that there is some entropy acceleration that humans are creating on earth (i.e., $\ddot{S} > 0$) which is greater than the background universal entropy acceleration rate if, indeed, such a metric exists. Terrestrially at least, the persistent entropization of the environment occurs either via the production of heat or by the creation of smaller molecules with less chemical potential to perform mechanical work [21,22].

Without the successive discovery or design of each of our technologies, beginning with the first sharp stones, we would not have climbed our own technological version of what Richard Dawkins refers to as "Mount Improbable," which, in fact, is the world as manifest today and which happened one genetic mutation at a time over the course of the lives of the

~10^{22} organisms that have lived in the 3.7 billion year history of this planet (~10^8 species multiplied by an average of 10^{14} organisms per species) [23]. The difference, of course, between biological evolution and technological evolution is that, while biology has to "wait" on mutations and material transfer of actual genetic material, technology can get away with being much more impatient by spreading memetically via whatever new technological media emerges. In the case of genetic algorithms used for design optimization, this happens at rates as great the speed of light as design algorithms are run on computers. Some of the fastest biological generations can occur on the order of minutes [24], but technological evolution can occur even faster. In fact, if you classify some of the algorithms written by hedge fund investors to destabilize and ultimately crash the markets in 2008 [25] as technologies, these have generation cycles on the order of milliseconds.

Just as Spencer Wells is working to trace every human alive [26] both temporally and spatially, every technology is traceable along a set of discrete steps all the way back to the first sharpened stones and the first flames fanned by humanity. Even though no such formal study has been conducted, presumably every technology could be placed into a specific location on a phylogenetic tree in a manner similar to that used in evolutionary biology studies [27,28]. A rough outline of what such a tree might look like is depicted in Figure 18.1.

Every technology exists in two forms: the purely abstract and the purely material. In the case of the airplane, the abstract form is the set of drawings, material specifications, and testing protocols required to build the airplane. An abstract form of the airplane existed in the notebooks of Leonardo da Vinci long before the first successful material form took flight at Kitty Hawk. And while the material form of any technology ultimately falls prey to entropic effects imposed by its environment, the ideal form of the airplane can effectively achieve immortality insofar as it lives at least as long as its creator. This idea is not completely new. It was first proposed by Pythia of Delphi nearly 3000 years ago when deliberating whether a boat that had been slowly patched and rebuilt until none of the original material remained was indeed the same boat. The instructions for building an airplane can be maintained

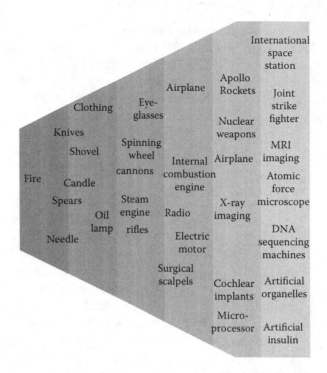

Figure 18.1 A phylogenetic tree of technology, with the past to the left and the present on the right. Larger technologies are at the top, and smaller ones are on the bottom. Conceivably, every technology has a predecessor, or in some cases, multiple "parents." An obvious example is the Swiss Army Knife, which is a combination of multiple tools or technologies. Technologies, however, unlike biologies, do not require direct transfer of matter from parent to child, and mutations are not bound by naturally dictated generation cycles. Actual lineages of inheritance have been left out of this figure. A more complete version would have lines connecting each technology, just as a biological phylogenetic tree would. Similar to a biological phylogenetic tree, specific genes may emerge at multiple times. Unlike a biological phylogenetic tree, individual technologies may have more than two "parents."

unmutated as long as proper backups are made on reliable media, be they paper, magnetic media, or stone. Each of these media will also ultimately fall prey to entropy, as long as the copying is robust, the airplane, or any technology, will remain immortal. Even as the material forms of all of our technologies undergo perpetual negative mutations, ultimately becoming a dispersed set of atoms, molecules, or fragments of metal, glass, and plastic with no apparent relation to its former structure,

with the proper maintenance of our oral and written tradi-
tions, our technologies can attain immortality. However, as
we have already transformed the planet's biosphere to a point
that it unfortunately will require additional vigilance, if not
maintenance, we will likely find little use for many of the technologies that were responsible for its transformation.

Information and Entropy

While composing this chapter, finding an author who had previously fully captured the central theme was difficult. There
have been a few attempts to unify the works of Shannon and
Carnot, but most have been unsatisfactory. However, the invo-
cation of the works of both of these engineers, along with contributions from Gibbs and Darwin, are necessary if we are
to develop a comprehensive framework for predicting future
technologies. Perhaps the closest echoing of this chapter's
central point was made recently by Matt Ridley in *Genome*.[29]
Ridley was quoted by *Dawkins in the Oxford Book of Modern
Science Writing*, stating that

> *Shannon's idea is that information and entropy are opposite faces of
> the same coin and that both have an intimate link with energy. The
> less entropy a system has, the more information it contains. A steam
> engine parcels out entropy to generate energy because of the informa-
> tion injected into it by the designer.* (p. 16)

These three sentences mention three key metrics for comparing technologies: energy, entropy, and information. Although
concise and thought provoking, these three sentences contain
at least one flaw as well as an oversight. The *flaw* is that an
engine does not generate energy, it transforms energy from
chemical to mechanical, which ultimately becomes thermal, a
concept not clearly understood or appreciated by a surprising
number of engineers and scientists. A steam engine has a combustion chamber as well as a closed system of interconnected
chambers for converting liquid water to steam and back. This
cycle of hot and cold exploits water's expansion and contraction as it moves through the engine. This expansion and

contraction then converts a change in gas volume to a linear displacement, which is then converted to a rotational motion used to drive wheels, gears, and so forth. The end result, of course, is the transportation of humans from one location to another, the plowing of a field, or the production of an electric field to supply computing power, motive power, chemical potential, and so forth [9]. Ridley's *oversight* is in "the less entropy a system has, the more information it contains." Of course, the reverse is true: more information is required to fully describe systems with more entropy. Greater entropy may be the result of either a greater quantity of matter to describe or a greater number of configurations available to a system in proportion to its temperature via the Boltzmann distribution (Box 18.1). What Ridley seems to be hinting at is the fact that Shannon entropy and Gibbs entropy may be expressed similarly. The difference, of course, is in the fact that Shannon entropy has dimensions of bits and Gibbs entropy has dimensions of joules per kelvin (Table 18.1).

BOX 18.1

Gibbs entropy, which is physical in nature, and has dimensions of energy per temperature is typically written as

$$S = k_B \ln(\Omega_i(E_i))$$

where k_B is Boltzmann's constant, E_i is the set of all possible energy states with weights Ω_i. *Shannon entropy,* or what we define here more unambiguously as *Shannon information*, is expressed as

$$I = -\Sigma \, p_i \log_2 p_i,$$

where p_i is the probability of a given symbol occurring within a message. Both entropy and information are scalar values and both are extensive (extrinsic) values, in contrast to metrics such as temperature which is intensive (intrinsic). However, just as a scalar such as temperature or pressure may be used to describe a multidimensional

---

physical system and just as a digitized three-dimensional (3D) computer-aided drawing may be sent as a dimensionless sequence of bits, both entropy and information can be used to characterize and thus evaluate disparate systems, thus making them ideal metrics for evaluating technologies via a single scalar, α, which will be defined later in the chapter. Again, the critical distinction to make and one that is frequently overlooked in the literature is that physical entropy, or more succinctly *entropy*, is always evaluated in units such as joules per kelvin. Information, which is dimensionless, has units of bits. Having made this distinction, there have been previous attempts to establish a relationship between entropy and information (John Avery, *Information Theory and Evolution, 2003*). For example, a physical system with less entropy takes less information to describe. A radio signal that is 100% "noise" takes more information to record, store, and reproduce than one transmitting a "pure" tone. A physical system with a greater mass, temperature, and number of chemical constituents is likely to be more entropic than a small amount of pure substance at low temperature. Commensurately, a greater amount of information is required to completely describe the larger, hotter, less pure system and less is required to describe the small cool pure one. Paradoxically though, it may require more energy to maintain the small, cool, pure system than the large hot multicomponent system depending upon how far from equilibrium the two systems are. Certainly there exists a unique minimal amount of information, I_m, required to reproduce a given technology. Furthermore, this amount of information is media-independent. For example, the information required to reproduce a given physical manifestation of a steam engine could just as readily be transferred via memory stick as it could via blueprints. However, normalizing the bit density on an energy, mass, or volume basis, the memory stick will likely be superior to that of the blueprint. Shannon also explored message fidelity on an energy basis when considering the "cost" of sending an error-free message with a given amount of signal power. So while the number of bits required to reproduce a given technology must have some minimum value, the efficiency of its reproduction will be media-dependent. It is also worth extending the discussion to the environmentally dependent entropy production rates of any heat engine. For example, the engine will have a greater Carnot efficiency when operating in a colder environment. If we assume that the engine operates at a given internal temperature of T_H and an environmental temperature of T_L, the Carnot efficiency is expressed as $\eta = (T_H - T_L)/T_H$. Thus, if we fix T_H at the safe upper limit for the engine, it will operate more efficiently in a cooler temperature. In principle, this would be absolute zero. However, practical constraints such as maintaining water in liquid form must also be considered. For example, a steam engine, or any heat engine, operating at a greater temperature will need to do more irreversible work and thus generate more entropy in order to overcome the same (i.e., gravitational) energy barrier than it would at a lesser temperature. In order to create a superior engine, specifically one that is capable of not melting at a greater temperature, requires additional time, insight, and without question, access to more information than was required to create the more archaic engines with lower maximum operating temperatures.

Thus, Ridley would be more correct in stating that "the greater a system's ability to partition entropy, the more information required to design it." This statement has held true from the very early Savery → Newcomen → Watt steam engine development that occurred 200 years ago to the improvements in microscale and nanoscale technologies that are occurring today. As Ridley points out, the system only remains successful if it continues to maintain a low entropy state (i.e., not exploding or corroding). But ironically, the most successful machine may be one that requires relatively little information to describe. The "information injected into it by the engineer"

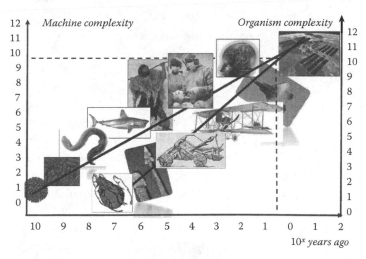

Figure 18.2 Qualitative representation of the relative rates of complexity acceleration in the biological and technological spheres. The abscissa is logarithmic time in years. The left ordinate represents a relative complexity metric for machine complexity, roughly quantified as the number of parts in a machine. The right ordinate represents a relative complexity metric for biological complexity, roughly quantified as the number of molecular interactions or potential to produce a wide spectrum of behavioral characteristics. The point of convergence is Kurzweil's singularity. (See Kurzweil, R., *The Singularity Is Near: When Humans Transcend Biology.* 2005, New York: Viking.)

is, of course, detained in all of the tried and failed attempts that were performed prior to arriving at the final embodied form. What remains to be seen, however, is if our technologies are capable of *both* reducing human corporeal entropy while simultaneously minimally affecting environmental entropy. To the author's knowledge, this has yet to be formally quantified [5]. Taking a long-sighted look at the confluence mentioned in the opening paragraphs, biological and technological evolution may be plotted qualitatively and approximately with general complexity as a function of time. This exponentially increasing complexity in both the technological and biological realms is depicted in Figure 18.2. Similar plots have been made by cosmic evolutionist Eric Chaisson [11] who has chosen specific energy throughput, ϕ, with dimensions of energy per time per mass as an ever-increasing variable with no apparent bound to describe how much energy passes through a system of a given mass per unit time. Nature constantly produces systems capable of greater amounts of specific energy throughput [12]. Chaisson attributes this propensity to the very expansion of the universe and to the thermal gradient that typically exists between radiant bodies and massive bodies (i.e., the sun is hotter than its planets; thus, thermal energy flows from the sun to the planets "down" the thermal gradient). Specifically, he states that as hot as the sun is, its power per unit mass is relatively paltry in comparison to a living organism or an energy-hungry technology. The earth essentially "feeds" upon the "waste" radiation from the sun. And just as the steam engine must maintain a specific configuration to continue to convert energy, so must photosystems I and II as they convert radiant energy to chemical energy [30]. In fact, as discussed in Chapter 15 on micro- and nanosolar technologies, we have a long way to go until we have achieved anywhere close to the efficacy of photosystem I (PSI) and photosystem II (PSII), with molecular weights of a few hundred kDa, dimensions of tens of nm, are found at a density of a few thousand per cubic micrometer and thus, have a global quantity of 10^{33} and convert radiant energy from photons to electrons at a rate of 5×10^{31} per second and at ambient temperatures. Meanwhile, our typical industrial-scale solar cells have efficiencies of 10% and typically generate substantial heat as a result of electron transfer. Micro- and nanosolar technologies that are capable of enhanced biomimicry may become an economic and sustainability imperative as we approach the "photosynthetic ceiling" [31,32]. For the reader not familiar with the photosynthetic ceiling, this is a concept introduced by Diamond in 2005 to quantify the ratio between human technological energy consumption, which stands near 16 terawatts (TW) and solar incident radiation, which is approximately 1.6 exawatts (EW). Although it is true that this "ceiling" neglects to include other nonsolar energies such as tidal, geothermal, and nuclear (either fission or fusion based), it does account for 85% to 90% of our current energy supply, namely fossil fuels, which are solar derived. The main point here is that in order power our future technologies, be they nano or macro, we will need to pay an increasing amount of attention to their energy consumption rates. Even if fossil fuel–based power generation diminishes (as it certainly must within two centuries based upon current reserve-to-consumption ratios), every technology emits heat; thus, we will have to deal with the thermal load of 500 exajoules (EJ) of thermal energy whether or not it is hot CO_2 or hot non-greenhouse gases.

Plots similar to those used by Chaisson have also been used by Coren to explain the saltatory nature of emergent life forms and technologies [13,14]. In this case, *saltatory*, similar to the term as used biologically where a process happens in jumps or leaps rather than continuous equal increments, implies that the "birth" of a new technology typically leads to a spawning of numerous other technologies. These then mature and eventually fizzle, and then a new burst occurs. Before Coren, Gould described evolution as saltatory [6]. In fact, Kurzweil has produced a metaplot of several studies similar to that of Coren's [3]. Of particular interest is the point in the near future where machine complexity reaches a level comparable to, or surpasses, that of the arguably most complex biological system, our own brain. This is roughly represented by the intersection of the biological and technological trajectories of Figure 18.2.

According to Kurzweil, this will happen before the close of this century, and the implications could be as simple as a permanently implanted neural prosthetics for accelerated cognition in a large fraction of the human population to a reality as complex as one where all humans achieve a sort of immortality through continuous upgrades to their consciousnesses. These "people" will have completely abandoned their own carbon-based cognition systems but will be fully "alive" and "aware" as "apps" on next-generation computers [2].

We have already seen early manifestations of embedded human–machine symbioses such as cochlear implants (Chapter 16) and direct brain interface devices (Chapter 16). Presently, these prosthetics are designed as an attempt to either restore or mimic normal human capabilities. However, there are other recent technologies such as exoskeleton projects [32], Lasik surgery [17], or of course anabolic steroids [17] that have had the effect of allowing a small subset of humans to exceed what even the most gifted would be able to achieve naturally. The other manifestation of what the future holds exists in the (typically physics violating) art of science fiction. The increased prevalence of movies such as *Terminator, Ironman, Spiderman,* and the *X-Men* are evidence of this trend.

Mechanoevolution

As a first attempt to cast a formal framework for mechanoevolution, let's classify all machines into four types, numbered I through IV (Table 18.2). Each of the four types may be classified along four bases. The four bases selected are not independent. The first basis is "entropy source." A designation of "no" indicates that once the machine or artifact has been manufactured, it is susceptible to thermal degradation just as all machines and biological systems are. However, in the absence of a human operator, it is incapable of generating entropy, but simply becomes a victim of environmentally induced entropic effects. This is why a Type I machine receives a designation of yes/no, rather than a clear "yes" as logic would seem to imply from the other fields in the table. A clear example of a Type I machine is a garden hoe left outside to weather. While working the soil, the hoe is dentropicizing the soil, raising lower layers to the top to create orderly rows, but once set aside, the hoe merely falls prey to entropy. If left alone long enough, the hoe becomes soil. We have lived with Type I machines for roughly two million years [33,34]. The surface rocks we used to form early tools are entropy sinks. Once sharpened, a surface rock becomes more susceptible to wear.

The triage criterion for a Type I machine versus a Type II machine is whether or not the machine is capable of accelerating the universal rate of entropy generation without the direct guidance of a human operator. The first Type I machines appeared during the Modern Stone Age (MSA), 300 kYa to 50 kYa, prior to the Early Stone Age (ESA) 2.5 MYa to 300 kYa, which produced stone artifacts but nothing resembling the honed tools of the MSA. The transition from ESA to MSA could roughly approximate the likely unabrupt transition from the predominate usage of tools with only a single part (i.e., the hand-axe of Galeria and Gran Dolina to the prevalent usage of spear points), which would have needed at least three parts: tip, shaft, and binding. This transition is located at roughly the "5" on the abscissa of Figure 18.2

TABLE 18.2

Categorization and Examples of the Four Basic Machine Types

Machine Type	Entropy Source	Motorized	Self-Operating	Self-Regenerating	Examples
I	yes/no	no	no	no	shovel, wheelbarrow
II	yes	yes	no	no	power lawnmower, power screwdriver
III	yes	yes	yes	no	auto-piloted airplane, computer server cluster
IV	yes	yes	yes	yes	self-replicating robot, artificial organelle

Clearly the bases "entropy relation" and "motorized" of Table 18.2 are nearly perfectly correlated and, thus, technically do not compose a mathematical basis. However, if we imagine a scenario where all tools are being used all of the time, Table 18.1 becomes one with "yes" along and below the diagonal and "no" above. Were we to be able to hit a "pause" button on all human activity and observe the resulting behavior of our technologies, the yes/no of the upper left corner would become a "no." If this imaginary pause button were to be held long enough, "no" would ultimately fill in the entire table. Thus, the basis with four axes is not a true mathematical basis because all four are not strictly independent. In other words, the presence or absence of a human operator is ultimately responsible for the immediate classification of the machine; a Type I machine becomes an entropy source rather than a sink when in the hands of its operator.

For better or worse, frequently, the advancement of weapons technology drives the advancement of our understanding of light and matter more than any other applied science. This is symbolized by the catapult of and continues through the development of TNT, napalm, and into the present with technologies, such as the F-35 Joint Strike Fighter (JSF), which has been touted as the most technologically sophisticated single machine conceived by humankind, (www.jsf.mil/). Other obvious examples of exceptionally complex machines include the International Space Station (ISS) and the Hubble Space Telescope (HST). Each of these three "big" technologies, the JSF, the ISS, and the HST, can be classified as bridges between Type II and Type III machines. Each has the ability to extract chemical or radiant energy from the environment, maintain a trajectory somewhat autonomously, and sense its environment in a semiautonomous manner. The general purpose of each of these advanced technologies, as with any technology, is twofold. The first is to give their operators access to information about the environment which those who do not possess the technology cannot access. The second, more subtle but implicit purpose, is to "shed entropy" via the maximum entropy production principle (MEPP) [20] onto those without access to the technology. The contrapositive to "entropy shedding" is "information shielding." Whereas entropy shedding onto one's enemy gives the shedder the upper hand, "information shielding" has the same end effect—greater probability of survival of the shedder and reduced probability of survival of the shedee. All three of these technologies, either by design or by their very nature, shield their information from people without the ability, access, or interest to comprehend it.

To a crude approximation, the JSF, the ISS, and the HST are leaf blowers. The owner of the leaf blower uses the blower, a Type II machine, to deentropicize his or her swath of the earth's surface. The typical purpose of this is to enhance the appearance of the landscape by removing leaves that would either be tracked into the house, clog the street gutter, or leave an impression of slovenliness to passersby. The resulting physical entropy spread from the blower's use may be quantified via the second law as

$$\Delta S = \Delta Q \cdot T^{-1} \tag{18.1}$$

where ΔS is the change in entropy of the environment, ΔQ is the chemical energy consumed by the engine, and T is the operating temperature of the machine. Gasoline has an energy density of roughly 50 MJ·kg^{-1}. Thus, if the blower is consuming chemical energy at a rate of 1 kW, then it is generating entropy at a rate of approximately 3.3 J·s^{-1}·K^{-1} or a mass-specific entropy generation rate of about 0.67 J·s^{-1}·K^{-1} kg^{-1}. More challenging is to quantify both the information embodied in the leaf blower as well as the information gained as a result of the leaves being blown. However, the amount is equivalent to the number of bits required to store the mechanical drawings used to manufacture each part (~200 Mb), the assembly instructions (~50 Mb), the operating instructions (~0.5 Mb), and the formulation for the fuel (~0.1 Mb). In the case of the leaf blower, there is no obvious resulting information gain to the user. And, in terms of its information processing, because it does not have an on-board computers, it essentially has an information processing rate of one bit per unit time that it is on since it is in only two states. The lawn may now be visible, but the information required to encode this image is likely less than that required to encode the image of the leaves. Turning back to our three more advanced technologies, the JSF, the ISS, and the HST, each of these have mass-specific entropy generation rates of 113, 0.45, and 0.07 W/K/kg, respectively.

The JSF, the ISS, and the HST consume energy at a rate of 34 MW, 500 kW, and 2250 W, respectively. The purpose of the JSF is to entropicize the enemy, the purpose of the HST is to collect information on the state of the universe, and the purpose of the ISS is to explore the potential for living beyond the confines of our planet. Notably, if a Kurzweilian future becomes manifest and we avoid asteroid impact, efforts of the ISS become moot. So how are three of the most advanced technologies like a leaf blower? Like the JSF, the leaf blower sprays entropy elsewhere, leaving the operator to enjoy an environment devoid of intrusive organic debris. Like the ISS, the leaf blower creates a habitat for its user that is devoid of organic particulate matter. Like the HST, the leaf blower allows the user to know where all foreign objects lie. In this regard, the leaf blower may be superior to the HST in that it can put the foreign objects, however inefficiently, into a pile. The same measurements may be made of any technology, regardless of scale. Micro- and nanoscale technologies of course have lower masses, and because most are typically designed for information processing (i.e., integrated circuits), they have greater specific information throughput rates as well as comparable specific power throughput rates and, thus, greater specific entropy generation rates with respect to the ambient environment than the larger technologies. In fact, the estimates for the information throughput rates for the large technologies discussed thus far were based primarily upon the number of processors that each has.

Entropy Generation Rate and Background Entropy

How does the entropy generation rate of a given technology compare to the overall background entropy generation rate of the universe? In other words, how far above the background rate of entropy increase is a given technology and is there a limit to this? Lloyd recently discussed the limit [36] but again did not make the distinction between entropy and information. According to Chaisson, the universe reached maximum entropy after only a small fraction of its current age [11]. Questions that remain unanswered from Chaisson's work, however, are the effects of the expansion of space. For example, is the universe becoming more entropic simply because there is more physical space and, thus, a longer ledger required to track all 10^{80} particles, or is the universe becoming more entropic because the various energy manifestations (i.e., material, electromagnetic, etc.) have yet to come to equilibrium? Also, there is continuing debate about the entropy at the event horizon of a black hole and the information thus required to describe its behavior (Grabbe 2006) [64]. Nevertheless, what is certainly clear of any terrestrial system, either biological or technological, is that both use their embodied information to partition entropy by exploiting the second law. As long as thermal gradients exist, such as the one between the earth and the sun, there will be "free" energy available to drive biological and technological engines. For six of the most prevalent technological engines such as the Atkinson, Brayton, Otto, Dual, Miller, and Diesel [37] defined maximization of the ecological function

$$E = P - T_0 \dot{S} \qquad (18.2)$$

to be the limit of technological "effectiveness." In Equation 18.2, E has dimensions of power per effectiveness, P is the power output of one cycle, T_0 is the temperature of the environment, and $\dot{S} = \sigma$ is the entropy generation rate with dimensions of energy per Kelvin. This approach has been used by other authors such as Angulo-Brown et al. [38] to maximize the effectiveness of the power generated by power plants. However, effectiveness has been left poorly or completely undefined. Efficacy is the more common term for converting between various units or dimensions. For example, a reading light is more efficacious if the reader can absorb more bits per joule.

We must now establish a working definition of the relationship between energy and information, and a new relationship between entropy and information emerges. The first definition that must be established is that *information has dimension of bits and is a purely abstract* (nonphysical) *entity,* and *entropy has dimensions of energy per temperature* and is physical. Frequently this distinction is not made. However, the two state variables may be related via

$$\dot{S} = \hat{\alpha}\dot{I} \qquad (18.3)$$

where α is a system-dependent coefficient defined as the rate at which information is generated proportional to the rate at which entropy is generated. But as mentioned above, the purpose of any technology, be it macroscale or nanoscale, is to partition entropy, by reducing it locally at the expense of increasing it environmentally. We thus rewrite Equation 18.3 as

$$\Delta\dot{S} = \alpha\dot{I} \qquad (18.4)$$

where the Δ represents the difference between the mEPP and the MEPP (i.e., $\Delta S = S_{\mathrm{MEPP}} - S_{\mathrm{mEPP}}$).

This relation is similar to Shannon's original work in the field of data transmission, specifically, how many bits of information can be reliably transmitted per energy consumed per unit time [7].

A distinction was made between the maximum entropy production rate (MEPP) and the minimum entropy production principle (mEPP) by [20]. These two curves represent the upper and lower curves of Figure 18.1.

The expression in Equation 18.3 must be applicable to all technological and biological systems. For example, as a large organism uses its sensory organs, it does so to gain access to information about its environment. Of particular interest is the location of potential predators or other threats to the organism's corporeal self. For humans, the power devoted to vision is on the order of 2 to 3 watts, and the rate of information throughput can range anywhere from a few dozen bits per second for a slow reader reading a newspaper to several gigabits per second for someone moving through a richer four-dimensional space such as an National Basketball Association (NBA) basketball player, a surgeon, or someone panning for gold. However, on this topic a more detailed analysis is warranted. In *The User Illusion*, Tor Norretranders discusses the "user illusion" in computing the desktop graphical user interface (GUI): the friendly, comprehensible illusion presented to the user to conceal all the bouncing bits and bytes that do the actual work [39].

Contemporary gene sequencing machines, such as those of 454 Life Sciences, Lynx, Solexa, and Illumina, the GS20 and the GS FLX Titanium series and, more recently, a Helicos Biosciences machine developed by Stephen Quake, consume energy at a rate of 1-10 kW and produce genetic information about organisms at a rate of 10 to 10,000 bits per second. At a mass of 100 to 1000 kg, this gives them an α of $0.003 \text{ J} \times \text{s}^{-1} \times \text{K}^{-1}$ and a mass-specific $\alpha_m = 5 \times 10^{-6}$. By comparison, the JSF, ISS, and HST have $\alpha_m = 1.2 \times 10^{-10}$, 9×10^{-13}, and 6.8×10^{-13}, respectively. As gene sequencing machines evolve, the information gain rate per energy expenditure rate and, thus, entropy generation rate will increase, driving up α itself, implying a nonlinear and potentially exponential between α and itself—that is,

$$\dot{\alpha} \propto \alpha \qquad\qquad (18.5)$$

In other words, the better a technology is at partition-ing entropy with minimal information (both embodied and throughput), the greater the probability that the particular technology will evolve an even greater ability to partition entropy. What we have not considered, however, and what will be left for future work, is to describe the relationships and symbioses among various technologies. As an example, consider a technology such as the manufacturing and usage of carbon nanotubes, which have already found their way into several commercial applications such as memory devices and structural materials as mentioned in Chapter 7. Currently, these are manufactured in vacuum furnaces at high tempera-ture and low pressure. A constant input of energy is required to maintain these gradients. Current prices of carbon nano-tubes are primarily driven by research and development costs as well as by the specific chemistry and morphology being produced, but as the research and development costs become absorbed by emerging markets, what remains to be seen is whether energetic costs will come into play in dictating the market value as appears to be the case with the manufactur-ing of silicon-based photovoltaic cells. Silicon, phosphorous, and boron are all cheap. The manufacturing time and manu-facturing energy required to essentially drive the entropy out of them by arranging them in a single-crystal form is not. The same is true of carbon nanotubes, carbon nanowires, and so forth. Each must be manufactured bottom-up, one atom at a time, and thus intimately linking a monetary cost function with the "negentropy" that essentially flows into an assembly of atoms in a highly unlikely, yet highly repetitive configura-tion. Just as a full description of a silicon crystal requires only the specification of the relative three-dimensional positions of fewer than 10 unit-cell atoms and the gross dimensions of the crystal requires relatively little information, a full descrip-tion of a multiwalled carbon nanotube requires relatively little information: chirality, number of walls, and length. The point here is that, typically, a relatively greater amount of time, temperature, and money is required to create a material with

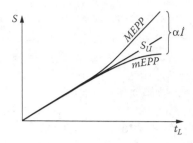

Figure 18.3 The relationship between the maximum entropy production principle (MEPP), the minimum entropy production principle (mEPP), and the rate of information throughput, $\dot{I}$. The coefficient, α, defines a given technology's ability to partition entropy. The greater a technology's α, the better that technology is at reducing its user's local entropy level against the universal rate of entropy increase $\dot{S}_U$, and maximizing entropy shed to the environment.

relatively little information content as quantified by the number of bits to specify the locations of all the atoms.

However, consider the case where there are a few impurities in the single-walled carbon nanotube. Not only does this configuration require a commensurately greater number of bits to describe exactly where these imperfections are, but now the technology that employs the nanotubes should, if our prediction of Figure 18.3 is correct, for the same, α, almost certainly result in a greater fraction of ΔS ending up above S_U. Specifically, let's say that a new membranous material composed of one continuous carbon nanotube has been developed and woven into an enormous sheet to replace the failed levies of New Orleans. If the sheet is indeed pure, one need only know the chemistry, the weave pattern, and the extent of the fabric. However, if impurities exist in the continuous tube, it will be weaker in proportion to the number of impurities [39–41]; thus, while it may be capable of maintaining a relatively low mEPP for some time, protecting those within the membrane from destructive pressure gradients, given a large enough load and enough time, the membrane will rupture, resulting not only in a shift in the environmental entropy above S_U, but will also eliminate the portion of ΔS represented by $S_U - S_{mEPP}$.

Carbon sequestration, which will increasingly become a by-product of microfabrication and nanofabrication, in large, pressurized concentrated regions is likely a bad idea because of its relevance to contemporary environmental concerns. Pumping a gas that is a toxin to much of life on earth, especially in pressurized tanks, virtually ensures that at some point these vaults will rupture, killing all nearby life in a blanket of suffocation. What was an ordered, low information content concentration of gas becomes a disorderly array of dying organisms.

Whether or not the universe became "fully entropicized" very early as Chaisson suggests, deserves further attention. However, what is clear is that without entropy gradients, energy would not flow and life would be incapable of tapping into "free energy" reserves. So as Gibbs, to some degree [21], and later Schrödinger pointed out [43], we thrive on entropy gradients and amplify them. Eventually, as with the carbon nanotube levy example, the boundaries fail and the gradients vanish into a more entropic state than at earlier times. The collapse of the Twin Towers represents the rapid diminishment of the entropy partitioning they were performing for over 30 years. The resulting pile of technological rubble and biological death that resulted greatly entropicized lower Manhattan and produced a drop in the Stock Market volume, which represents reduced information partitioning. The closing of this entropy partition is depicted as the collapsing bubble in the upper right of Figure 18.4.

However, in Ian Morris' recent work, *Why the West Rules for Now* [65], he included a cofactor that he terms *energy capture* as a metric for discrepancies in the relative success of Western versus Eastern societies. His basic argument is that Western societies have either had greater access to agricultural energy either via biological, climatological, or geographical disparities. This brings up the fundamental question of what the best metric might be for measuring the success of a society. Is it merely a society's ability to exploit natural resources, be they material or energetic? Is a society's success measured by its financial resources? Certainly these two contribute, but the more fundamental metric for measuring a society's success is certainly the quality of the physical and mental, health of its citizens. Frequently, greater access to material, energetic, and financial resources results in superior physical and mental health. However, when material extraction (i.e., mining or energy consumption [i.e., carbon-based combustion], result in

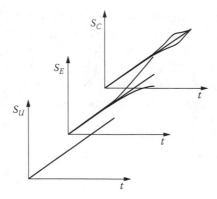

Figure 18.4 Three possible scenarios for entropy production in the universe. In the foreground, in black is the overall entropy increase of the universe, S_U. Just above this in dark gray is depicted a region of the universe such as the earth where life has used its information base to partition entropy, S_E. At the top of the figure in light gray is depicted a region of the universe where entropy partitioning began but collapsed as perhaps by war or overexploitation of the environment, S_C.

"excessive entropization" of the environment [i.e., watershed pollution and anthropogenic climate change], then, surely, the measure of success is diminished when the society ends up "drowning in its own entropy").

If you are reading this chapter on a computer or other microelectronic device, energy is flowing through the machine at Chaisson's energy throughput rate, ϕ, and information, I, is also flowing from it into your visual cortex and then through the language processing regions of your brain to then be transformed and stored in your neural circuits. In this specific example, the energy throughput rate is approximately 25 W·kg^{-1}. A laptop runs at approximately 100 W and weighs approximately 4 kg. Reading rates vary, but range from 100 to 1000 words per minute. This is the equivalent of 200 bits per second. Thus, α for reading on a laptop is approximately

$$\alpha = \frac{\dot{S}}{\dot{I}} = 0.0016 \text{ J b}^{-1} \text{ K}^{-1}. \tag{18.6}$$

New microtechnologies such as microengineered lithium-ion battery technologies and microcapacitive screen technologies have enabled information to flow from these devices nearly as rapidly as from a newspaper and with less environmental entropy produced per bit.

An increasing fraction of our technological primary energy consumption is being funneled into power micro-devices and in researching nanodevices. Already, we are within four orders of magnitude of Diamond's photosyn-thetic limit [31]. In other words, we use the equivalent of 1% of 1% of the sun's incident energy for heating, cooling, transportation, manufacturing, and now computing. This occurs on a continuous basis. We do not, in fact, harvest 1% of 1% of solar energy directly for purposes such as water heating, electricity generation, or transportation. But we do consume as a species approximately 200 megajoules (MJ) of technological energy per day, which is the equivalent to the solar energy received by 10 to 20 square meters at typical location in a day. With conversion efficiencies of 10%, this requirement raises to 100 to 200 square meters. It also does not account for metabolic energy, which is 10 MJ per day and has an efficiency of a, fraction of 1%. All of our tech-nological and metabolic energy consumption is consumed for the purpose of maintaining our corporeal selves, a large fraction of which is devoted to brain maintenance. So, argu-ably, whatever fraction of metabolic energy is not spent on genetic reproduction is spent on memetic production, repro-duction, and consumption. Computers are the preferred media for spreading memes.

Sustainability

Already computing and the Internet allow us to share infor-mation at a rate that well exceeds our collective abilities to process it. What remains to be seen, however, is whether a greater number of bits per joule will result in a decrease or an increase in joules per capita per unit time. Obviously, in order to become sustainable, new microtechnologies and nano-technologies must be capable of delivering to their users infor-mation cost-effectively in order to avoid energy-hungry and entropy-intensive consequences (in other words, devices with a large $\dot{I}$ and a small ϕ and $\dot{S}$, or equivalently a small α). A few simple examples are the new remote home power monitoring systems, small remote seismographs for regions with poor infrastructure, or portable health-care devices. Respectively, these represent a saved trip home to turn off the furnace, an expensive search and rescue mission, or emergency trips to the hospital. The overall goal of sustainability as we move to adopt a greater number of microdevices and nanodevices is thus threefold:

1. Reduce the entropy of our corporeal selves for the purpose of sustaining our own lives or enhancing the probability of our own genes, stored in the DNA of our children, or enhancing the probability of survivability of our memes through what is typically referred to as the "grandmother hypothesis" [44].
2. Increase the entropy level in the environment of competitors or enemies. This is carried out constantly through direct warfare such as the 9/11 bombings, the use of radar jamming equipment, or the use of chemicals to kill plants and animals.
3. Increase the access level to pure information, which serves the purpose of driving points 1 and 2. This is done on an individual organism level, a societal level, and likely at a genetic level with individual genes within a single organism competing as well as cooperating for expression levels.

It may even be fair to conclude that when the majority of human technological artifacts are engineered at the atomic scale, they will be fully capable of converging with the extant biologies. In many ways, they could become fully and inextricably symbiotic with us at a molecular level, just as mitochondria became inextricably symbiotic with a separate discrete cell early in the history of life.

We live in what Martin Rees calls the mesoscale [45]. In fact, it is the specific ratio, N, between the magnitude of the force of gravity at our planet's surface and the magnitude of electrostatic forces between the molecules within living organisms on our planet that allow for the rich diversity of life that we see. Each human is composed of between 10^{28} and 10^{29} atoms. If the smallest object that humans typically manipulate is a pin and the largest object is a sofa, then we manipulate agglomerations of matter that are between 100 mg and 100 kg, or between 10^{21} and 10^{29} atoms, with our bare hands. With the advent of optical microscopy, the lower end of this range drops by one or two orders of magnitude. With the advent of the internal combustion engine and conventional rocketry, we can move masses on the order of 500 million kilograms or between 10^{34} and 10^{35} atoms (many more times this much mass was moved during some of the largest atomic weapons detonations, but not in an orderly manner). A distinction is necessary between mass moved and mass transported, because modern nuclear weapons tests can be sensed from across the planet, but the energy released by some of the largest nuclear weapons, approximately 200 gigajoules (GJ) is enough energy to give every human on the planet a trip into space of 70 km, or to give one fifth of the planet's population a trip to the International Space Station.

Paradoxically, it is our continued ability to manipulate matter at the microscale and nanoscale that has enabled the movement of such large quantities of mass and energy. Also somewhat paradoxically, some of the largest energy-consuming scientific instruments are required to probe some of the smallest and transient particles in the universe. And it is likely that advances in computing will further enable this. Moore's law has held true for over four decades and has been written about extensively. In fact, in a recent paper by Seth Lloyd [46], he predicts that the ultimate limit of a computer with a mass of 1 kg and a volume of 1 L is capable of performing 5.4258×10^{50} calculations per second, which falls well beyond Kurzweil's singularity, and which itself occurs when computers operate at 10^{14} to 10^{15} calculations per second. Unfortunately, in the Lloyd ultimate computer, all of the mass turns to energy, so it is not clear whether the "information" generated will be useful, or purely entropic.

Another idea that is germane to our discussion of the evolutionary trajectory that technologies might take is Gould's "left wall hypothesis" [47], namely, that most living organisms exist at some average level of complexity with fewer complex organisms being capable of living far out to the right on the complexity scale, and those that live too far to the left are incapable of survival and are thus consumed. This is certainly true of technologies as well. For example, some of the most primitive and simple technologies, such as knives, utensils, and other hand tools demonstrate utility over entire lifetimes and frequently over many generations, whereas complex technologies, such as personal computers, automobiles, and cellular phones, are considered old well before their first decade.

In 1943, Salvador Luria demonstrated that when exposed to environmental challenges or opportunities, bacteria both retool their metabolic and defensive molecular machinery to enhance their probability of survival, t_L, by simultaneously maintaining their energy throughput, ϕ [48]. Telomerase and its embodied information plays a similar role in maintaining the material integrity of the distal-most ends of ends of our precious chromosomes. Remarkably, but not surprisingly, the ability of these molecular machines, designed through natural selection to simultaneously deentropicize a cell by cobbling together stray nucleic acids and chemically welding them back onto the fraying split ends of the double helix, manifests at the organism scale. In fact, it appears that, literally, more information as embodied in the length of chromosomes increases longevity [45]. This is true on an organismal basis, not as a cross-species comparison. An organism with a longer genome is not necessarily likely to live longer than one with a short genome. However, if during cell division, genetic information is lost during division, the chance of mortality of healthy daughter cells, and thus the organism, in general, diminishes. For specific cancers, there may be exceptions whereby the loss or mutation of inherited DNA may render a cell immortal yet diminish the life span of the organism that carries the resulting tumor. By doing so, telomerase reduces the degrees of freedom that the cell has by reducing the total number of molecules in the cell and thus reducing the entropy. It also adds to the amount of information contained by the cell. The result of this is that the cell and its progeny are more likely to persist longer into the future. Thus, we see that, with the emergence of molecular machines such as telomerase, they seal

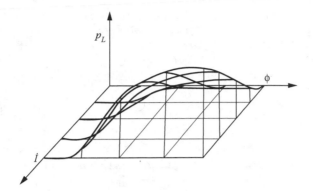

Figure 18.5 All technologies or molecular machines may be placed at a point on these three axes. P is the extensive power flowing through the machine and is directly proportional to the entropy emanating from it, $\dot{I}$ is the rate of information flowing through the machine, and t_L is the lifetime of the machine. This plot is intended to suggest that entities that are "smart" and powerful have a greater probability of survival.

their fate as invaluable arrangements of matter that propel their host cells farther along the $\dot{I}$, ϕ, p_L axes (Figure 18.5). One way to define the sophistication of a machine is to do so on a triaxial basis consisting of the ability to funnel energy at a given rate, ϕ, the ability to process information quickly, $\dot{I}$, and its propensity toward longevity, t_L.

Another broader perspective worth exploring on the relationship between technological evolution and society is the complex relationship among scientific thought: the ability to program a computer, the ability of other talented people to hire computer programmers, and the resulting flow of abstract money through machines. Only a tiny fraction of financial transactions involve hard currency. Now, most financial transactions are attached to some thread, either through the financial records kept of a check written, the issuance of a stock or bond, or the numerical signature left on a broker's disk. Money has memory [50]. Perhaps the most poignant statement in Lanier's article is

> There's an old cliché that goes "If you want to make money in gambling, own a casino." The new version is "If you want to make money on a financial network, own the server." If you own the fastest computers with the most access to everyone's information, you can just search for money and it will appear. A clear example of this principle

is the rise of "high-frequency trading," in which the clan that owns the server gets to pull money out of a market before nonmembers can even try to make trades. (p. 127)

How does Lanier's observation apply to the assertion of this chapter that the human–machine symbioses that will control the future will be ones with the greatest values along the $\dot{I}$, ϕ, p_L axes? We have already seen that each of these three metrics, namely the ability to funnel energy, ϕ, and the ability to funnel information, $\dot{I}$, can determine the ability to do this for a long time, p_L. There are other situations where the three are independent. For example, a machine can be very powerful but have a very short lifetime and almost no information processing ability. An example is an improvised explosive device (IED). A machine may have a very low power requirement but have the ability to process a relatively large amount of information such as a PDA. Or, a machine may persist for a very long time, have no power requirement, and have modest information processing capacity, such as an abacus.

Human–Machine Symbiosis

The fallacy of the common misconception that humans are "more evolved" than our close genetic relatives the chimpanzees implies that all contemporary living organisms may be "equally evolved" [18,50]. It is also a misnomer to state that a particular machine or a particular human–machine symbiosis is more evolved than another. Consider the sophistication of a system consisting of a camera being developed for special needs children that will allow them to take photographs and share them with friends. These students would be nearly nonambulatory without their wheelchairs, crutches, and walkers. This symbiosis that is developing between the students and their ambulatory prosthetics is facilitated by a highly trained staff that serves to prevent the students from getting injured and to enrich their environments so that their brains stay stimulated and engaged. Thus, the system consists of camera, student, and trainer. This is much more sophisticated of a system than any single "normal" human. In earlier societies, children with these disabilities would not likely have lived beyond 1 or 2 years. Thus, even though these people as individuals may be "evolutionarily challenged," in a purely Darwinian sense, through their technological symbioses, they are "more human" as were Gates, Glenn, and Gordon. It is, of course, the charge of academic leaders at our institutions of advanced education to enrich the experiences of their students. It is the responsibility of our best aging athletes to provide role models for the next generation. It is also the role of our best intellectuals to challenge the wise-cracking 18-year-old in the lecture hall and the job of the most far-thinking technologists to challenge the minds and hands of burgeoning and aspiring young students to bring into reality the next generation of machines that will propel our species to the stars, unravel the molecular mechanisms that make life possible, and to conceive of the next generation of micro- and nanotechnologies that will make the Kurzweilian dream of human immortality a reality.

Some of the most profound changes to humanity will come from what Kurzweil refers to as the NGR revolution: "N" for nanotechnology, "G" for genetics, and "R" for robotics. The idea is that once each of these fields of study matures, they will reach a crescendo where nanoscopic robots are able to both read and write genetic code in its native language. Some test-tube scale experiments have yielded preliminary results [51–53]. However, none of these use nanobots per se. Some attempts have been made at scales approaching the nanoscale (i.e., [54,55]), but there is always an issue with actuation. We are, thus, typically constrained to dealing with small batches using conventional bottom-up techniques or frequently painstaking top-down techniques (e.g., [56]). For example, in order to manipulate even a few femtoliters (1 μm^3) of matter to extract a single parameter such as Young's modulus, which can be represented with 64 bits of data, from a nanopipette can take nearly 1 kW of power and several hours [57]. So, in this case, $\alpha = 750$, or a relatively poor return on information per joule. Most of Shakespeare's works contain approximately 30,000 words. Someone reading at 500 words per minute and 25 bits per word, has an information processing rate of 208 bits per second, resulting in $\alpha = 0.0016$. From the local maximum that emerges in Figure 18.6 wherein α is plotted as a function of volume at the mesoscale of about 1 cubic meter, at least for the technologies discussed, there is a relatively poor return on bits per joule. Or, restated, unless there is substantial information processing occurring at a relatively low energy consumption rate, the α for a given technology is poor. The ISS, JSF, and HST do well because of their on-board computing power. In general, most other advanced imaging systems are going to have relatively large α values because extracting information from very small volumes of space requires more energy than does extraction at the human scale. Two extreme α values to find would be that of the Search for Extra-Terrestrial Intelligence (SETI) experiment, which is essentially being conducted with a large number of computers over a large fraction of the universe as well as advanced high field gradient magnetic resonance imaging techniques or atomic

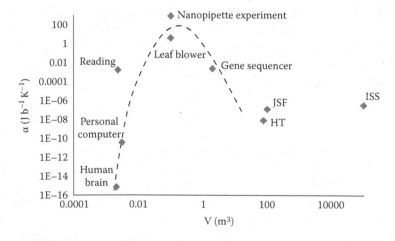

Figure 18.6 The technologies discussed in this chapter appear to follow a trend, whereby mesoscale technologies have poorer bit return per joule invested. The goal of any microtechnology or nanotechnology should logically demonstrate an economy that has a large bit return per joule invested or bit return per entropy unit (J/K). Here, the volume of a personal computer includes the volume of the tower, screen, keyboard, and so forth. The single data point indicates a rough estimate. The number of personal computers that are smaller than human brains is large and growing. The point indicating "reading" represents the actual act of serial reading one word at a time. Consequently, the bit rate is lower and is greater than that required of the visual cortex when engaging in an activity such as basketball.

force microscopy experiments. The difficulty is in defining the boundaries of the experiment. The overall goal, however, in any advanced experiment or technology is to have a high return on information for a low investment in energy and subsequently entropy. We will also develop machines that read DNA on a desktop [58,59] that will surpass the gene sequencer tag, essentially tunneling from the right side of the curve to the left as size diminishes and information production rates rise. Also worth noting are still-evolving nanodevices. Many of the devices described in Freitas's Nanomedicine [60] are yet to materialize, but these, too, will likely, in principle, have very high bit to joule production ratios.

As an additional example of what we discussed in the introductory paragraphs, consider the portable scanners now being made available to remote regions of the world with almost no technological infrastructure. A portable scanner such as the GE LOGIQ Book XP with a few tens of watts of power is capable of collecting information about a pregnant mother's health status and that of her baby at a rate of several million bits per second. These form the images on the monitor, all of which then become condensed in the mind of the technician working locally and the trained expert working remotely. But what value of information throughput is the proper metric to use here? The number of bits flowing through the machine or the binary decision being made by the doctor or clinician: administer drug or not, admit to emergency room or not, and so forth. In other words, the definition of α may be relativistic in the sense that it is different for different observers. For the technician reading the output from the scanner, billions of bits move from the scanner's screen to the technician's retinas, there to be processed at a rate of terahertz in the technician's brain. However, from the patient's point of few, there is only a single bit of information to consider: C-section or vaginal birth?

How will we cool our computing devices? The author recently received a request to estimate the cooling power of Lake Michigan for a server farm. This practice is already common for nuclear technology and the steel industry, which severely alters local ecosystems. Global engineering issues such as this and the growing concern over the politics of energy (e.g., [61]) led a paper on the topic wherein energy densities and their associated monetary and societal costs were discussed [31]. The calculation for the server farm is simple. The heat thermal capacitance of water is 4.2 joules per gram per Kelvin, the volume of Lake Michigan is 5000 km^3, its mass is 5000 km^3 $\times$ 1000^3 m^3 km^{-3} $\times$ 1000 kg m^{-3} $\times$ 1000 g kg^{-1} = 5 $\times$ 10^{18} g. For a cluster running at 100 kW for a year, the heat generated is 100,000 $\times$ 356 $\times$ 24 $\times$ 60 $\times$ 60 = 3 $\times$ 10^{12} J. So one cluster of this size being cooled by Lake Michigan would raise its temperature by 3 $\times$ 10^{12} J $_{,}$ 5 $\times$ 10^{18} g $_{,}$ 4.2 J g^{-1}K^{-1} = 0.15 $\times$ 10^{-6} K, or about a sixth of a millionth of a degree in 1 year. Six thousand such clusters using a natural body of water to shed thermal energy would thus raise the temperature of Lake Michigan by one one-thousanth of a degree in a year. It would take an entire eon for these six thousand clusters (that's the equivalent of one every quarter mile) to raise the temperature of the lake 1°K, and that's assuming that the lake does not dissipate this thermal energy into the atmosphere or earth. A similar order of magnitude calculation yields that the 425 $\times$ 10^{18} J that we consume annually heats the atmosphere 0.1 K per year [31].

As stated in the opening paragraph, smaller technologies typically have greater "information payback" than larger ones. For example, a cell phone, which consumes energy at a rate of less than 1 watt, is capable of processing information at a rate of a few kbps, comparable to, or in many cases superior to a PC. How does this small technology already redirect human metabolic energy? For example, the cell phone can help the user find the nearest gas station, the nearest restaurant with the best menu, his favorite movie at a local theatre, or could be used to tell the user that his doctor's appointment has been canceled. It could also be used as a monitor for a home energy monitoring system or as an early storm evacuation warning. In each of these examples, information provided to the end user via the phone allows the user to minimize his or her own path to a rich source of information or energy. Place the smart cell phone with its embedded micro- and nanotechnologies in contrast with large technologies, such as earth-moving equipment, military equipment, commercial aircraft, or oil refineries. If we first consider large technologies such as these and strip them of all of their embedded small technologies, which are typically used for control, their embodied information or the information required to reproduce them (i.e., the information in the blueprints) is likely equal to that required to create the cell phone in the previous example. Obviously, in each of these cases, the gross energy throughput is much greater than that of the small technology. If the large commercial aircraft is not equipped with gadgets such as radar, radio, or other sophisticated telecommunication or control equipment, but merely its fuselage, seats, engines, and a simple power-assist manual control (basically only what is required for transportation of its passengers and crew), examining the same metrics that we did for the cell phone, the plane consumes energy at a rate of 50 MW, is capable of processing information at whatever rate the pilot is capable, say 10^{13} b s^{-1}, and thus has $\phi = 440$ W kg^{-1}, and thus, has $\alpha = 1.7 \times 10^{-8}$. If we were to add in all of the computer technology and equip all passengers with laptops, this number becomes 2×10^{-11}, comparable to a single computer. After having read this paragraph, the reader may wonder, what about the energy being consumed by the infrastructure to support the function of the cell phone? As with any thermodynamics problem, a boundary must be clearly defined. Certainly the cell phone is no more a singular discrete entity than the brain would be without motor, sensory, and communication organs. Thus, it becomes necessary to carefully define the boundaries. For example, in the definition of the machine types, the boundary of a Type II machine would enclose both the technology and the machine.

It may be worth considering an α_m and an α_M, the first of which gives the ability to stay under the background entropy generation rate, and α_M which defines the amount by which it is exceeded. For example, $\dot{S}_{MEPP} - \dot{S}_U = \dot{\alpha}_M \dot{I}_M$ and $\dot{S}_U - \dot{S}_{mEPP} = \dot{\alpha}_m \dot{I}_m$.

It has been suggested that natural physical entropy, S, is not a function of time, but that our observation of increasing entropy is responsible for the perception of time [62]. This is consistent with the fact that subatomic particles at temperatures we experience on earth have lifetimes essentially equal to that of the planet and thus do not "age." At high energy levels, neutron decay has a half-life of 17 minutes, but some heavy nuclei remain stable for billions of years, and carbon 14 takes 5700 years to decay into nitrogen, an electron, and an electron antineutrino. The proton itself is stable for at least 6.6×10^{33} years. Large molecules, of course, are more likely to fall victim to entropic events and thus have shorter lifetimes, yet some may have lifetimes that exceed that of the organism they serve. A poignant example is collagen, which was recently extracted from a *Tyrannosaurus rex* bone [63] revealing it to be closely related to birds. DNA that is tens of thousands of years old has also been found in *Homo* remains. If we then allow ourselves to consider that time is not an independent variable in the Newtonian sense or even a relativistic variable in the Einsteinian sense but a dependent variable that is a function of an entropy partitioning clock that is related to the information processing prowess of a biological or technological machine, then arguably, this allows less-sophisticated systems from the "past" and more-sophisticated systems of the "future" to coexist in the "present." In fact, the idea of coexistence is implicit in Einstein's general relativity. The question of whether our observed "arrow of time" is a result of the universe's expansion and the measureable increases in entropy we have discussed is still open for debate. Most physical theories either predict the absence of time or the reversibility of time at the quantum level, concluding that entropy measures are merely the result of statistical compilations of the numerous ways in which matter organizes itself, and the fact that the past may be known but not affected and the future affected but not known leads to a human perception of time.

Summary

In conclusion, the micro- and nanotechnologies that are most likely to persist and thrive will be ones that proffer the greatest selective advantage to their respective users. This selective advantage may be generally quantified as the human–technology symbioses that provide the owner/user with the greatest amount of usable information. To reiterate, the "usable information" is information that gives its user the ability to partition entropy: reduced entropy internally at the expense of above-background entropy acceleration environmentally. Usable information will emerge at multiple scales: individual RNA expression levels, basic blood chemistry, economic markets, weather and traffic patterns, and perhaps even geophysical and cosmological events.

Acknowledgments

This material is based upon work partially supported by the National Science Foundation under Grant No. 0900101. "Any opinions, findings, and conclusions or recommendations expressed in this material are those of the author(s) and do not necessarily reflect the views of the National Science Foundation." This work was also conducted with funds from the Keck Foundation for the Drexel Center for Attofluidics, the United States Department of Agriculture, USDA/CREES NRI grant 2008-35100-04413 and a grant from NetScientific. Opinions expressed by the author do not necessarily reflect those of these funding agencies.

Bad entropy day.

References

1. Kurzweil, R., *The Age of Intelligent Machines*. 1990, Cambridge, MA: MIT Press. xiii, 565 p.
2. Kurzweil, R., *The Age of Spiritual Machines: When Computers Exceed Human Intelligence*. 1999, New York: Viking. xii, 388 p.
3. Kurzweil, R., *The Singularity Is Near: When Humans Transcend Biology*. 2005, New York: Viking. xvii, 652 p.
4. Layton, B.E., Recent patents in bionanotechnologies: nanolithography, bionanocomposites, cell-based computing and entropy production. *Recent Patents in Nanotechnol.*, 2008. 2(2): 72–83.
5. Layton, B.E., *Mechanoevolution: New Insightes into the Connection Between Information Theory and the Second Law*, in *ASME International Mechanical Engineering Conference and Exposition*. 2010: Vancouver, Canada.
6. Gould, S.J., *The Book of Life*. 1993, New York: W.W. Norton. 256 p.
7. Shannon, C.E., A mathematical theory of communication. *Bell System Tech. J.*, 1948. 27: 623–656.
8. Carnot, S., E. Clapeyron, and R. Clausius, *Reflections on the Motive Power of Fire*. 1960, New York: Dover Publications. xxii, 152 p.
9. Boyle, G., B. Everett, and J. Ramage, *Energy Systems and Sustainability: Power for a Sustainable Future*. 2003, New York: Oxford University Press.
10. Dawkins, R., *The Selfish Gene*. 1976, New York: Oxford University Press. xi, 224 p.
11. Chaisson, E., *Cosmic Evolution: The Rise of Complexity in Nature*. 2001, Cambridge, MA: Harvard University Press. xii, 274 p.
12. Chaisson, E., Energy rate density as a complexity metric and evolutionary driver. *Complexity*, 2011, 16(3): 27–40.
13. Coren, R.L., *The Evolutionary Trajectory: The Growth of Information in the History and Future of Earth: The World Futures General Evolution Studies,* vol. 11, 1998, Amsterdam, Netherlands: Gordon and Breach Publishers. xiv, 220 p.
14. Coren, R.L., Empirical evidence for a law of information growth. *Entropy*, 2001. 3: 259–272.
15. Moore, G.E., Cramming more components onto integrated circuits. *Electronics*, 1965, 38(8): 114–117.
16. Berger, M., Nanotechnology, transhumanism and the bionic man. *Nanowerk*, 2008, May 28.

17. Roco, M.C., and W.S. Bainbridge, *Converging Technologies for Improving Human Performance Nanotechnology, Biotechnology, Information Technology and Cognitive Science*. 2002, Arlington, VA National Science Foundation.
18. Bakewell, M.A., P. Shi, and J.M. Zhang, More genes underwent positive selection in chimpanzee evolution than in human evolution. *Proc. Natl. Acad. Sci. USA*, 2007. 104(18): 7489–7494.
19. Petroski, H., *The Evolution of Useful Things*. 1st Vintage Books ed. 1994, New York: Vintage Books. xi, 288 p.
20. Martyushev, L.M., and V.D. Seleznev, Maximum entropy production principle in physics, chemistry and biology. *Phys. Rep.*, 2006. 426: 1–45.
21. Bumstead, H.A., and R.G. Van Name, *The Scientific Papers of J. Willard Gibbs Ph.D. LL.D.* 1907, London: Longmans, Green and Co.
22. Carnot, S., *Reflections on the Motive Power of Fire*. 1824, Paris.
23. Dawkins, R., *Climbing Mount Improbable*. First American ed. 1996, New York: Norton. xii, 340 p.
24. Luria, S., and M. Delbruck, Mutations of bacteria from virus sensitivity to virus resistance. *Genetics*, 1943. 28: 491–511.
25. Hodgson, G.M., The great crash of 2008 and the reform of economics. *Cambridge J. Econ.*, 2009. 33: 1205–1221.
26. Wells, S., *Deep ancestry: Inside the Genographic Project*. 2006, Washington, DC: National Geographic. 247 p.
27. Altschul, S. et al., Basic local alignment search tool. *J. Mol. Biol.*, 1990. 215: 403–410.
28. Layton, B. et al., Collagen's triglycine repeat number and phylogeny suggest an interdomain transfer event from a devonian or silurian organism into *Trichodesmium erythraeum*. *J. Mol. Evol.*, 2008. 6: 539–554.
29. Ridley, Matt. 1999. Genome: *The Autobiography of a Species in 23 Chapters*. New York: HarperCollins.
30. Blankenship, R.E., *Molecular Mechanisms of Photosynthesis*. 2002, London: Blackwell.
31. Diamond, J.M., *Collapse: How Societies Choose to Fail or Succeed*. 2005, New York: Viking. xi, 575 p., [24] p. of plates.
32. Layton, B.E., A comparison of energy densities of prevalent energy sources in units of joules per cubic meter. *Intl. J. Green Energy*, 2008. 5(6): 438–455.
33. Ferris, D., The exoskeletons are here. *J. Neuroeng. Rehabil.*, 2009. (6): 17.
34. Whittaker, J.C., and G. McCall, Handaxe-hurling hominids: An unlikely story. *Curr. Anthropol.*, 2001. 42(4): 566–572.

35. Shea, J., The origins of lithic projectile point technology: Evidence from Africa, the Levant, and Europe. *J. Archaeological Sci.*, 2006. 33(6): 823–846.
36. Lloyd, S., Ultimate physical limits to computation. *Nature*, 2000. 406: 1047–1054.
37. Chen, L., W. Zhang, and F. Sun, Power, efficiency, entropy-generation rate and ecological optimization for a class of generalized irreversible universal heat-engine cycles. *Appl. Energy*, 2007. 84: 512–525.
38. Angulo-Brown, F., G.A. de Parga, and L.A. Arias-Hernandez, A variational approach to ecological-type optimization criteria for finite-time thermal engine models. *J. Phys. D: Appl. Phys.*, 2002. 35: 1089–1093.
39. Daniels, H.E., The statistical theory of the strength of bundles of threads I. *Proc. Roy. Soc. London Series A-Math. Phys. Sci.*, 1945. 183(995): 405–435.
40. Nørretranders, T., The User Illusion: Cutting Consciousness Down to Size. 1998. New York: Viking.
41. Harlow, D.G., and S.L. Phoenix, Chain-of-bundles probability model for strength of fibrous materials. 2. Numerical study of convergence. *J. Comp. Mater.*, 1978. 12(JUL): 314–334.
42. Harlow, D.G., and S.L. Phoenix, Chain-of-bundles probability model for strength of fibrous materials. 1. Analysis and conjectures. *J. Comp. Mater.*, 1978. 12(APR): 195–214.
43. Schrödinger, E., *What Is Life? The Physical Aspect of the Living Cell and Mind and Matter.* 1967, Cambridge: Cambridge University Press. 178 p.
44. Alvarez, H.P., Grandmother hypothesis and primate life histories. *Am. J. Phys. Anthropol.*, 2000. 113(3): 435–450.
45. Rees, M. J. (2000). *Just Six Numbers: The Deep Forces That Shape the Universe.* New York, Basic Books.
46. Lloyd, S., Ultimate physical limits to computation. *Nature*, 2000. 406(6799): 1047–1054.
47. Gould, S.J., The evolution of life on the earth. *Sci. Am.*, 1994. 271(4): 84–91.
48. Blackburn, E.H., C.W. Greider, and J.W. Szostak, Telomeres and telomerase: The path from maize, Tetrahymena and yeast to human cancer and aging. *Nat. Med.*, 2006. 12(10): 1133–1138.
49. Lanier, J., Of Money and Memory. *Playboy*, 2010. 57(2): 127–128.
50. Diamond, J.M., *The Third Chimpanzee: The Evolution and Future of the Human Animal.* 1st ed. 1992, New York: HarperCollins. viii, 407 p.
51. Brenner, S., The Role of Planet Earth in Life's Origins. In *Origins of Life Initiative.* 2009. Cambridge, MA.

52. Liedl, T., T.L. Sobey, and F.C. Simmel, DNA-based nanodevices. *Nano Today*, 2007. 2(2): 36–41.
53. Mao, C., W. Sun, and N.C. Seeman, Assembly of Borromean rings from DNA. *Nature*, 1997. 386(6621): 137–138.
54. Zhou, Y., B.J. Nelson, and B. Vikramaditya, Integrating optical force sensing with visual servoing for microassembly. *J. Intell. Robotic Syst.*, 2000. 28(3): 259–276.
55. Tan, J.L. et al., Simple approach to micropattern cells on common culture substrates by tuning substrate wettability. *Tissue Eng.*, 2004. 10(5–6): 865–872.
56. Patel, R.A. et al., Parameterization of a Piezoelectric Nanomanipulation Device. In *8th Biennial ASME Conference on Engineering Systems Design and Analysis ESDA*. 2006. Turin, Italy.
57. Warren, J., Y. Gogotsi, and B.E. Layton. Mechanical characterization of carbon based injection nanopipettes. In *Proceedings of ASME 2010 First Global Congress on NanoEngineering for Medicine and Biology*. 2009. Houston, TX.
58. Meller, A., L. Nivon, and D. Branton, Voltage-driven DNA translocations through a nanopore. *Phys. Rev. Lett.*, 2001. 86: 3435–3438.
59. Storm, A.J., C. Storm, J. Chen, H. Zandbergen, J. Joanny, and C. Dekker, Fast DNA translocation through a solid-state nanopore. *Nano Lett.*, 2005. 5: 1193–1197.
60. Freitas, R.A., *Nanomedicine*. 1999, Austin, TX: Landes Bioscience.
61. Clark, W.R., *Petrodollar Warfare: Oil, Iraq and the Future of the Dollar*. 2005, Gabriola Island, BC: New Society. xix, 265 p.
62. Bizo, L.A. et al., The failure of Weber's law in time perception and production. *Behav. Processes*, 2006. 71(2–3): 201–210.
63. Organ, C.L. et al., Molecular phylogenetics of Mastodon and Tyrannosaurus rex. *Science*, 2008. 320(5875): 499.
64. Grabbe, J.O., 2006. "The holographic principle is a simple consequence of the divergence theorem." billstclair.com/grabbe/holographic_universe.pdf.
65. Morris, I., 2010. Why the West Rules—for Now: The Patterns of History, and What They Reveal About the Future. New York: Farrar, Straus, and Giroux.

Index

A

J

T